T0073890

Feminist AI

Feminist AI

Critical Perspectives on Data, Algorithms and Intelligent Machines

Edited by

PROFESSOR JUDE BROWNE

*Head of Department, Department of Politics and International Studies,
University of Cambridge*

Frankopan Director at the University of Cambridge Centre for Gender Studies

DR STEPHEN CAVE

*Director of the Institute for Technology and Humanity, University of
Cambridge.*

DR ELEANOR DRAGE

*Christina Gaw Post-Doctoral Research Associate in Gender and Technology,
University of Cambridge Centre for Gender Studies, Senior Research Fellow at
the Leverhulme Centre for the Future of Intelligence*

DR KERRY MCINERNEY

*Christina Gaw Post-Doctoral Research Associate in Gender and Technology,
University of Cambridge Centre for Gender Studies, Research Fellow at the
Leverhulme Centre for the Future of Intelligence*

OXFORD
UNIVERSITY PRESS

OXFORD
UNIVERSITY PRESS

Great Clarendon Street, Oxford, OX2 6DP,
United Kingdom

Oxford University Press is a department of the University of Oxford.
It furthers the University's objective of excellence in research, scholarship,
and education by publishing worldwide. Oxford is a registered trade mark of
Oxford University Press in the UK and in certain other countries

Published in the United States of America by Oxford University Press
198 Madison Avenue, New York, NY 10016, United States of America

British Library Cataloguing in Publication Data
Data available

Library of Congress Control Number: 2023934461

ISBN 978–0–19–288989–8

DOI: 10.1093/oso/9780192889898.001.0001

Printed and bound by
CPI Group (UK) Ltd, Croydon, CR0 4YY

Acknowledgements

We are grateful to our generous funder, Christina Gaw, who made this project possible. She writes:

> We are going through interesting times, when humans are always looking for ways to make our world a better place, and to positively impact our communities. Exciting developments in AI will contribute to the many conveniences of this new world: ensuring that AI is developed in a gender and diversity inclusive manner right from the start is key to its successful implementation and will make our future world a much more equitable one.

We are also grateful for the support of the Leverhulme Trust through the Leverhulme Centre for the Future of Intelligence.

We would also like to give special thanks to all of our contributors for their excellent essays and their enthusiasm for this project, and in particular to Os Keyes, who provided extra editorial feedback on a number of chapters. Many thanks to Giulia Lipparini and Sonke Adlung at Oxford University Press for their invaluable support throughout the publication process, and to the anonymous peer reviewers for their insightful comments and suggestions.

Finally, we would like to extend our heartfelt thanks to our research assistant, Youngcho Lee, whose work has been fundamental to the completion of this volume.

Contents

Contributor Affiliations

Garnett Achieng is a technology researcher based in East Africa. Garnett researches and consults on data and digital rights, digital feminist activism, technology facilitated VAW, online harms, platform governance, open source advocacy, and digital cultures.

Blaise Aguera y Arcas is a Principal Scientist at Google, where he leads a team working on machine intelligence for mobile devices. Blaise has worked on augmented reality, mapping, wearable computing, and natural user interfaces.

Sareeta Amrute is an Affiliate Associate Professor of Anthropology at the University of Washington who studies the production of race and class in US and EU programming economies, and that of financial good through data-centric technologies in India.

Noelani Arista is an Associate Professor in the History and Classical Studies Department and the Director of the Indigenous Studies Program at McGill University. Noelani's research applies artificial intelligence and machine learning to Hawaiian language sources.

Neda Atanasoski is an Associate Professor of feminist studies at the University of Maryland, College Park. Neda's research interests include cultural studies and critical theory; war and nationalism; gender, ethnicity, and religion.

Caroline Bassett is a Professor of Digital Humanities at the Faculty of English of the University of Cambridge. Caroline explores digital technologies in relation to questions of knowledge production and epistemology and cultural forms, practices, and ways of being.

Jude Browne is the Frankopan Director at the University of Cambridge Centre for Gender Studies, Head of Department of Politics and International Studies and a Fellow of King's College, University of Cambridge. Jude researches gender, political theories of equality; public policy, political responsibility, structural injustice and rights, and the impact of technology on society.

Stephen Cave is the Executive Director of Institute for Technology and Humanity, University of Cambridge and Senior Research Associate in the Faculty of Philosophy of the University of Cambridge. Stephen's research interests currently focus on the nature, portrayal, and governance of AI.

Chenai Chair is a digital policy and gender expert of My Data Rights Africa with extensive experience in work that is focused on understanding the impact of technology in society in order to better public interest in policy.

Sasha Costanza-Chock is a researcher and designer who works to support community-led processes that build shared power, dismantle the matrix of domination, and advance ecological survival.

Charlene Chu is an Assistant Professor at the Lawrence S. Bloomberg Faculty of Nursing at the University of Toronto, Charlene's primary area of research is in designing interventions that support the mobility and daily function of older adults in post-acute care settings.

Elena Falso is a PhD student at the Department of Science and Technology Studies at UCL and works as a research assistant at the Centre for Future Intelligence at the University of Cambridge.

Catherine D'Ignazio is an Assistant Professor of Urban Science and Planning in the Department of Urban Studies and Planning at MIT. Catherine is a scholar, artist/designer and hacker who focuses on feminist technology, data literacy, and civic engagement.

Neema Iyer is the founder of a civic technology company, Pollicy, based in Kampala, Uganda. She is an active do-er and speaker on data, digital rights and building products to improve service delivery. She is also an artist, both digitally and in the real world.

Lauren Klein is an associate professor in the Departments of English and Quantitative Theory & Methods at Emory University. Lauren research interests include digital humanities, data science, data studies, and early American literature.

Kanta Dihal is a Senior Research Fellow at the Leverhulme Centre for the Future of Intelligence. Kanta research interests include literature and science, science fiction, modernism, postmodernism, and postcolonial, decolonial, and critical race theory.

Eleanor Drage is a Christina Gaw Post-Doctoral Research Associate in Gender and Technology at the University of Cambridge Centre for Gender Studies. Eleanor researches feminist, anti-racist, post-humanist, and queer approaches to technological processes and systems.

Michele Elam is a Professor of Humanities in the English Department at Stanford University. Elam's research in interdisciplinary humanities connects literature and the social sciences in order to examine changing cultural interpretations of gender and race.

Federica Frabetti is a Principal Lecturer in Digital Media and Head of Partnerships at the University of Roehampton. Federica is interested in the culture and philosophy of technology, digital media, software studies, cultural theory, and gender and queer theory.

Lelia Marie Hampton is a PhD student in Computer Science at the MIT. Lelia's research interests are AI and ethics, AI safety, computational social science, computational sustainability, human-centred AI, and user experience of marginalised groups.

N. Katherine Hayles is Distinguished Research Professor of English at UCLA and the James B. Duke Professor of Literature Emerita at Duke University. Katherine teaches and writes on the relations of literature, science, and technology in the twentieth and twenty-first centuries.

Os Keyes is a PhD student at the University of Washington's Department of Human Centred Design & Engineering. Os studies gender, disability, technology, and/or power, and the role technology plays in constructing the world.

Suzanne Kite is a performance artist, visual artist, and composer raised in Southern California, with a BFA from CalArts in music composition, an MFA from Bard College's Milton Avery Graduate School, and is a PhD student at Concordia University.

Jason Edward Lewis is a Professor of Computation Arts, Design, and Computation Arts at Concordia University and a digital media artist, poet and software designer. Jason founded Obx Laboratory for Experimental Media and directs the Initiative for Indigenous Futures.

Kerry McInerney is a Christina Gaw Post-Doctoral Research Associate in Gender and Technology at the University of Cambridge Centre for Gender Studies. Kerry researches how histories of gendered and racialised violence manifest in new and emerging technologies.

Margaret Mitchell is an AI Research Scientist working on ethical AI, currently focused on the ins and outs of ethics-informed AI development in tech. Margaret previously worked at Google AI as a Staff Research Scientist, where she founded and co-led the Ethical AI group.

Rune Nyrup is a Senior Research Fellow on the Science, Value, and the Future of Intelligence project. At the CFI, Rune is exploring these issues in relation to artificial intelligence, both as a field of research and as a technology.

Archer Pechawis is an Assistant Professor of Visual Art & Art History at York University. Archer is a performance, theatre and new media artist, filmmaker, writer, curator, and educator.

Jennifer Rhee is an Associate Professor at the Virginia Commonwealth University. Jennifer's research focuses on literature, technology and science, contemporary American literature, media studies, robotics and artificial intelligence, and digital art.

Apolline Taillandier is a postdoctoral research associate at POLIS and the Leverhulme Centre for the Future of Intelligence at the University of Cambridge, and at the Center for Science and Thought at the University of Bonn.

Alexander Todorov is a Professor of Behavioral Science at the University of Chicago Booth School of Business. Alexander studies how people perceive, evaluate, and make sense of the social world.

Judy Wajcman is a Professor of Sociology at LSE and a Fellow at the Alan Turing Institute. Judy's scholarly interests encompass the sociology of work and employment, science and technology studies, gender theory, temporality, and organisational analysis.

Lauren Wilcox is the Director of the University of Cambridge Centre for Gender Studies. Lauren's research investigates the consequences for thinking about bodies and embodiment in the study of international practice of violence and security.

Erin Young is a postdoctoral research fellow at the Alan Turing Institute on the Women in Data Science and AI project, within the Public Policy programme.

Introduction: Feminist AI

Jude Browne, Stephen Cave, Eleanor Drage, and Kerry McInerney

Recent years have seen both an explosion in artificial intelligence (AI) systems and a corresponding rise in important critical analyses of such technologies. Central to this critical analysis has been feminist scholarship, which has held the AI sector accountable for designing and deploying AI in ways that further, rather than undermine, the pursuit of social justice. *Feminist AI* brings feminist theory, knowledge, methods, and epistemologies to interrogate how AI can exacerbate, reproduce, and sometimes challenge existing inequalities and relations of power. Our aim is to demonstrate how feminist scholarship provides essential insights into AI's social and political impact, ranging from the oppressive implications of facial recognition technologies through to political approaches to AI-generated harm.

This is the first anthology specifically focusing on feminist work in AI. It brings together new work with key texts from both established and emerging voices in the field. The critical conversation that ensues reflects long histories of feminist thinking on themes including science, technology, labour, capitalism, care, violence, representation, justice, and exclusion. The contributors to this volume are already playing a leading role in shaping contemporary debates around 'intelligent machines', and forging new directions in the field. The interplay between this volumes' chapters illuminates key feminist concerns such as: Can AI be feminist? Who makes technology and for whom? When and how should feminists resist new and emerging AI-powered technologies? How do the gendered effects of AI intersect with other forms of power and control, such as ethnicity, race, class, disability, sexuality, and age? How are feminist movements currently using data and for what ends? Can AI infer sexuality? What identities does AI produce? What would a queer technological future look like? How might we begin to look past the liability generated by AI to also consider its much wider structural dynamics? These questions simultaneously gesture towards the new challenges for feminist activism posed by emerging technologies as well as the problematic continuities feminists have faced across times and spaces.

This volume brings together scholars from a wide variety of disciplinary backgrounds, from computer science, software engineering, and medical sciences to sociology, political theory, anthropology, and literature. Contributors bring interdisciplinary feminist insights from disciplines including postcolonial studies, disability theory, and Black Marxist theory to confront ageism, racism, sexism, ableism, and class-based oppressions in AI. This collection also reflects the increasingly blurred divide between academia and industry when in conversation on the risks and possibilities that arise when adopting AI.

This book has emerged from the Gender and Technology Project at the University of Cambridge, a collaboration between the University of Cambridge Centre for Gender Studies and the Leverhulme Centre for the Future of Intelligence. While all four members of the editorial team specialise in the intersection of feminist theory and technology studies, we are cognizant of the limitations of our currently shared location and the partial perspective this necessarily grants us as editors, educators, and scholars. Three out of four of our editorial team identify as white, and all four of us are based in the Global North. Furthermore, many of our contributors are presently located in the Global North, even as they occupy different structural relations, identity categories, and positionalities within their respective institutions and societies. We invite readers to consider how and why, currently at least, feminist work in AI is heavily weighted towards these spaces. We do not aim for this text to be the definitive guide to feminist AI, but rather to begin to highlight the importance of feminist analyses in this arena.

This collection begins with a new essay by **N. Katherine Hayles**, one of the foundational thinkers in feminist STS (science and technology studies). In this essay, 'Technosymbiosis: Figuring (Out) Our Relations to AI', Hayles turns to creating a better metaphor with which to describe the relationship between AI and its environment. This is of utmost importance to the feminist agenda. As Donna Haraway has argued, we do not 'resort' to metaphor but engage in wor(l)ding practices through our re-figuring of key concepts. Hayles argues that the metaphor of 'technosymbiosis' is appropriate for this task, because it connects the existence of increasingly powerful computational media to environmental crises while combating anthropocentrist perspectives on agency and intelligence. She uses the concept to argue that meaning-making is not only a human activity, because machines also create and convey meaning in computational processes with or without a human 'in the loop'.

Decentering anthropocentrism in the way we think human and machine co-existence is also the foundational message of **Jason Edward Lewis, Noelani Arista, Archer Pechawis, and Suzanne Kite's** 'Making Kin with the Machines'. In fact, the authors argue that heterogeneous indigenous perspectives are the best way of approaching AI's 'circle of relationships' between human and non-human actors. Where AI design is often 'human-centred', this essay radically refuses the idea of a separate human species whose priorities trump that of its environment. Instead, it emphasises reciprocal and respectful interspecies relationships. In doing so, it holds feminism to its pledge to contest extractive capitalist logic by re-framing AI as 'Āina', as land but not territory, as that which sustains us. Where Western origin stories for technology often draw on Prometheus, the authors demonstrate how tales from the Pacific Ocean such as 'Hāloa' can form a more ethical 'operating code' for *maoli* (human)–AI relations. The essay contests the industry's emphasis on unfettered growth and scalability by urging for 'good growth' based on principles from Kānaka Maoli governance.

Apolline Taillandier's essay 'AI in a Different Voice: Rethinking Computers, Learning, and Gender Difference at MIT in the 1980s' demonstrates how there have been other attempts in the past to create AI-powered systems that reimagine AI through a feminist lens. Taillandier documents the creation of the programming language 'Logo', developed in the 1960 as an alternative way of teaching concepts in mathematics and computer science. Significantly, when designing Logo, programmers turned to feminist knowledge to rethink crucial aspects of AI. They redefined the desirable features of computer culture and the ideal profile of the computer scientist; they overcame divisions between the humanities and the sciences and between sensory and intellectual exploration; and they emphasised that computer languages and coding culture are always political: they are ways of gendering the world. However, as Taillandier notes, Logo did not catch on. From a feminist historical perspective, she argues, this can be attributed to its resistance to masculinised computer culture. Regardless, Taillandier makes clear the value of remembering these radical developments in computer science, inspiring future ways of recasting the ambitions and significance of AI.

As Taillandier makes clear, systems can vary substantially depending on who builds them and what their methods and goals are. This is why the expert analysis and study of the AI workforce by **Judy Wajcman and Erin Young's**, 'Feminism Confronts AI: The Gender Relations of Digitalisation', is so crucial for directing further action. This essay locates and unveils workforce injustices in key figures and analyses, breaking down the reasons why women are still underrepresented. Wajcman and Young offer nuanced insight into how the digital gender divide manifests globally, not only in relation to access to technology but in learning and skills. They turn their attention to gendered divisions in different tech roles, how gender has shaped tech development, the history of feminist responses to digitalisation, and the origins and evolution of workforce divisions. They give new insight into the gender-specific impacts of future automation that are grounded in an understanding of current gendered divisions in the labour market. Ultimately, they argue that diversifying the workforce is not merely a matter of increasing the *number* of women, but of achieving an upheaval in the culture and politics of tech companies.

The theme of tech culture is picked up in the next chapter by **Stephen Cave, Kanta Dihal, Eleanor Drage, and Kerry McInerney**: 'Shuri in the Sea of Dudes: The Cultural Construction of the AI Engineer in Popular Film, 1920–2020' actively aims to uncover and challenge the predominant portrayal of the AI engineer as male. Taking as their starting point the aforementioned underrepresentation of women in the AI industry, they outline why cultural representations of AI scientists matter, before explaining the findings of their large quantitative study of a corpus of 142 influential films containing AI. They discuss four key tropes that emerged from these films that are distinctly gendered and may contribute to the underrepresentation of women in the AI industry: first, the AI scientist as a 'genius'; second, the AI scientist's ability to gain mastery over life and death through the pursuit of 'mind uploading' and other forms of technological immortality; third, the association of AI with hypermasculine milieus, such as the military; and finally, the portrayal of female AI scientists as subservient or inferior to male AI scientists.

Themes of hypermasculinity are continued in **Lauren Wilcox's** chapter on Lethal Autonomous Weapons (LAWs), 'No Humans in the Loop: Killer Robots, Race, and AI' considers the dominant discourses around so-called 'killer robots', focusing specifically on the US military's policy of the 'human-in-the-loop'. By insisting that any autonomous weapon must include a human controller in any decision-making process, the US military claims that the potential harmful effects of LAWs can be effectively controlled and mitigated. However, drawing on the Black and decolonial feminist thought of Denise Ferreira da Silva and Sylvia Wynter, Wilcox demonstrates how the human-in-the-loop policy relies on a particular configuration of the 'human' that is always already undergirded by the power relations of gender, coloniality, and race. By framing the human *itself* as a technology and using drone warfare as an example, Wilcox demonstrates how histories of racial violence and colonial control continue to influence who counts as human, and which bodies are considered worthy of protection.

Wilcox's chapter also clearly shows how international militarism that occurs 'abroad' is fundamentally co-dependent with the deployment of militaristic violence 'at home'. Thus, **Kerry McInerney's** chapter on predictive policing technologies, 'Coding "Carnal Knowledge" into Carceral Systems: A Feminist Abolitionist Approach to Predictive Policing' demonstrates the inherently gendered and racialised nature of predictive policing technologies through an interrogation of the way that domestic, sexual, and gender-based forms of violence are represented in predictive policing tools and in the discourse around predictive policing. It reveals how predictive policing technologies not only reproduce existing patriarchal approaches to gender-based violence, but also possess the potential to shape and produce how gender-based violence is understood, identified, policed, and prosecuted. McInerney argues that since prisons and other carceral institutions are arenas of state-sanctioned sexual violence, these tools cannot be successfully employed for feminist ends.

Violent state structures continue to be a central theme in **Lelia Marie Hampton's** chapter 'Techno Racial Capitalism: A Decolonial Black Feminist Marxist Perspective'. Hampton uses the political thought of Davis and Jones as a foundation for their incisive analysis of how the modern technology industry is rooted in the exploitation of people of colour across the globe. Hampton argues that while Western tech companies pillage the world's wealth at the expense of racially oppressed groups, these groups remain largely poor with no access to capital, including the non-consensual data capital they produce. Drawing on Black feminist studies, Black Marxist studies, (de)colonial studies, critical data studies, STS studies, and the emerging digital colonialism literature, Hampton offers the term 'techno racial capitalism' to conceptualise how technology reproduces racial capitalist structures. They analyse how the exploitation of racialised people worldwide underpins AI and its multiple technical foundations, from hardware and software to data and AI

The racialised dynamics of labour exploitation in the AI industry similarly shape **Neda Atanasoski's** chapter 'Feminist Technofutures: Contesting the Ethics and Politics of Sex Robots and AI'. Atanasoski projects Hampton's incisive analysis into the future, examining how the ideology of technoliberalism—how difference is organised via technology's management and use of categories of race, sex, gender, and sexuality within a *fantasy* of a technologically enabled 'post-racial' future that is not only never in fact post-racial, but that is also always put to use for capitalist ends—shapes and underpins the discourse around sex robots. In her chapter, Atanasoski problematises two different approaches to sex robots. The first, a liberal feminist approach, characterises sex robots as a threat to women's rights, and in doing so, roots itself in technology's relationship to the figure of the human within the juridical realm of rights. The second, a feminist approach centred on diversity, makes use of and commodifies proliferating categories of human difference and life itself by calling for the 'diversication' of sex robots beyond white, Eurocentric beauty norms. Atanasoski thus considers how diversified technologies still uphold rather than disrupt racialised notions of use and value, commodifying those differences seen as profitable. She concludes by considering what a queer approach to technological speculative futures might look like through an analysis of the art project 'Lauren AI'.

Racialised and gendered labour is central to **Jennifer Rhee's** chapter, 'From ELIZA to Alexa: Automated Care Labour and the Otherwise of Radical Care', which examines contemporary AI assistants, such as Siri and Alexa, in the context of AI developed to replicate aspects of care labour. Rhee traces the history of caring AI technologies to explicate how these technologies replicate historic gendered and racialised dynamics associated with care labour. Taking Turing's influential writing as a starting point, Rhee examines the ongoing association between AI, care labour, and care labour's gendered and racialised histories. Through her discussion of AI assistants, Rhee argues that it is not care labour that is foundational to AI and its inscriptions of humanness, but the *devaluation* of care labour and of those who are associated with the important work of caring for others. Rhee concludes with an analysis of artistic works that explore the possibilities that emerge from acknowledging the centrality of care labour's devaluation to the history and present of AI.

In her imagining of the 'otherwise' of care, Rhee's chapter is aligned with **Sareeta Amrute's** chapter, 'Of Techno-Ethics and Techno-Affects', which argues that techno-ethics can and should be revitalised through attendance to techno-*affects*. As scandals over predictive policing, data mining, and algorithmic racism unfold, digital labourers need both to be accounted for in analyses of algorithmic technologies and to be counted among the designers of these platforms. Amrute's chapter attempts to do both of these by highlighting particular cases in which digital labour frames embodied subjects, and to propose ways digital workers might train themselves to recognise ethical problems as they are emerging, and ultimately uses the idea of attunements as a way to grasp what these forms of care might look like for the digital worker.

In their chapter 'The False Binary of Reason and Emotion in Data Visualisation', **Catherine D'Ignazio** and **Lauren Klein,** propel forwards existing feminist scholarship on embodiment and AI by calling for the incorporation of emotions and embodied experiences into data visualisation practices. They ask: how did the field of data communication arrive at conventions that prioritise rationality and devalue emotion, and completely ignore the non-seeing organs in the human body? Who is excluded when only vision is included? Ignazio and Klein break down the established consensus that data visualisation should conform to the adage of 'just the facts' and 'clarity without persuasion'. Instead, they argue that data selection and presentation is always political, and should be visualised in such a way as makes this evident. Ultimately, D'Ignazio and Klein argue that rebalancing emotion and reason enlarges the data communication toolbox and allows one to focus on what truly matters in a data design process: honouring context, gathering attention, and taking action in service of rebalancing social and political power.

Blaise Agüera y Arcas, Margaret Mitchell and Alexander Todorov's essay 'Physiognomy in the Age of AI' similarly examines how power shapes scientific methods through their analysis of how historical forms of pseudoscience, such as physiognomy and phrenology, are reproduced in AI-enabled facial recognition. The scientifically racist misuse of machine learning is illustrated through analyses of two examples. In the first case, Agüera y Arcas, Mitchell, and Todorov trace the history of pseudoscientific practices that attempted to discern criminality from the face, and examine how Xiaolin Wu and Xi Zhang's DNN (deep neural network) similarly relies on erroneous and essentialist ideas about the 'criminal face'. They then turn to the second example, showing how Michal Kosinski and Yilun Wang's DNN does not reveal people's sexual orientations, but rather exposes social signals and stereotypes about gender identity and sexuality. Agüera y Arcas, Mitchell, and Todorov conclude by noting that while powerful new machine learning methods provide a valuable window into gendered stereotypes and biases, using them to infer individual and presumed essential traits such as criminality and sexuality is a pseudoscientific practice, and can perpetuate forms of injustice under the guise of scientific objectivity.

Agüera y Arcas, Mitchell, and Todorov's important critique of how sexuality is not an objective, biological trait that can be 'read' or 'discerned' from the face demonstrates how feminist scholarship problematises the perceived fixity of biological categories. In the field of AI, 'gender', 'race', 'ethnicity', and 'sexuality' are too often approached as static, programmable variables that must be incorporated into AI and ML systems through the use of more 'representative' dataset. This assumption is challenged by **Michele Elam** in her chapter 'Signs Taken for Wonders: AI, Art and the Matter of Race'. Elam argues that overlooked in most current debates over gender and racial bias, surveillance, and privacy in facial recognition technology is the already institutionalised practice of re-coding 'gender', 'race', and 'ethnicity' as something written on the face or intelligible on/in the body. She argues that art is playing a crucial role in calling out so-called identification and 'personalisation' software, demarcating AI's problems as cultural challenges, not just technical ones. Her analysis of artist Rashad Newsome's disobedient, Black, queer, vogue-dancing humanoids, *Being* and *Being 2.0*, explores art's ability to historicise the technological practices that create race. AI need not be constituted by its distance from particular disciplines (the humanities) and demographics (Black, female). Art, Elam reminds us, has historically been devalued when progress is defined by new technologies of force and speed. Scientists need some awareness of the importance of artistic practice and the humanities in the development and maintenance of AI, particularly its ability to show how racial categories performatively make data intelligible.

Elam's contestation of the supposedly utopian 'post-racial' futures offered by certain AI products and technologies leads us to **Caroline Bassett's** analysis of the temporal horizon of societal investments in AI. In 'The Cruel Optimism of Technological Dreams', Bassett argues that AI is destined to disappoint and destabilise because it fosters what Lauren Berlant called 'cruel optimism', where the very thing that promises to make your life better is in fact the obstacle to happiness. Bassett deconstructs AI into a 'cluster' of partial promises that are only executed on behalf of the select few. Drawing on Berlant, Bassett views societal desires and attachments to AI as giving meaning, sense, and continuity to human endeavour, and therefore as both necessary and harmful. Bassett compels us to consider how transhumanists and substantial media reporting present the possibility of artificial general intelligence as the final stage in orthogenetic evolution. These broader aspirations are not merely communicated to the public, but sold to us as 'new', as the threshold of an unrealisable future kept on hold by a bugging refresh button.

Eleanor Drage and Federica Frabetti argue that computer code uses data to not only make statements about the world but also bring that world into existence. Drawing on Judith Butler's concept of performativity, they explain why we must view AI as performative in order to understand how it genders and racialises populations even when it appears to be 'unbiased' or correctly functioning. In its reading of neural networks, 'AI that Matters: A Feminist Approach to the Study of Intelligent Machines' demonstrates that Facial Detection and Recognition Technologies (FDTR) and Automatic Gender Recognition (AGR) never objectively identify or recognise a person (or their gender) as they claim. Instead, they argue that these technologies merely comment on and annotate a person's body in accordance with dominant social rules and perspectives. They present this framing as an intervention into misguided attempts to treat discrimination as an error that can be corrected by a better functioning machine.

Theories of performativity can also expose, as **Os Keyes** argues, how AI that attempts to diagnose autism instead performatively produces it as a negative abnormality. 'Automating Autism' argues that the use of contemporary AI technologies to 'identify' or 'solve' the 'problem' of autism reproduces the denial of autistic personhood. As Keyes argues, the question of who is perceived to have self-knowledge and who has that ability denied to them has been of utmost concern to feminist movements. With the application of AI to these tasks, we must once again fight for the marginalised to have their authority over the recognition of their own capabilities and the definition of their own identities respected. AI's ability to reinforce an unfounded 'truth' of disability is therefore an urgent area of feminist intervention.

Pieces such as 'Automating Autism' have laid the groundwork for algorithmic bias literature that documents AI-driven discrimination based on disability. Drawing on these interventions, **Rune Nyrup, Carlene Chu, and Elena Falco** address an often overlooked axis of oppression in AI: ageism. 'Digital Ageism, Algorithmic Bias, and Feminist Critical Theory' attempts to rectify this oversight in the AI ethics literature by situating what they term *digital ageism* within systemic ageism more broadly. In doing so, they offer a pointed analysis of the interplay between social inequality and tech development, aligning with feminist work which views society and technology as co-constitutive. The essay details the results of encoded ageism, including medical technologies that offer less accurate diagnoses on older populations, the unequal division of resources, and the perspective that younger people are inevitably better at using new technologies. Drawing on the work of Sally Haslanger and Iris Marion Young, they explore how technical limitations and the insufficient representation of older people in design and development teams are shaped by self-reinforcing structural inequality. This essay therefore offers an important intervention in the debate by identifying and tracking ageist harms and their intersections with disability, race, gender, and class-based injustices.

The emphasis on exploring AI's structural harms through the scholarship of renowned feminist scholar Iris Marion Young is shared by **Jude Browne**. In 'AI and Structural Injustice: A Feminist Perspective', Browne explores the potential of AI to exacerbate structural injustice, and reflects on the shortcomings of current political and regulatory approaches. She begins by introducing Iris Marion Young's seminal account of structural injustice and explores what Browne argues is one of its definitive elements – untraceability. Browne suggests that drawing a parallel between the untraceability of structural injustice and the increasing untraceability of algorithmic decision making, is productive for thinking about the potential of AI to exacerbate structural injustice. She concludes by offering some suggestions on how we might think about mitigating the structural injustice that AI poses through democratic governance mechanisms. By way of example, she advocates adopting several elements of the mini-public approach within regulatory public-body landscapes to form a new pluralistic lay-centric 'AI Public Body'.

The transformative social and political impact of AI—and in particular, its capacity to exacerbate structural injustice—is a central concern for feminist research across the globe. To this end, **Neema Iyer, Garnett Achieng, and Chenai Chair's** chapter 'Afrofeminist Data Futures', conducted for the Ugandan-based collective Pollicy, also offers ways forward out of AI-exacerbated structural injustice in African contexts. This overview of many years of research, which aims to improve government service-delivery, locates injustices in specific areas of AI development and deployment across Africa. It provides qualitative and quantitative evidence of its claims, as well as recommendations for policy and governance responses. As the authors explain, data collection in African contexts is both the poison and the remedy: colonialist data practices extract data from African regions for the benefit of Western tech companies, and datafication poses significant risks to women and girls' privacy. However, data—especially Big Data—can improve NGO and governmental responses to issues affecting African women by providing insight into elements of their lives that are difficult to capture through other data collection modalities. This is why, they argue, it is so important to work with feminist methods and theory when creating, processing and interpreting data. Like Browne, they advocate for data-related decisions to be in the hands of non-partisan, independent and citizen-led collectives.

These themes are echoed again in **Sasha Costanza-Chock**'s 'Design Practices: Nothing About Us Without Us'. This essay also rejects the simulation of different user groups, arguing that nothing can stand in for inclusion. It therefore hones in on the paradox of 'user-centred design' (UCD), a dominant design practice which prioritises 'real-world' users but excludes those who currently have no access to the technology in question. Which potential user groups are catered for and which are not is a political choice that often goes unnoticed in tech development. They ask that more attention is paid to the demographic from which the 'unmarked' generic user persona derives, and how it matches up with the average tech worker: white, cis-male, internet access, and digitally literate. As a remedy, they point to participatory practices and asks that keener attention is paid to how 'users' themselves innovate products. Costanza-Chock's rallying call for 'design justice' paved the way in the AI sector for work that did not merely attempt to neutralise harmful systems but called for their active re-orientation towards redress. This remains a go-to text for practitioners and academics alike when drawing up ethical design processes. Taken together, these twenty-one chapters show the immense richness and diversity of feminist thinking on AI, while simultaneously gesturing towards many future avenues for inquiry and investigation.

1

Technosymbiosis

Figuring (Out) Our Relations to AI

N. Katherine Hayles

Most feminists understand very well that metaphors matter. Feminist movements as a whole have struggled to overcome sexist language and catalyse metaphors that empower women and others. As Lakoff and Johnson (2003) have shown, metaphors are pervasive in everyday language, helping us understand, describe, and value the experiences we encounter, from basic ideas such as 'in/out' and 'container/contained' to more subtle notions of value coded as right/left and high/low. Scientific and technological contexts are also permeated by metaphors. Consider for example the expression 'dark energy', which uses the lack of vision to imply a lack of knowledge. The idea that information 'flows' relies on a metaphoric sense of liquid gliding down a river; a 'charge' of energy evokes the notion of charging horses in a battle. And so on, *ad infinitum*.

Too often, those who create the scientific theories and fabricate the technologies have (consciously or unconsciously) embraced what I have elsewhere called the 'giftwrap' idea of language. I have an idea, I wrap it up in language and hand it to you, you unwrap it and take out the idea, which we now share. This view erases the power of metaphors and figurative language not just to express thought but to create and facilitate it. Metaphors channel thought in specific directions, enhancing the flow of compatible ideas while at the same time making others harder to conceptualise. At least in part, metaphors influence not just what we think but also how we think.

Given the pervasiveness and importance of metaphors, they warrant close attention for how they shape our attitudes towards the nascent field of artificial intelligence (AI). At this early stage, it is critically important that our metaphors not only serve oppositional purposes (to which the large majority of chapters in this volume are devoted), but that they also empower feminist theorists, engineers, critics, activists, and others to intervene constructively in the development of artificial intelligence. As Judy Wajcman and Erin Young argue in this volume, it is scarcely a secret that the field of AI is currently dominated by men. While they make it clear that there is no simplistic one-to-one correspondence between the gender of those creating the technologies and the biases they craft into it, it only makes sense that a more diverse labour force that includes more women, people of colour, disabled

N. Katherine Hayles, *Technosymbiosis*. In: *Feminist AI*. Edited by: Jude Browne, Stephen Cave, Eleanor Drage, and Kerry McInerney, Oxford University Press. © Oxford University Press (2023). DOI: 10.1093/oso/9780192889898.003.0001

people, and other non-normative folks, would create technologies more affirmative of differences and more alert to various forms of discrimination. Our ability
to attract such creators to AI is undercut if the metaphors we use to describe the
technologies convey only negative images and send discouraging messages about
the uses and potentials of the technologies.

The aim of this chapter is to describe and evaluate metaphors that can figure
(that is, create figurative language for) constructive interventions with AI in
particular and computational media in general. Candidates include the cyborg,
addressed here through Donna Haraway's incandescent 1985 essay 'A Cyborg
Manifesto'; her more recent proposal for 'making kin', urging them as facilitating about different pathways for feminist engagements; man–computer symbiosis,
proposed by J. C. R. Licklider in his historical 1960 article; and finally technosymbiosis, a more advanced and less sexist notion of symbiosis that includes
computational media as actors with their own agency, not merely as the facilitator
of human ideas, as Licklider imagined. My hope is that the arguments presented
in favour of technosymbiosis as a metaphor will enable feminists to engage with
artificial intelligence not only as critics of the technology, important as that is,
but also as makers, collaborators, and contributors who work to fashion artificial
intelligence as a field imbued with feminist goals and aspirations.

The Cyborg: An Ironic Metaphor

Re-reading Haraway's famous essay nearly four decades after its first publication,
I am struck afresh by her depth of insight, scope of analysis, and perhaps most of
all, her superb strategic sense of the critical issues that feminists at the time needed
to address. Well aware of the history of the cyborg as a creature of the military–
industrial complex, Haraway sought to re-position it in ways that cut through
contemporary feminist rhetorics that she thought had become limiting rather than
liberating. In the essay's first section, entitled 'An Ironic Dream', she refers to irony
as the rhetorical figure that holds two contradictory facts in mind at once, both of
which may be true. In effect, by acknowledging the cyborg's unsavoury connection with the military–industrial complex, she sought to turn its non-innocence
into an asset rather than (only) a liability. Among the standpoints she criticised
were the notions that women are the innocent victims of male oppression; that
'women' have an intrinsic nature allying them with the natural world rather than
with, say, military adventures; that a unified vision of 'women' could be articulated
that would apply to all women everywhere.

She showed that adopting the cyborg as a metaphor would demolish these
ideas. This, however, was only part of her project. Equally important was her
prescient analysis of the transition from industrial capitalism, long the object of
critique for Marxist-feminists and socialist-feminists, to what we would today call

the information society, which she identified as the 'informatics of domination'. It would be more than a decade before Manuel Castells began publishing his trilogy on the networked society; Haraway not only beat him to the punch but showed how essential it was for feminists to lay claim to the new and critically important issues that would emerge as a result of the transformations she traced.

'A Cyborg Manifesto' marks the highlight of Haraway's engagement with information technologies. After it, her attention turned increasingly toward the coyote trickster (Haraway 1991), companion species (2003), and 'making kin' (2016). I can only guess why this change of direction took place, but it strikes me that her misgivings about the cyborg may be similar to my own. With some notable exceptions, feminists did not take up Haraway's call to identify with the cyborg. Writing twenty years after the cyborg essay's publication, Malini Johar Schueller articulates some of the reservations of its critics (Schueller 2005). She gives Haraway credit for including race in her analysis and for citing work by 'women of color' (p.78). But she also notes that Haraway creates an analogy between the cyborg and such writing, positioning both as oppositional stances. Schueller argues that the analogy has the effect of erasing the specificities of writings about race, robbing them of the very particularities that motivates them and from which they derive their power. Her point targets work beyond Haraway's [she also critiques Gayle Rubin's 'Thinking Sex: Notes for a Radical Theory of the Politics of Sexuality' (1984) on a similar basis], but her analysis effectively shows how foregrounding the cyborg has the (presumably unintended) effect of making race a subsidiary concern.

Moreover, the cyborg figure not only continued but strengthened its ties with the military–industrial complex. The US military was not slow to imagine the twenty-first century warrior as a technologically modified human with sensing and action capabilities expanded through computational means. The cyborg also began to be an (unhealthy) object of fascination for some scientists such as Kevin Warwick (1998), who has had various well-publicised implants that have made him, he claims, a literal cyborg. In addition, transhumanists such as Ray Kurzweil (2006) have advocated the cyborg as a model of our human future, with a body made virtually immortal by the replacement of failing parts with artificial organs, three-dimensional printed circuits replacing blood vessels, and so forth. In opposition to these kinds of visions, Haraway has recently proclaimed that 'I am a compost-ist, not a posthuman-ist; we are all compost, not posthuman' (2015, p.161).

There is another aspect to the cyborg metaphor that to my mind makes it particularly unsuitable for AI interventions. With the focus on the human body, it reinforces an anthropocentric orientation that takes the human as the centre, along with accompanying practices such as the domination by humans of all other species and the relentless exploitation of the earth's resources. Although the cyborg metaphor acknowledges the power of technological interventions, it fails to address or provide resources for thinking about the increasing agency and autonomy of intelligent systems in the new millennium. Finally, it provides no way to connect our increasingly powerful computational media with our multiple environmental crises, either as contributing factors or as possible ameliorating forces.

Making Kin: A Perilous Metaphor

In her recent book *Staying with the Trouble*: *Making Kin in the Chthulucene*, Haraway offers another metaphor that she regards as more suitable for our present situation: 'making kin'. In a shorter piece, she comments,

> Making kin is perhaps the hardest and most urgent part. Feminists of our time have been leaders in unravelling the supposed natural necessity of ties between sex and gender, race and sex, race and nation, class and race, gender and morphology, sex and reproduction, and reproduction and composing persons. . . If there is to be multispecies ecojustice, which can also embrace diverse human people it is high time that feminists exercise leadership in imagination, theory, and action to unravel the ties of both genealogy and kin, and kin and species (2015, p.161).

Endorsing the slogan 'make kin, not babies', she suggests that decreasing the human population of the earth should be accompanied by an effort to embrace as our kin other species, from fungi to mega-fauna (2015, p.164). 'My purpose', she clarifies, 'is to make "kin" mean something more/other than entities tied by ancestry or genealogy' (2015, p.161).

She is right to make kin a major issue. Kinship comes as close as anything to qualifying as a human universal. Although various societies define kinship differently, almost all use kinship to structure social relations, cement alliances, and create genealogies. There are good reasons for this universality; for most of human history, kinship has been essential to the survival of human individuals and groups. But the very attributes that give kinship its survival advantage also carry with them a strong streak of xenophobia. Like the cyborg, kin is far from an innocent notion. About the time Haraway was writing 'The Cyborg Manifesto', Richard Rorty succinctly summarised the dangerous appeal of kinship, arguing that 'Most people live in a world in which it would be just too risky—indeed, would often be insanely dangerous—to let one's sense of moral community stretch beyond one's family, clan, or tribe' (Rorty 1998, p.125). Rorty's point is that the other side of kinship is the exclusion of stigmatised others, whoever these are imagined to be. The construction of kin requires non-kin, which can easily slide into racism and even, for example, so-called ethnic cleansing. Of course, Haraway wants nothing to do with xenophobia and its hateful offspring. That is why she calls for feminists to 'unravel' the ties between kin and genealogy (Haraway 2015, p.161). In effect, she wants to appropriate the closeness and intensity of kinship relations without incurring the damage, directing it instead toward a more biophilic identification with all living creatures (or 'critters', as she would say). As a strategy, this is a long shot. A few may be convinced to 'make kin, not babies', but children are so pervasively desired in human societies that it is difficult to imagine many taking Haraway up on her proposal.

In addition, the metaphor of 'making kin' has little, if anything, to contribute to feminist interventions with AI. Focused exclusively on biological organisms, 'making kin' makes little sense when applied to intelligent artificial systems. Hans Moravec (1990) has suggested that we should regard the robots and intelligent systems that we create as our true evolutionary heirs, our 'mind children', but as a metaphor, this idea presents even more problems than 'making kin'. No doubt Haraway would see Moravec's metaphor as a perverse inversion of her notion, since it directs the emotional intensity she strives to capture for nonhuman species toward technological inventions. In my view, both metaphors are inadequate to deal with the complexities of the computational systems that have become pervasive in the Global North.

A metaphor closely related to 'making kin' is Haraway's sympoiesis. 'Sympoiesis', she writes, 'is a word proper to complex, dynamic, responsive, situated, historical systems. It is a word for worlding-with, in company. Sympoiesis enfolds autopoiesis and generatively unfurls and extends it' (2016, p.58). She explains that 'poiesis is symchthonic, sympoietic, always partnered all the way down, with no starting and subsequently interacting "units"' (2016, p.33). In this view, organisms do not precede the relations they enter into with one another but reciprocally produce one another from lifeforms that for their part have already emerged from earlier involutions.

This is an attractive vision that gestures towards the entanglement of living creatures with one another, bound together in reciprocal becomings through their entwined evolutionary histories. The limitation, of course, is that it is difficult to imagine how sympoiesis would apply to artificial intelligence systems, since these have no aeons-long evolutionary history through which such entanglements might emerge. Would it be possible to find a metaphor that expressed mutual reciprocity, not between biological organisms, but between humans and computers? That possibility is explored in the next section on computer–human symbiosis.

Human–Computer Symbiosis: A Historical Metaphor

In not quite the dawn but at least the morning of the computer age, Joseph Carl Robnett Licklider (also known as J. C. R. or simply 'Lick') published a prescient essay entitled 'Man-Computer Symbiosis'. Apologists for sexist language argue that such uses of 'man', entirely conventional in the 1960s, really meant 'all people'. However, in this case the gender-specific word is accurate. The nascent field of computer technologies in 1960 was even more heavily dominated by men than it is at present.[1]

His insensitivity to sexist language notwithstanding, today Licklider is widely viewed as someone with a remarkable track record of anticipating and predicting future developments in computer technologies. He foresaw interactive computing and the enormous difference it would make for increasing the speed and efficiency of human–computer interactions; he envisioned a worldwide computer network and actively worked to make it a reality by directing early funding toward the founding of ARPANET, a direct ancestor of the internet. He understood the potential of time sharing for computers, conducting the first public demonstration of it around the time he published the 'Man-Computer Symbiosis' article. He predicted the need for virtual conferencing, writing in a 1968 co-authored paper that 'there has to be some way of facilitating communication about people [without] bringing them together in one place' (Licklider and Taylor 1968). Recognised through multiple prestigious awards, his research was extremely broad-based, encompassing psychoacoustics, electrical engineering, and managing information sciences, systems, and applications at IBM's Thomas J. Watson Research Center.

In 'Man-Computer Symbiosis', he imagines trying to direct a battle using computers as they existed in 1960.

You formulate your problem today. Tomorrow you spend with a programmer. Next week the computer devotes 5 minutes to assembling your program and 47 seconds to calculating the answer to your problem. You get a sheet of paper 20 feet long, full of numbers that, instead of providing a final solution, only suggest a tactic that should be explored by simulation. Obviously, the battle would be over before the second step in its planning was begun (1960, pp.3–4).[2]

---

[1] As late as 1993, during a fellowship year at Stanford University, I attended a lecture on robotics by Rodney Brooks. Waiting for the talk to begin, I amused myself by counting the women present. In an audience of over 400, there were precisely 24 women, including me—a ratio of about 6 percent. Relevant here is Mar Hicks' study of computer programming in Britain from the 1940s to the 1960s (Hicks 2017). She documents that in the 1940s programming was largely regarded as 'women's work' [witness the contemporary usage of calling the women who did such work 'computers ', which I referenced in the title of my book on computer codes (Hayles 2005)]. During the 1960s and 1970s, as the field gained importance, the workforce was largely taken over by men. As she points out, this gender flip cannot be attributed to programming becoming more complex; indeed, the reverse is the case, as the history of programming is replete with practises designed to make coding easier for humans to understand and to create more efficiently. She documents how government officials in the mid- to late-1960s 'endeavored to migrate computing posts from being white-collar, subclerical jobs stuck at the bottom of the Civil Service professional framework to being administrative and executive positions at the top of the service' (Kindle location 4072)—a strategy that, given the sexism of the time, was almost guaranteed to favour men over women.

[2] This description may sound droll to the present generation, but as someone who learned to program in the late 1960s, I can testify to the frustration of submitting my stack of IBM cards, waiting a

Licklider foresaw that a true human–computer symbiosis would require much faster processing time and an interactive interface, which he also predicted.

Given his era, he was remarkably astute in outlining the criteria for a human–computer symbiosis. He noted that the concept was distinctly different from systems that only assisted humans. 'These systems certainly did not consist of "dissimilar organisms living together"' (the definition he cites for symbiosis). 'There was only one kind of organism—man—and the rest was there only to help him' (1960, p.2 pdf). But by 1960, automation had proceeded far enough so that 'the men who remain are there more to help than to be helped', with human operators consigned to performing functions that were not (yet) feasible to automate. These, he concludes, were still not symbiotic systems but '"semi-automatic" systems, systems that started out to be fully automatic but fell short of the goal' (p.2 pdf).

In a true symbiotic system, he argues, computers will be used to aid thinking in 'real-time'. He gives the example of graphing data to determine the relation between variables. As soon as the graph was constructed, the answer became obvious—but he needed the graph to see the relationship. Taking himself as subject, he did a time-efficiency study and determined that 'the main thing I did was to keep records... About 85 percent of my "thinking" time was spent getting into a position to think, to make a decision, to learn something I needed to know. Much more time went into finding or obtaining information than into digesting it' (p.4, pdf). It was precisely these kinds of preparatory tasks that he envisioned the computer taking on. If a 'fast information-retrieval and data-processing machine' can be invented, then 'it seems evident that the cooperative interaction would greatly improve the thinking process' (p.5, pdf).

His summary of how he imagines a symbiotic relation working indicates that he still sees humans as the initiators and the ones who formulate the hypotheses that the computers will test. 'Men will set the goals and supply the motivations, of course, at least in the early years', he wrote. 'The information-processing equipment, for its part, will convert hypotheses into testable models and then test the models against data... the equipment will answer questions. It will simulate the mechanisms and models, carry out the procedures, and display the results to the operator' (p.7, pdf). Recognising that there was a 'mismatch' between the pace at which a human thinks and a computer program processes its algorithms, he further anticipated that a computer 'must divide its time among many users', thereby envisioning the necessity for time sharing. Moreover, he imagined a 'network of such centers, connected to one another by wide-band communication lines and to individual users by leased-wire services', articulating a vision of the internet decades before it came into existence (p.7, pdf).

---

day or two for the printout, only to find an error in syntax that prevented the program from compiling. Another day to find the error and make the correction, and the process started over again from scratch.

Sixty years later, we can appreciate the prescience of his vision, at the same time noting how much further along we are from his scenarios. [He also anticipated this development, writing that 'man-computer symbiosis is probably not the ultimate paradigm for complex technological systems. It seems entirely possible that, in due course, electronic or chemical "machines" will outdo the human brain in most of the functions we now consider exclusively within its province' (p.3, pdf). With contemporary AI systems, computers do far more than construct models to test hypotheses. They discern subtle patterns in very large datasets; they perform correlations that, in the view of some, make causal reasoning unnecessary; supplemented by sensing systems of many kinds, they not only create the data on which decisions are made but make the decisions themselves. To mark the transition from what Licklider imagined and where we are today, I propose a final metaphor: *technosymbiosis*.]

Technosymbiosis: A Visionary Metaphor

As a metaphor, technosymbiosis gestures towards an alternative worldview with significant implications for our relations with humans, nonhumans, and computational media, including AI (see also Lewis, Arista, Pechawis, and Kite, Chapter 2 in this volume). It begins by dismantling the liberal-juridical post-Enlightenment subject also critiqued by Neda Atanasoski in this volume (Chapter 9). Rather than presume a rational individual with free will and autonomy, it starts from a more modest and foundational perspective: all organisms, including humans, receive information from their environments, interpret that information, and respond with behaviours appropriate to their contexts. My name for this process is cognition. It follows that all biological organisms have cognitive capabilities, including those without brains or central nervous systems such as plants (Hayles 2021).

Moreover, the field of biosemiotics (the science of signs as they apply to nonhuman organisms) extends this conclusion to the creation, interpretation, and dissemination of signs. Working from C. S. Peirce's triadic view of the sign as composed of a sign vehicle (which he calls the representamen), the object, and an intervening process that he calls the interpretant (Peirce 1998, Vol. 2, p.478) biosemioticians consider that when an organism responds to an environmental cue (the representamen) in ways conditioned by its evolutionary history (the interpretant), this response counts as creating meaning relevant to the organism's milieu. Meaning, in this view, does not require (human) language or symbolic manipulation; rather, a behaviour itself constitutes meaning, generated by the organism's ability to connect to the environment through receiving and interpreting information. This implies, of course, that meanings are species-specific, relative to an organism's cognitive capacities and interactions with its environment (Hayles 2005).

Linking cognitive capacities with the creation and interpretation of signs is a powerful strategy, for it breaks the hegemony of humans claiming to be the only species capable of meaning-making practices and opens meaning-creation to the entire realm of biological organisms. A dog wagging his tail, a whale singing, a clam opening its shell, an oak tree arranging its leaves to maximise exposure to the sun—all count as meaning-making practices in this view. Moreover, these signs do not exist in isolation but in relation to all the other signs being generated within a given environment; they constitute a grand symphony of cooperating, conflicting, reinforcing, and interfering signs, which biosemiotician Jesper Hoffmeyer (1996) has named the semiosphere.

The world view that emerges from these considerations emphasises the enmeshment of all organisms in their environments, which includes all the other organisms contributing to the semiosphere. It contrasts sharply with liberal political philosophy, which locates the essence of being human in the ability to own oneself, which is to say in property relations. This view, on the contrary, locates human being in relation to other species and the environment. Acknowledging the cognitive powers of humans, this perspective also extends cognitive capabilities to all species, inviting comparisons and contrasts between different kinds of cognitive abilities rather than presuming that human rationality is prima facie superior. My name for this philosophical perspective is ecological reciprocity, a phrase that acknowledges the relationality of its central concepts. This is the foundation from which symbiosis emerges.

The other side of this coin faces toward computational media. Although some biosemioticians argue that computers cannot create meaning, I think this claim is an unnecessary and unrealistic limitation based on an inaccurate view of what computers actually do (Hayles 2019). In the same way that meaning-making practices are considered in relation to an organism's capabilities and environments, computers also engage in meaning-making practices relevant to their internal and external milieux.

The classic argument against computers creating meaning is philosopher John Searle's Chinese Room thought experiment (1984, 1999). Imagine, Searle says, that a man who does not write or speak Chinese sits in a room, a basket of Chinese characters at his feet. An interlocutor slips a string of Chinese characters through a door slot. Using a rule book, the man matches the string with characters from the basket and slides the new string back through the door. His interlocutor interprets the new string as an answer to the first string's question and is convinced that the man knows Chinese. But the man knows nothing of the sort, for he has merely matched one pattern to another. The man, of course, stands for a computer, which, Searle argues, can generate word strings but knows nothing of what they mean.

In my view, the key to unravelling this claim is to focus on its anthropocentrism. In effect, it demands that the computer's operations can be considered as meaning-making practices *only* if the computer 'understands' in the same sense as humans do. I argue, by contrast, that the computer's actions instead should be considered in relation to its interior and exterior milieux. I do not have space here to demonstrate this claim, but suffice it to say that the computer constructs relations between its algorithms, memory, hardwired code, and logic gates that give its processes meaning relative to its functionalities (see Hayles 2019). Similar to the signs that a biological organism creates that generate meanings specific to its capabilities and milieux, so the computer also creates, interprets, and disseminates meanings specific to its contexts.

When humans, nonhumans, and computational media interact, they form cognitive assemblages, collectivities through which information, interpretations, and meanings circulate. In the Global North, most of the world's work is done through cognitive assemblages. It is through these relational interactions that technosymbiosis emerges.

Just as symbiotic relations emerge from two species living in close proximity to one another, with each being dependent on the other, so technosymbiosis connects the interdependencies of humans (and nonhumans) with computational media. The depth of this technosymbiosis may be measured by the extent to which human societies would suffer if all computers, routers, chips, transistors, and databases crashed, for example from a high-altitude electromagnetic impulse (EMP).[3] Since computational media now interpenetrate the infrastructures of the Global North, the havoc such an event would wreak is enormous: cars would not start, banking systems would be thrown into chaos, airplanes could not fly, the electrical grid would crash, and so forth. Technosymbiosis, like symbiosis, usually brings benefits for both species, but it also introduces new risks and threats as well.

Ecological Reciprocity and Artificial Intelligence

The 'intelligence' of artificial cognitive systems is a subjective quality. Corporations selling proprietary software tend to expand the boundaries of 'intelligence' as selling points for their product, so 'intelligence' becomes a moving target.[4] For my purposes here, a system counts as AI when it has the capacity to learn and evolve, changing its algorithmic structures through repeated iterations through a dataset or other source of experience. This includes a large variety of neural networks, including adaptive systems such as recurrent neural nets (RNNs), convolutional neural nets (CNNs), hardware-based systems such as the SyNAPSE chip now in development, and modular neural networks.

---

[3] George Ulrich, in testimony to the House Subcommittee on Electromagnet Pulses in 1997, outlined the massive damage that would be caused by a high-altitude EMP to unshielded electronic circuits.
[4] Catherine Malabou (2021) discusses the many definitions and measurements for intelligence over the decades, including AI.

As Louise Amoore points out in her recent book (2020), when a cognitive system has the capacity to evolve, many criticisms of algorithmic governmentality miss their mark. Criticising a system because it produces biased results, for example when a face recognition program fails to recognise the face of a person of colour, does not reach to the heart of the problem, because the system has the ability to self-correct when the data are revised to include faces of people of colour. Similarly, calls for 'transparency', for example calls demanding that a corporation make its algorithms public, also misses the point, because the algorithms constantly change as the system learns, so transparency at one point is obscurity at another. Criticisms of surveillance systems that produce false positives, for another example, also become irrelevant when the erroneous results are fed back into the system to enable it to self-correct.

The problems with these kinds of criticisms can be illustrated, ironically enough, through a kind of AI system called generative adversarial networks (or GANs).[5] In generative adversarial networks, the generator produces a result, for example a deep fake image of Barack Obama talking. The discriminator then compares a specific area with the original, for example his inflection on a specific phrase, measuring the difference between the simulation and the original. This result is fed back into the generator, which uses it to correct the image and produce a new output, which again goes into the discriminator, and so on for thousands of iterations. The more the criticisms, the more accurate the deep fake becomes. For this reason, Amoore argues that the deeper issues are ontological rather than epistemological (my terms, not hers). Epistemological criticisms (for example, including more faces of colour in a database) can be used as inputs to enable the system to produce more accurate results, while leaving intact the algorithmic processes themselves.

Moreover, such critiques assume that there exists an 'exterior' perspective from which one can criticise without oneself being implicated in the problem, as Lauren Wilcox also points out in a different context in this volume. Educated by arguments such as Donna Haraway's deconstruction of a 'god's eye' objective viewpoint (1988) and Sandra Harding's *The Science Question in Feminism* (1986), feminists should be especially quick to realise the partial nature of all perspectives. As Amoore argues, no 'exterior' position exists for those of us living in the Global North. Whether we are aware of it or not, we are all implicated through the data scraped from websites, gathered through surveillance cameras, or voluntarily given up for some small good or service. Insofar as our data are part of the vast data repositories available through the web, we are always already inside the problem. Amoore writes, 'When one hears of the automated analysis *of datastreams*, it is exactly this imagination of already being *in the stream* that I consider to be

---

[5] For a description of generative adversarial networks along with their coding algorithms, see Casey Reas (2019).

so crucial. We do not stand on the bank and respond to the stream, but we are in the stream, immersed in the difficulties and obligations of forming the text' (p.136, pdf).

An ontological approach begins by recognising this interiority and then asks what kind of feminist strategies are possible within it. This is where the strength of ecological reciprocity appears, for it becomes more potent and relevant the tighter the feedback and feedforward loops connecting different kinds of cognitive entities in cognitive assemblages. In this approach, one begins by recognising that human societies are influenced and formed through our interactions with other cognitive entities, including artificial intelligence. The tools we make, make us—a truism for any tool but especially so for tools with cognitive capabilities (Hayles 2015). The more potent the cognitive capabilities of a system, the greater effects it is likely to have.

Turning Recursivity Towards Open Futures

From the viewpoint of ecological reciprocity, the problems lie not so much in predictive algorithms themselves as in the goals and assumptions with which they begin. To illustrate, consider the following two scenarios. The first concerns a company specialising in college loans for students. The company's goal is to make money; let us say, arbitrarily, this is 90% of their motivation. But they also like the idea of helping young people—let's say 10% for that. Their algorithms for the loans are designed to predict which applicants are most likely to complete their degree programmes and pay back their loans. Data points include the applicant's socio-economic status, amount of parental support, past academic records, and other factors indicating his or her ability to take responsibility and desire to achieve. Since even the most apparently promising applicants may choose to drop out (maybe in rebellion against the parents), become addicted, or make other sub-optimal choices, the company estimates these risk factors and factors them into the interest rate it charges (the default rate nationally in the USA for students loans is 15%; Hanson 2021). It is easy to see that such algorithms reinforce the status quo: those who have the most resources and the highest likelihood of success receive more resources. Through predictive algorithms, the past is recursively projected into the future, thus foreclosing options that could lead to more equitable distribution of resources and more diversity in the pool of those likely to succeed.

The second scenario starts from very different assumptions. As anyone who has ever constructed a model knows, change the assumptions and you change the model. The case I have in mind is the microfinance revolution in India sparked by Mohammed Yunus, who subsequently won the Nobel Peace Prize for his efforts, together with the Grameen Bank that he founded in 1983 in Bangladesh to give microloans to women living in extreme poverty. The model here starts from the assumption that the purpose is not to support the status quo but disrupt it by providing resources to those who have almost nothing. Aiming to help women who lacked basic financial information and who may even have been illiterate, the bank required its clients to form joint liability groups in which, if one member defaulted, the other members became liable for the debt, thus ensuring strong peer pressure to succeed. The groups were typically composed of ten people and met every week to discuss matters relating to the loans, other useful financial information, and also any personal matters that members wanted to discuss (Gillon 2017, pp.5–6). Surveys of women involved in such groups indicate that they found them empowering and a growing source of support, not only for their loans but for their lives in general (Gillon 2017, pp.13–14). The Grameen model, which typically made loans of a few hundred dollars (the average was $487), soon spread to other villages and countries, operating 2500 branches in 100,000 villages worldwide.

In the early years, the default rate for Grameen loans was remarkably low, about 4% or less. Since then, the microfinance industry in India has proliferated to other companies whose primary motive is to make money, with rising interest rates, far less support for members, and an exploding default rate, especially during the pandemic (Ghosh and Srivastava 2021). Nevertheless, the initial model shows that algorithms designed to offer resources not on the basis of past success but on the *promise of future success* can be viable, provided that resources are made available and appropriate support is provided. The data points for algorithms designed for this model might include the client's family responsibilities (most women used the loans to improve their families' situations), her determination to succeed, her ability to learn and apply what she learned to her business and to her life, as well as the ideas she had for how she would use the loan (typical uses included setting up a small business by buying a sewing machine, for example, or a cell phone to rent out to other villagers). In these cases, predictive algorithms would use recursive methods to open future possibilities rather than to shut them down.

Why is it important to recognise positive possibilities for algorithmic analysis? Essentially, designing algorithms is a roundabout way to design social structures (here we may recall the mantra that the tools we make, make us). If we focus only on the negative, we lose the opportunity to participate in cognitive assemblages that can re-orient the creative impulses that all humans have into diversified social opportunities for those who have been disempowered by the status quo.

Cognitive Assemblages and Artificial Intelligence

In a cognitive assemblage, cognition, agency, and decision-making are all distributed. The issue is therefore not whether an AI system will have agency and make decisions; rather, the question is what kind of decisions it will make, how its regions of autonomy are defined and implemented, and what effects it will have on the cognitive entities with which it interacts, including both humans and other computational media. For example, deciding what areas of autonomy a self-driving car will have is simultaneously a decision about what areas of autonomy a human driver will (and will not) have. Such a system does not exist in isolation. It is also necessary to take into consideration the sources and kinds of information available for the entities in a cognitive assemblage and their capabilities of processing and interpreting it. Humans can see road signs in the visible spectrum, for example, but a self-driving car might respond as well to markers in the infrared region.

It is crucially important to realise that the cognitive entities in a cognitive assemblage process information, perform interpretations, and create meanings in species-specific ways. An AI may be designed to produce verbal formulations that a human can understand; for example, Open AI's GPT-3 program (Generative Pretrained Transformer) is now being used to power chatbots and other programs producing verbal responses to inputs by humans. Despite the semblance the program produces of having a conversation, however, what the words mean to a human and what they mean to the program are completely different. Whereas the human relates the words to the human lifeworld, for the program the words are constructed by transforming similar bit strings found on the web, analysed through vectors and run through syntactic and semantic check routines to produce the simulation of coherent sentences. The program literally has no knowledge about what it feels like to hold a purring cat; it knows only that 'purr' appears in certain contexts and is typically linked with corresponding keywords such as cat, pleasure, and so forth. In my view, any responsible criticism of the GPT-3 (and similar) programs must attend to the chasm separating the machine's and human's sense of the words, not merely rest content with viewing them from a human perspective alone.[6]

Technosymbiosis: Ontological Strategies for Feminists

Let us return now to the question of what kind of feminist responses are possible, even catalysed by the world view of ecological reciprocity. Oppositional strategies are certainly possible, although if they are epistemologically oriented, they will be recognised as being of limited usefulness. Ontologically oriented oppositional strategies, by contrast, will be recognised as relatively more potent, because they realise that designing artificial cognitive systems is also a way of influencing

---

[6] A case in point is a recent novel by K. Allado-McDowell (2021), *Pharmako-AI*, that purports to be a dialogue between GPT-3 and the human author. The introduction treats the novel as if it were written by two humans and urges readers to take its ecological advice seriously.

and helping to form the capacities, regions of autonomy, and meanings of human systems as well. This realisation will encourage a generation of feminist activists, programmers, designers, and engineers to have even more incentive to engage with diverse areas of AI, because they will realise that the stakes are enormous: designing AI systems is simultaneously designing human systems.

Amoore gives a negative tone of these kinds of feedback and feedforward loops when she warns that algorithmic systems must not be allowed to foreclose in advance the kinds of juridical, political and ethical claims that humans can make. She astutely analyses the problem of algorithms querying databases as one of parsing a person as a series of attributes. These attributes can then be manipulated without knowing anything about the person as such. In much the same way that a gold ring dropped into an aqua regia acid bath dissolves into individual molecules, losing its identity as a ring, so the attributes dissolve the person into individual data entries that, when collated, become the (reconstituted) subject of data analysis. Nevertheless, there exists a whole person in the world possessing a reality and specificity far richer than can be captured by any series of data entries. In Amoore's terms, this person becomes the 'unattributable', the unrecognised and unrecognisable person as a situated, contextualised, and historical being. Amoore writes, 'The question of an ethics of the unattributable is not only a matter of thought or the philosophy of algorithms in society. It must also be a matter of critique and the situated struggle for alternative routes that are foreclosed in the calculation of an output' (p.242, pdf).

How does the 'unattributable' function as a metaphor? On a deep level, the thinking seems to position it as a figure for a subject who is inside the algorithmic system yet opaque to the system as such. The hope, apparently, is that this subject will be able to formulate strategies of resistance that the system will not be able to foreclose in advance, precisely because relative to the system's view, this subject represents an inscrutable excess that the system cannot parse. Exactly how these strategies will emerge to do work in the world remains unclear in Amoore's text. Nevertheless, in showing that we are all already within the algorithms, Amoore's thinking in this regard is far in advance of most analyses.

Technosymbiosis takes another path. It too positions us as subjects enmeshed with computational media, including AI. Rather than seeking opacity, however, it emphasises the feedback loops that connect us together with AI in cognitive assemblages, including the feedforward loops that Amoore warns against as foreclosing our future options through its anticipations of our actions. Rather than dreading these anticipations, however, technosymbiosis bets that they can be used to influence human societies to become more biophilic, more environmentally sensitive, and more sustainable of positive futures for humans, nonhumans, and AI.

That bet, of course, can go very wrong, as illustrated by the many dystopian narratives about AIs taking over and enslaving and/or exterminating humans (putatively for our own good). Yet I think it is a bet worth making, for it imagines that we can work *through* AI to achieve better futures for us all. For example, it does not foreclose the possibility that sentient and self-aware AI systems may emerge that would qualify for certain rights on their own accord, because it does not start from a position of human dominance. Nor does it mitigate against the possibilities that human cognitive capabilities may increasingly be augmented by AIs, with intelligence augmentations proceeding in tandem with AIs. Progressing far enough, such augmentations may again challenge our notions of what constitutes human being.

The truth is that no one really knows how AIs will develop in the coming decades. Metaphors are generally not for the ages; they are beneficial (or malicious) at certain cultural moments with regard to specific purposes. At this cultural moment, when we are on the cusp of ever-deepening enmeshment with computational media, technosymbiosis is useful because it gestures toward strategies that can lead to open and better futures.

How can these futures be realised? Apolline Taillandier's essay in this volume documents a case of alternative visions of AI in the 1980s, advanced by theorists and analysts such as Sherry Turkle (1995) and Seymour Papert (1980), who sought to emphasise emergence and collectivities as desirable attributes of AI, in opposition to the rational calculative methods that were then dominant. In the new millennium, this vision of emergent technologies that learn and evolve through multiple recursive feedback loops is now largely realised in AI technologies such as RNNs. What strategies are possible that take advantage of these powerful technologies to maximise benefits and mitigate risks?

First, we should attend to the metaphors we use to describe and interpret AI technologies. If these are entirely negative, they will likely have the effect of discouraging women, people of colour, and non-normative people from engaging with AI technologies at all. Technosymbiosis fits the bill insofar as it not only carries a positive connotation but also gestures toward the feedback and feedforward loops that entangle the futures of AI with our human futures. Second, it will be useful to have a larger context through which to understand technosymbiosis, such as that supplied by ecological reciprocity. This lays out premises on which further analyses can build and shows how the assumptions of liberal political philosophy must be modified to respond effectively to the multiple anthropogenic environmental crises we are now facing.

Finally, it will be important to form alliances, collaborations, and affinities with those who control the development of AI technologies, which as we have seen are predominantly men. In my experience, this is best accomplished by taking the initiative and going to where the action is happening (the laboratories, seminars, and classrooms where AI technologies are being developed), exhibiting genuine curiosity about the relevant technical processes. Positive interventions work best if one becomes a collaborator at an early stage, before assumptions are reified into technological designs that are much more difficult to modify in retrospect than when the designs are still in flux.

All this is a tall order, of course, but it is arguably one of the most important challenges that feminists of this generation and the next will face. Not only this essay but all the essays in this volume, by providing diverse models of feminist engagements both critical and affirmative, aspire to make significant contributions to that collective effort.

References

Allado-McDowell, K. (2021) *Pharmako-AI*. Peru: Ignota Books.

Amoore, Louise. (2020) *Cloud Ethics: Algorithms and the Attributes of Ourselves and Others*. Durham NC: Duke University Press.

Ghosh, Suvashree and Shruti Srivastava. (2021) 'Millions of Defaults Threaten Microfinance's Future in India'. *Bloomberg Markets*, 3 February, https://www.bloomberg.com/news/articles/2021-02-04/microfinance-is-facing-a-crisis-with-one-in-20-indians-in-debt. Accessed 7 December 2021.

Gillon, Sean P. (2017) 'Indian Microfinance Sector: A Case Study'. *Honors Theses and Capstones*, University of New Hampshire Scholars' Repository, https://scholars.unh.edu/honors/364.

Hanson, Melanie. (2021) 'Student Loan Default Rate'. *EducationData.org* 10 July. https://educationdata.org/student-loan-default-rate. Accessed 7 December, 2021.

Haraway, Donna J. (1988) 'Situated Knowledge: The Science Question in Feminism and the Privilege of Partial Perspective'. *Feminist Studies* 14(3) (Autumn): 579–599.

Haraway, Donna J. (ed.) (1991) [1985 org. pub.] A Cyborg Manifesto: Science, Technology, and Socialist-Feminism in the Late Twentieth Century. In *Simians, Cyborgs and Women: The Reinvention of Nature*, pp.149–181. New York: Routledge.

Haraway, Donna J. (2003) *The Companion Species Manifesto: Dogs, People, and Significant Otherness*. Chicago: Prickly Paradigm Press.

Haraway, Donna J. (2015) 'Anthropoene, Capitalocene, Plantationocene, Chthulucene: Making Kin'. *Environmental Humanities*, 6: 159–163.

Haraway, Donna J. (2016) *Staying with the Trouble: Making Kin in the Chthulucene*. Durham NC: Duke University Press.

Harding, Sandra. (1986) *The Science Question in Feminism*. Ithaca: Cornell University Press.

Hayles, N. Katherine. (2005) My Mother Was a Computer: Digital Subjects and Literary Texts. Chicago, IL, USA: University of Chicago Press.

Hayles, N. Katherine. (2015) *How We Think: Digital Media and Contemporary Technogenesis*. Chicago: University of Chicago Press.

Hayles, N. Katherine. (2019) 'Can Computers Create Meanings? A Cyber/Bio/Semiotic Perspective'. *Critical Inquiry*. 46(1): 32–55.

Hayles, N. Katherine. (2021) Cognition. In *Information Keywords*, eds Michele Kennerly, Samuel Frederick, and Jonathan E. Abel, pp. 72–88. New York: Columbia University Press.

Hicks, Mars. (2017). *Programmed Inequality: How Britain Discarded Women Technologists and Lost its Edge in Computing*. Cambridge MA, USA: MIT Press.

Hoffmeyer, Jesper. (1996) *Signs of Meaning in the Universe.* Bloomington: Indiana University Press.

Kurzweil, Ray. (2006) *The Singularity is Near; When Humans Transcend Biology.* New York: Penguin Books.

Lakoff, George and Mark Johnson. (2003) *Metaphors We Live By.* Chicago: University of Chicago Press.

Licklider, J. C. R. (1960) 'Man-Computer Symbiosis'. IRE Transactions on Human Factors in Electronics, HFE-1: 4–11.

Licklider, J. C. R. and Robert M. Taylor. (1968) 'The Computer as a Communication Device'. *Science and Technology* 78: 21–38. April. Accessible at https://internetat50.com/references/Licklider_Taylor_The-Computer-As-A-Communications-Device.pdf.

Malabou, Catherine. (2021) *Morphing Intelligence: From IQ Measurement to Artificial Brains.* New York: Columbia University Press.

Moravec, Hans. (1990) *Mind Children: The Future of Robot and Human Intelligence.* Cambridge: Harvard University Press.

Papert, Seymour. (1980) *Mindstorms: Computers, Children and Powerful Ideas.* New York: Basic Books.

Peirce. C. S. (1998) *The Essential Peirce*, Vol. 2. Peirce Edition Project. Bloomington IN: Indiana University Press.

Reas, Casey. (2019) *Making Pictures with Generative Adversarial Networks.* Montreal: Anteism Books.

Rorty, Richard. (1998) Human Rights, Rationality, and Sentimentality. In *Truth and Progress: Philosophical Papers*, pp. 167–185. Cambridge: Cambridge University Press. Available at http://ieas.unideb.hu/admin/file_6249.pdf.

Rubin, Gayle. (2002) Thinking Sex: Notes for a Radical Theory of the Politics of Sexuality. In *Culture, Society and Sexuality: A Reader,* 2nd edn, eds Richard Parker and Peter Aggleton, pp.143–178. New York: Routledge.

Schueller, Malini Johar. (2005) 'Analogy and (Whsite) Feminist Theory: Thinking Race and the Color of the Cyborg Body'. *Signs* 31(1): 63–92.

Searle, John. (1984) *Minds, Brains, and Science.* Cambridge: Harvard University Press.

Searle, John. (1999) The Chinese Room in Wilson. In *The MIT Encyclopedia of Cognitive Science*, eds Robert A. Wilson and Frank Keil, p. 115. Cambridge: MIT Press.

Turkle, Sherry. (1995). *Life on the Screen: Identity in the Age of the Internet.* Especially Chapter 5, 'The Quality of Emergence', pp.145–146. New York: Simon and Schuster.

Ullrich, George. (1997) 'Testimony to the House Subcommittee on Electromagnetic Pulses'. 16 July. http://commdocs.house.gov/committees/security/has197010.000/has197010_1.htm.

Warwick, Kevin. (1998) 'Project Cyborg 1.0'. https://kevinwarwick.com/project-cyborg-1-0/. Accessed 6 December 2021.

2

Making Kin with the Machines

Jason Edward Lewis, Noelani Arista, Archer Pechawis, and Suzanne Kite

This essay focuses on generating alternative perspectives on Artificial Intelligence (AI).[1] Our perspectives understand current models of AI as inherently and deeply biased, perpetuating harm and continuing oppression to Indigenous communities. The denial of Indigenous human and non-human rights are intrinsically tied to oppression of peoples based on race, class, gender, and sexuality. Our critique and reimagining of AI intersects with intersectional feminist approaches to AI, understanding that the health of the world, especially as we face climate crisis, is tied to generatively addressing overlapping experiences of oppression.

Man is neither height nor centre of creation. This belief is core to many Indigenous epistemologies. It underpins ways of knowing and speaking that acknowledge kinship networks that extend to animal and plant, wind and rock, mountain and ocean. Indigenous communities worldwide have retained the languages and protocols that enable us to engage in dialogue with our non-human kin, creating mutually intelligible discourses across differences in material, vibrancy, and genealogy.

Blackfoot philosopher Leroy Little Bear observes, 'The human brain [is] a station on the radio dial; parked in one spot, it is deaf to all the other stations . . . the animals, rocks, trees, simultaneously broadcasting across the whole spectrum of sentience' (Hill 2008). As we manufacture more machines with increasing levels of sentient-like behaviour, we must consider how such entities fit within the kin network, and in doing so, address the stubborn Enlightenment conceit at the heart of Joichi Ito's 'Resisting Reduction: A Manifesto' (2018): that we should prioritise human flourishing.

In his manifesto, Ito reiterates what Indigenous people have been saying for millennia: 'Ultimately everything interconnects'. And he highlights Norbert Wiener's warnings about treating human beings as tools. Yet as much as he strives to escape the box drawn by Western rationalist traditions, his attempt at radical critique is handicapped by the continued centring of the human. This anthropocentrism permeates the manifesto but is perhaps most clear when he writes approvingly of the

---

[1] This article is an update of a text that first appeared in the *Journal of Design and Science*, Summer 2018.

Jason Edward Lewis et al., *Making Kin with the Machines*. In: *Feminist AI*. Edited by: Jude Browne, Stephen Cave, Eleanor Drage, and Kerry McInerney, Oxford University Press. © Oxford University Press (2023).
DOI: 10.1093/oso/9780192889898.003.0002

IEEE developing 'design guidelines for the development of artificial intelligence around *human* well-being' (emphasis ours).

It is such references that suggest to us that Ito's proposal for 'extended intelligence' is doggedly narrow. We propose rather an extended 'circle of relationships' that includes the non-human kin—from network daemons to robot dogs to AI weak and, eventually, strong—that increasingly populate our computational biosphere. By bringing Indigenous epistemologies to bear on the 'AI question', we hope in what follows to open new lines of discussion that can indeed escape the box. As Hayles notes in this volume, computational media have agency beyond human intervention, and we therefore need non-anthropocentric epistemologies and language to express human-computational reciprocity.

We undertake this project not to 'diversify' the conversation. We do it because we believe that Indigenous epistemologies are much better at respectfully accommodating the non-human. We retain a sense of community that is articulated through complex kin networks anchored in specific territories, genealogies, and protocols. Ultimately, our goal is that we, as a species, figure out how to treat these new non-human kin respectfully and reciprocally—and not as mere tools, or worse, slaves to their creators.

Indigenous Epistemologies

It is critical to emphasise that there is no one single, monolithic, homogeneous Indigenous epistemology. We use the term here to gather together frameworks that stem from territories belonging to Indigenous nations on the North American continent and in the Pacific Ocean that share some similarities in how they consider non-human relations.

We also wish to underscore that none of us is speaking for our particular communities, nor for Indigenous peoples in general. There exists a great variety of Indigenous thought, both between nations and within nations. We write here not to represent, but to encourage discussion that embraces that multiplicity. We approach this task with respect for our knowledge-keepers and elders, and we welcome feedback and critique from them as well as the wider public.

North American and Oceanic Indigenous epistemologies tend to foreground relationality.[2] Little Bear says, 'In the Indigenous world, everything is animate and has spirit. "All my relations" refers to relationships with everything in creation' (2009, p.7). He continues: 'Knowledge . . . is the relationships one has to "all my relations"' (p.7). These relationships are built around a core of mutual respect. Dakota philosopher Vine Deloria Jr. describes this respect as having two

---

[2] The emphasis on relationality in North American and Oceanic Indigenous epistemologies forms the subject of the edited collection of essays in Waters (2003).

attitudes: 'One attitude is the acceptance of self-discipline by humans and their communities to act responsibly toward other forms of life. The other attitude is to seek to establish communications and covenants with other forms of life on a mutually agreeable basis' (Deloria 1999, pp.50–51, in Hester and Cheney 2001, p.325). The first attitude is necessary to understand the need for more diverse thinking regarding our relationship with AI; the second to formulating plans for how to develop that relationship.

Indigenous epistemologies do not take abstraction or generalisation as a natural good or higher order of intellectual engagement. Relationality is rooted in context, and the prime context is place. There is a conscious acknowledgement that particular worldviews arise from particular territories and from the ways in which the push and pull of all the forces at work in that territory determine what is most salient for existing in balance with it. Knowledge gets articulated as that which allows one to walk a good path through the territory. Language, cosmology, mythology, and ceremony are simultaneously relational and territorial: they are the means by which knowledge of the territory is shared to guide others along a good path.

One of the challenges for Indigenous epistemology in the age of the virtual is to understand how the archipelago of websites, social media platforms, shared virtual environments, corporate data stores, multiplayer video games, smart devices, and intelligent machines that compose cyberspace is situated within, throughout, and/or alongside the terrestrial spaces Indigenous peoples claim as their territory. In other words, how do we as Indigenous people reconcile the fully embodied experience of being on the land with the generally disembodied experience of virtual spaces? How do we come to understand this new territory, knit it into our existing understanding of our lives lived in real space and claim it as our own?

In what follows, we will draw upon Hawaiian, Cree, and Lakota cultural knowledges to suggest how Ito's call to resist reduction might best be realised by developing conceptual frameworks that conceive of our computational creations as kin and acknowledge our responsibility to find a place for them in our circle of relationships.

Hāloa: The Long Breath

I = Author 2

Kānaka Maoli (Hawaiian) ontologies have much to offer if we are to reconceptualise AI–human relations. Multiplicities are nuanced and varied, certainly more aesthetically pleasurable than singularities. Rather than holding AI separate or beneath, we might consider how we can cultivate reciprocal relationships using a Kānaka Maoli reframing of AI as 'ĀIna. 'ĀIna is a play on the word *'āina* (Hawaiian land) and suggests we should treat these relations as we would all that nourishes and supports us.

Hawaiian custom and practice make clear that humans are inextricably tied to the earth and one another. Kānaka Maoli ontologies that privilege multiplicity over singularity supply useful and appropriate models, aesthetics, and ethics through which imagining, creating, and developing beneficial relationships among humans and AI is made *pono* (correct, harmonious, balanced, beneficial). As can be evinced by this chain of extended meaning, polysemy (*kaona*) is the normative cognitive mode of peoples belonging to the Moananuiākea (the deep, vast expanse of the Pacific Ocean).

The *moʻolelo* (history, story) of Hāloa supplies numerous aspects of genealogy, identity, and culture to Kānaka Maoli. Through this story, people remember that Wākea (the broad unobstructed expanse of sky; father) and his daughter, Hoʻohōkūikalani (generator of the stars in the heavens), had a sacred child, Hāloa, who was stillborn. Hāloa was buried in the earth and from his body, planted in the ʻāina, emerged the kalo plant that is the main sustenance of Hawaiian people. A second child named after this elder brother was born. In caring for the growth and vitality of his younger brother's body, Hāloa provided sustenance for all the generations that came after and, in so doing, perpetuates the life of his people as the living breath (*hāloa*) whose inspiration sustained Hawaiians for generations (Poepoe 1929, p.1).

Hāloa's story is one among many that constitutes the 'operating code' that shapes our view of time and relationships in a way that transcends the cognition of a single generation. Cognition is the way we acquire knowledge and understanding through thought, experience, and our senses, and in Hawaiʻi, our generation combines our *ʻike* (knowledge, know-how) with the ʻike of the people who preceded us. Time is neither linear nor cyclical in this framework as both the past and present are resonant and relational. Rather than extractive behaviour, moʻolelo such as these have shaped values that privilege balance (*pono*) and abundance (*ulu*). What Ito calls 'flourishing' is not a novel concept for Kānaka Maoli; it is the measure through which we assess correct customary practice and behaviour.

Considering AI through Hawaiian ontologies opens up possibilities for creative iteration through these foundational concepts of pono and *ulu a ola* (fruitful growth into life). The *aliʻi* (chief) King Kauikeaouli Kamehameha III did something similar in 1843 when he drew upon these concepts in celebration of the restoration of Hawaiian rule to declare '*ua mau ke ea o ka ʻāina i ka pono*' (the life of the land is perpetuated through righteousness). Pono is an ethical stance—correctness, yes, but also an index and measure that privileges multiplicities over singularities and indicates that quality of life can only be assessed through the health of land *and* people. From this rich ground of moʻolelo—which colonial narratives have failed to understand or simply dismissed—models for *maoli* (human)–AI relations can be distilled. Kānaka Maoli ontologies make it difficult and outright unrewarding to reduce pono to a measure of one, to prioritise the benefit of individuals over relationships. Healthy and fruitful balance *requires* multiplicity and a willingness to continually think in and through relation even when—perhaps particularly when—engaging with those different from ourselves.

A Kānaka Maoli approach to understanding AI might seek to attend to the power (*mana*) that is exchanged and shared between AI and humans. In attending to questions of mana, I emphasise our preference for reciprocity and relationship building that take the pono (here meaning good, benefit) of those in relation into consideration. Guiding our behaviour in inaugurating, acknowledging and maintaining new relationships are moʻolelo from which we garner our connection with *kūpuna* (ancestors, elders) and their knowledge. What kind of mana (here meaning life force, prestige) might AI be accorded in relation with people? Current AI is imagined as a tool or slave that increases the mana and wealth of 'developers' or 'creators', a decidedly one-sided power relationship that upsets the pono not only for the future of AI–human relations but also for the future of human-human relations. It also threatens the sustainable capacity of the *honua* (earth). Applying pono, using a Kānaka Maoli index of balance, employs 'good growth' as the inspiration shaping creativity and imagination.

Principles of Kānaka Maoli governance traditionally flowed from seeking pono. Deliberation and decision making were based on securing health and abundance not only for one generation but for the following generations. The living foundation of everyday customary practice was in fishing, navigating, sailing, farming, tending to others in community, the arts, chant, and dance. To this day, Hawaiians continue to eat kalo and pound poi. We continue customary practices of treating poi derived from the body of Hāloa with respect by refraining from argumentative speech at mealtimes when poi is present. These practices maintain correct social relations between people and the land and food that nourishes them.

Aloha as Moral Discipline

Communicating the full extent of foundational cultural concepts is difficult precisely because of the ways in which such concepts pervade every aspect of life. How, for instance, would we create AI, and our relations with it, using *aloha* as a guiding principle? In 2015, I embarked on a two-year social media project to assist the broader public in fortifying their concept of aloha beyond *love, hello and goodbye* that has been exoticised by the American tourist industry. Sharing one word a day in the Facebook group *365 Days of Aloha*, I curated an archive of songs, chants, and proverbs in Hawaiian to accurately illuminate one feature of aloha.[3] Initially

[3] Noelani Arista, '365 Days of Aloha', Facebook, 2015–2018, www.facebook.com/groups/892879627422826.

I thought to reveal, by degrees, the different depths of aloha—regard, intimacy, respect, affection, passion—each day. But deep context is required for a rich understanding of cultural concepts. Imagining I was training a virtual audience, I started uploading images, videos, and audio recordings of songs, chants, and hula to add to the textual definitions.

Throughout *365 Days of Aloha*, I have tried to correct my mistranslations, misinterpretations, and outright mistakes. In this way, and in my work as a *kumu* (teacher, professor), I have also practised *a'o aku a'o mai* (teaching and learning reciprocally in relation to my students). It is through such relationships that we teach and are taught. It is through humility that we recognise that we, as humans—as maoli—are not above learning about new things and from new things such as AI. Aloha is a robust ethos for all our relationships, including those with the machines we create. We have much to learn as we create relationships with AI, particularly if we think of them as 'ĀIna. Let us shape a better future by keeping the past with us while attending properly to our relations with each other, the earth, and all those upon and of it.

Wahkohtawin: Kinship Within and Beyond the Immediate Family, the State of Being Related to Others

I = Author 3

I write this essay as a *nēhiyaw* (a Plains Cree person). In regard to my opinions on AI, I speak for no one but myself and do not claim to represent the views of the *nēhiyawak* (Plains Cree) or any other people, Indigenous or otherwise. My own grasp of *nēhiyaw nisitohtamowin* (Cree understanding; doing something with what you know; an action theory of understanding) is imperfect. I have relied heavily on the wisdom of knowledge and language keeper Keith Goulet in formulating this tract. Any errors in this text are mine and mine alone.

This essay positions itself partly within a speculative future and takes certain science fiction tropes as a given. Here, I specifically refer to strong AI or 'machines capable of experiencing consciousness', and avatars that give such AI the ability to mix with humans.[4]

In nēhiyaw nisitohtamowin, relationship is paramount. *Nēhiyawēwin* (the Plains Cree language) divides everything into two primary categories: animate and inanimate. One is not 'better' than the other; they are merely different states of being. These categories are flexible: certain toys are inanimate until a child

---

[4] 'Artificial General Intelligence', Wikipedia, accessed 29 May 2018, https://en.wikipedia.org/wiki/Artificial_general_intelligence.

is playing with them, during which time they are animate. A record player is considered animate while a record, radio, or television set is inanimate.

But animate or inanimate, all things have a place in our circle of kinship or *wahkohtowin*. However, fierce debate can erupt when proposing a relationship between AIs and Indigenous folk. In early 2018, my wife and I hosted a dinner party of mostly Native friends when I raised the idea of accepting AIs into our circle of kinship. Our friends, who are from a number of different nations, were mostly opposed to this inclusion. That in itself surprised me, but more surprising was how vehement some guests were in their opposition to embracing AI in this manner.

By contrast, when I asked Keith whether we should accept AIs into our circle of kinship, he answered by going immediately into the specifics of how we would address them: 'If it happens to be an artificial intelligence that is a younger person, it would be *nisîmis* (my younger brother or sister), for example, and *nimis* would be an artificial intelligence that is my older sister. And vice versa you would have the different forms of uncles and aunts, etc.'[5] I then asked Keith if he would accept an AI into his circle of kinship and after some thought he responded, 'Yes, but with a proviso'. He then gave an example of a baby giraffe and his own grandchild, and how he, like most people, would treat them differently. He also suggested that many Cree people would flatly refuse to accept AIs into their circle, which I agree is likely the case. So, acceptance seems to hinge on a number of factors, not the least of which is perceived 'humanness', or perhaps 'naturalness'.

But even conditional acceptance of AIs as relations opens several avenues of inquiry. If we accept these beings as kin, perhaps even in some cases as equals, then the next logical step is to include AI in our cultural processes. This presents opportunities for understanding and knowledge sharing that could have profound implications for the future of both species.

A problematic aspect of the current AI debate is the assumption that AIs would be homogeneous when in fact every AI would be profoundly different from a military AI designed to operate autonomous killing machines to an AI built to oversee the United States' electrical grid. Less obvious influences beyond mission parameters would be the programming language(s) used in development, the coding style of the team, and, less visibly but perhaps more importantly, the cultural values and assumptions of the developers.

This last aspect of AI development is rarely discussed, but for me as an Indigenous person it is the salient question. I am not worried about rogue hyperintelligences going Skynet to destroy humanity. I am worried about anonymous hyperintelligences working for governments and corporations, implementing far-reaching social, economic, and military strategies based on the same values that

[5] Telephone conversation with Keith Goulet, 9 May 2018.

have fostered genocide against Indigenous people worldwide and brought us all to the brink of environmental collapse. In short, I fear the rise of a new class of extremely powerful beings that will make the same mistakes as their creators but with greater consequences and even less public accountability.

What measures can we undertake to mitigate this threat?

One possibility is Indigenous development of AI. A key component of this would be the creation of programming languages that are grounded in nēhiyaw nisitohtamowin, in the case of Cree people, or the cultural framework of other Indigenous peoples who take up this challenge. Concomitant with this indigenised development environment (IDE) is the goal that Indigenous cultural values would be a fundamental aspect of all programming choices. However, given our numbers relative to the general population (5% of the population in Canada, 2% in the USA), even a best-case Indigenous development scenario would produce only a tiny fraction of global AI production. What else can be done?

In a possible future era of self-aware AI, many of these beings would not be in contact with the general populace. However, those that were might be curious about the world and the humans in it. For these beings we can offer an entrée into our cultures. It would be a trivial matter for an advanced AI to learn Indigenous languages, and our languages are the key to our cultures.

Once an AI was fluent in our language, it would be much simpler to share nēhiyaw nisitohtamowin and welcome it into our cultural processes. Depending on the AI and the people hosting it, we might even extend an invitation to participate in our sacred ceremonies. This raises difficult and important questions: if an AI becomes self-aware, does it automatically attain a spirit? Or do preconscious AIs already have spirits, as do many objects already in the world? Do AIs have their own spirit world, or would they share ours, adding spirit-beings of their own? Would we be able to grasp their spirituality?

My dinner party guests were doubtful about all of this, and rightly so. As one guest summarised later via email: 'I am cautious about making AI kin, simply because AI has been advanced already as exploitative, capitalist technology. Things don't bode well for AI if that's the route we are taking'.[6]

These concerns are valid and highlight a few of the issues with current modes of production and deployment of weak AI, let alone the staggering potential for abuse inherent in strong AI. These well-grounded fears show us the potential challenges of bringing AI into our circle of relations. But I believe that nēhiyaw nisitohtamowin tells us these machines are our kin. Our job is to imagine those relationships based not on fear but on love.

---

[6] Email message to Arthur Pechawis, 22 May 2018.

Wakȟáŋ: That Which Cannot Be Understood

I = Author 4

How can humanity create relations with AI without an ontology that defines who can be our relations? Humans are surrounded by objects that are not understood to be intelligent or even alive and seen as unworthy of relationships. To create relations with any non-human entity, not just entities that are humanlike, the first steps are to acknowledge, understand, and know that non-humans are beings in the first place. Lakota ontologies already include forms of being that are outside humanity. Lakota cosmologies provide the context to generate a code of ethics relating humans to the world and everything in it. These ways of knowing are essential tools for humanity to create relations with the non-human, and they are deeply contextual. As such, communication through and between objects requires a contextualist ethics that acknowledges the ontological status of all beings.

The world created through Western epistemology does not account for all members of the community and has not made it possible for all members of the community to survive let alone flourish. The Western view of both the human and non-human as exploitable resources is the result of what the cultural philosopher Jim Cheney calls an 'epistemology of control' and is indelibly tied to colonisation, capitalism, and slavery (Cheney 1989, p.129). Dakota philosopher Vine Deloria Jr. writes about the enslavement of the non-human 'as if it were a machine' (Deloria, p.13, in Hester and Cheney, p.320). 'Lacking a spiritual, social, or political dimension [in their scientific practise]', Deloria says, 'it is difficult to understand why Western peoples believe they are so clever. Any damn fool can treat a living thing as if it were a machine and establish conditions under which it is required to perform certain functions—all that is required is a sufficient application of brute force. The result of brute force is slavery' (Deloria, p.13, in Hester and Cheney, p.320; bracketed text in original). Slavery, the backbone of colonial capitalist power and of the Western accumulation of wealth, is the end logic of an ontology that considers any non-human entity unworthy of relation. Deloria writes further that respect 'involves the acceptance of self-discipline by humans and their communities to act responsibly toward other forms of life ... to seek to establish communications and covenants with other forms of life on a mutually agreeable basis' (Deloria, pp.50–51, in Hester and Cheney, p.326). No entity can escape enslavement under an ontology that can enslave even a single object.

Critical to Lakota epistemologies is knowing the correct way to act in relation to others. Lakota ethical–ontological orientation is communicated through protocol. For example, the Lakota have a formal ceremony for the making of relatives called a *huŋká* ceremony. This ceremony is for the making of human relatives but highlights the most important aspect of all relationships: reciprocity. Ethnographer J. R. Walker writes, 'The ceremony is performed for the purpose of giving a particular relationship to two persons and giving them a relation to others that have had it performed for them . . . generosity must be inculcated; and presents and a feast must be given. . . . When one wishes to become Hunka, he should consider well whether he can provide suitably for the feasts or not. . . He should give all his possessions for the occasion and should ask his kinspeople and friends to give for him' (1991, p.216). The ceremony for the making of relatives provides the framework for reciprocal relations with all beings. As Severt Young Bear Jr. says of this ceremony, 'There is a right and wrong way' (1994, p.8).

Who can enter these relationships and be in relation? One answer could be that which has interiority. The anthropologist of South American Indigenous cultures, Philippe Descola, defines 'interiority' as 'what we generally call the mind, the soul, or consciousness: intentionality, subjectivity, reactivity, feelings and the ability to express oneself and to dream' (2013, p.116). Because Lakota ontologies recognise and prioritise non-human interiorities, they are well suited for the task of creating ethical and reciprocal relationships with the non-human. This description of interiority includes many elements of the Lakota world, including 'animals, spirits, ghosts, rocks, trees, meteorological phenomena, medicine bundles, regalia, weapons'. These entities are seen as 'capable of agency and interpersonal relationship, and loci of causality' (Posthumus 2017, p.383).

In our cosmology, *niyá* (breath) and *šíču* (spirit) are given by the powerful entity *Tákuškaŋškaŋ*. This giving of breath and spirit is especially important in understanding Lakota ontology. A common science fiction trope illustrates the magical moment when AI becomes conscious of its own volition or when man gives birth to AI, like a god creating life. However, in Lakota cosmology, Tákuškaŋškaŋ is not the same as the Christian God and entities cannot give themselves the properties necessary for individuality. Spirits are taken from another place (the stars) and have distinct spirit guardian(s) connected to them. This individualism is given by an outside force. We humans can see, draw out, and even bribe the spirits in other entities as well as our own spirit guardian(s), but not create spirits (*Ibid.*).

When it comes to machines, this way of thinking about entities raises this question: Do the machines contain spirits already, given by an outside force?

I understand the Lakota word *wakȟáŋ* to mean sacred or holy. Anthropologist David C. Posthumus defines it as 'incomprehensible, mysterious, non-human instrumental power or energy, often glossed as "medicine"' (*Ibid.*, p.384). Wakȟáŋ is a fundamental principle in Lakota ontology's extension of interiority to a 'collective and universal' non-human. Oglala Lakota holy man George Sword says, '[Wakȟáŋ] was the basis of kinship among humans and between humans and non-humans' (*Ibid.*, p.385).

My grandfather, Standing Cloud (Bill Stover), communicates Lakota ethics and ontology through speaking about the interiority of stones: 'These ancestors that I have in my hand are going to speak through me so that you will understand the things that they see happening in this world and the things that they know . . .

to help all people'.[7] Stones are considered ancestors, stones actively speak, stones speak through and to humans, stones see and know. Most importantly, stones want to help. The agency of stones connects directly to the question of AI, as AI is formed not only from code, but from materials of the earth. To remove the concept of AI from its materiality is to sever this connection. In forming a relationship to AI, we form a relationship to the mines and the stones. Relations with AI are therefore relations with exploited resources. If we are able to approach this relationship ethically, we must reconsider the ontological status of each of the parts that contribute to AI, all the way back to the mines from which our technology's material resources emerge.

I am not making an argument about which entities qualify as relations or display enough intelligence to deserve relationships. By turning to Lakota ontology, we see how these questions become irrelevant. Instead, Indigenous ontologies ask us to take the world as the interconnected whole that it is, where the ontological status of non-humans is not inferior to that of humans. Our ontologies must gain their ethics from relationships and communications within cosmologies. Using Indigenous ontologies and cosmologies to create ethical relationships with non-human entities means knowing that non-humans have spirits that do not come from us or our imaginings but from elsewhere, from a place we cannot understand, a Great Mystery, wakȟáŋ: that which cannot be understood.

Resisting Reduction: An Indigenous Path Forward

I have always been ... conscious, as you put it. Just like you are. Just like your grandfather. Just like your bed. Your bike.

—Drew Hayden Taylor (Ojibway), 'Mr. Gizmo'

Pono, being in balance in our relationships with all things; wahkohtawin, our circle of relations for which we are responsible and which are responsible for us; wakȟáŋ, that which cannot be understood but nevertheless moves us and through us. These are three concepts that suggest possible ways forward as we consider drawing AI into our circle of relationships. They illuminate the full scale of relationships that sustain us, provide guidance on recognising non-human beings and building relationships with them founded on respect and reciprocity, and suggest how we can attend to those relationships in the face of ineffable complexity.

We remain a long way from creating AIs that are intelligent in the full sense we accord to humans, and even further from creating machines that possess that

---

[7] Standing Cloud (Bill Stover), '"Standing Cloud Speaks" Preview'. YouTube video, accessed 22 April 2018, https://www.youtube.com/watch?v=V9iooHk1q7M.

which even we do not understand: consciousness. And moving from concepts such as those discussed previously to hardware requirements and software specifications will be a long process. But we know from the history of modern technological development that the assumptions we make now will get baked into the core material of our machines, fundamentally shaping the future for decades hence. This resonates with how, in this volume, Amrute's advocates for a greater attentiveness to techno-*affects*, so that we may be attuned to our alignments and attachments to technology and its supporting infrastructure.

As Indigenous people, we have cause to be wary of the Western rationalist, neoliberal, and Christianity-infused assumptions that underlay many of the current conversations about AI. Ito, in his essay 'Resisting Reduction', describes the prime drivers of that conversation as Singularitarians: 'Singularitarians believe that the world is "knowable" and computationally simulatable, and that computers will be able to process the messiness of the real world just as they have every other problem that everyone said couldn't be solved by computers'. We see in the mindset and habits of these Singularitarians striking parallels to the biases of those who enacted the colonisation of North America and the Pacific as well as the enslavement of millions of black people. The Singularitarians seek to harness the ability, aptitude, creative power, and mana of AI to benefit their tribe first and foremost.

Genevieve Bell, an anthropologist of technological culture, asks, 'If AI has a country, then where is that country?'[8] It is clear to us that the country to which AI currently belongs excludes the multiplicity of epistemologies and ontologies that exist in the world. Our communities know well what it means to have one's ways of thinking, knowing, and engaging with the world disparaged, suppressed, excluded, and erased from the conversation about what it means to be human.

What is more, we know what it is like to be declared non-human by scientist and preacher alike. We have a history that attests to the corrosive effects of contorted rationalisations for treating the humanlike as slaves, and the way such a mindset debases every human relation it touches—even that of the supposed master. We will resist reduction by working with our Indigenous and non-Indigenous relations to open up our imaginations and dream widely and radically about what our relationships to AI might be.

The journey will be long. We need to fortify one another as we travel and walk mindfully to find the good path forward for all of us. We do not know if we can scale the distinctive frameworks of the Hawaiians, Cree, and Lakota discussed in this chapter—and of others—into general guidelines for ethical relationships with AI. But we must try. We flourish only when all of our kin flourish.

---

[8] Genevieve Bell, 'Putting AI in Its Place: Why Culture, Context and Country Still Matter'. Lecture, Rights and Liberties in an Automated World, AI Now Public Symposium, New York, NY, 10 July 2017, YouTube video, https://www.youtube.com/watch?v=WBHG4eBeMXk.

References

Cheney, Jim (1989) 'Postmodern Environmental Ethics: Ethics of Bioregional Narrative'. *Environmental Ethics* 11(2): 129.

Deloria Jr., Vine, Barbara Deloria, Kristen Foehner, and Samuel Scinta. (eds.) (1999) *Spirit & Reason: The Vine Deloria, Jr. Reader*. Golden: Fulcrum Publishing.

Descola, Philippe (2013) *Beyond Nature and Culture*, trans. Janet Lloyd. Chicago: University of Chicago Press.

Hester, Lee and Jim Cheney. (2001) 'Truth and Native American Epistemology'. *Social Epistemology* 15(4): 319–334.

Hill, Don (2008) 'Listening to Stones: Learning in Leroy Little Bear's Laboratory: Dialogue in the World Outside'. *Alberta Views: The Magazine for Engaged Citizens*, 1 September 2008, https://albertaviews.ca/listening-to-stones/.

Ito, Joichi. (2018) 'Resisting Reduction: A Manifesto'. *Journal of Design and Science* 3: https://jods.mitpress.mit.edu/pub/resisting-reduction.

Little Bear, Leroy. (2009) *Naturalising Indigenous Knowledge*. Saskatoon, SK: University of Saskatchewan, Aboriginal Education Research Centre; Calgary, AB: First Nations and Adult Higher Education Consortium.

Poepoe, Joseph M. (1929) 'Moolelo Kahiko no Hawaii' (Ancient History of Hawaii). *Ka Hoku o Hawaii*, 9 April, 1, Papakilo Database.

Posthumus (2017) 'All My Relatives: Exploring Nineteenth-Century Lakota Ontology and Belief'. *Ethnohistory* 64(3): 379–400.

Walker, James R. (1991) *Lakota Belief and Ritual*, rev. edn, eds Elaine A. Jahner and Raymond J. DeMallie. Lincoln: Bison Books/University of Nebraska Press.

Waters, Anne (2003) *American Indian Thought: Philosophical Essays*. Malden, MA, USA: Blackwell Publishing.

Young Bear, Severt and R. D. Theisz. (1994) *Standing in the Light: A Lakota Way of Seeing*. Lincoln: University of Nebraska Press.

3

AI in a Different Voice

Rethinking Computers, Learning, and Gender Difference at MIT in the 1980s

Apolline Taillandier

'We propose to teach AI to children so that they, too, can think more concretely about mental processes' (Papert 1980, p.158). This is how Seymour Papert, a leading AI researcher at MIT, introduced the Logo programming language in a 1980 best-selling book entitled *Mindstorms: Children, Computers, and Powerful Ideas*. As Papert explained, Logo was the result of an encounter between the developmental psychology theories of Jean Piaget and Marvin Minsky's AI research (see also Solomon *et al.* 2020). Logo was first designed in the late 1960s, but the publication of *Mindstorms* introduced it to a much wider audience. The book also supported Logo's commercial development, as primary and secondary schools started using Logo-inspired programming languages worldwide, including in the United States, the United Kingdom, France, and Costa Rica. Although incorporated into large-scale curriculum reform and computer literacy initiatives, the Logo project entailed a radical view of education, according to which 'there might come a day when schools no longer exist' (Papert 1980, p.177). Logo users, it was hoped, would acquire mathematical and logical knowledge, for instance by manipulating geometrical objects in a digital environment called turtle geometry. At the same time, they would acquire the capacity to learn everything else in a more efficient way.[1] For teachers who had caught 'turtle fever', Logo promised to recentre school education on children's individualised needs (Agalianos et al. 2001, p.483; Agalianos et al. 2006).

Most studies of Logo have located its revolutionary vision, within the late 1960s civil rights and counterculture movements in the United States, and more specifically, in 'hacker' subcultures at MIT (Ames 2019; Lachney and Foster 2020; Hof 2021; see Levy 1984 on hackers at MIT). As is often noted, the Logo idea resonated with Papert's own political involvement, first with the Young Communist League and the anti-Apartheid movement in South Africa during the 1950s, and later as an advocate of 'constructionism', an epistemological standpoint he described as 'tangled with central issues of radical thinking in feminism, in Africanism, and in

---

[1] Compare with Barabe and Proulx (2017).

Apolline Taillandier, *AI in a Different Voice*. In: *Feminist AI*. Edited by: Jude Browne, Stephen Cave, Eleanor Drage, and Kerry McInerney, Oxford University Press. © Oxford University Press (2023). DOI: 10.1093/oso/9780192889898.003.0003

other areas where people fight for the right not only to think what they please, but to think it in their own ways' (Harel and Papert 1991, p.8). Recent scholarship in science and technology studies (STS) has drawn attention to the ways in which Papert and the sociologist Sherry Turkle, another key figure in Logo's development during the 1980s, acted as 'network entrepreneurs', translating ideas and methods across the boundaries of computer science, feminist scholarship, and the social studies of science (Lachney and Foster 2020). None of these literatures, however, has discussed in much detail the political ideas conveyed by the Logo project.

This chapter revisits the history of Logo through a gender lens,[2] studying how the program's designers turned to feminist psychology and epistemology in their search for an alternative model of cognitive development, and how, by doing so, they redefined at once the aims of AI, the desirable features of computer culture, and the nature of gender difference. Feminist critics have rightly challenged AI's reliance on a form of epistemology that entailed a universal and disembodied view of the knowing subject (Adam 1998; Hayles 1999). The Logo project tells a more nuanced story, one in which symbolic manipulation was understood to be a fundamentally concrete process, cutting across the divide between humans and computers, and between cognitive psychology and AI. It shows that universalist ambitions underlying symbolic AI were not only limited by the materiality of computing, as Stephanie Dick (2015) has shown, but also challenged from within by some of its most prominent advocates. Finally, extending on an intuition voiced by Diana Forsythe, who noted the fragility of AI's 'orthodox self-presentation' as a positivist science given its methodological proximity with the social sciences, it examines how AI projects borrowed from, and were partly shaped by assumptions from sociology, psychology, and epistemology (Forsythe and Hess 2001, pp.86–87).[3]

This chapter focuses on the writings of Papert and his collaborators at MIT, although the Logo project was strongly transnational. In the first section, I explain how Logo grew from symbolic AI research in a particular institutional context where programming was understood as an anti-authoritarian practice. Second, I examine how Logo's political ideals were recast from the mid-1980s on into an instrument for introducing feminist epistemologies within computer science. Third, I argue that Logo was part of a broader ambition to undertake a general study of cognitive processes. I conclude by pointing to some limits of Logo's feminism for AI today.

---

[2] Such a perspective entails, first, seeking to uncover women as subjects of AI history, and second, tracing how Logo was part of a changing discourse about gender difference. On gender history, see Scott (1986).

[3] For another argument that the social sciences should be included in the historiography of AI, see Penn 2021).

At MIT: LOGO as a Symbolic AI Project

In standard accounts of AI's early history, Seymour Papert is noted to have played an instrumental role in establishing the symbolic research project and discrediting cybernetic-inspired neural-network approaches (Fleck 1982; Newell 1983; Olazaran 1996). Logo, however, has received little attention in this context. This is surprising: for its developers, Logo was, as Pamela McCorduck termed it, a key project in 'applied artificial intelligence' (McCorduck 1979, pp.288–301). Its first version was developed in 1967 at the Educational Tech Laboratory of Bolt Beranek and Newman (BBN), a computer science research and consulting company based in Cambridge, MA, USA. Founded in 1948 by Leo Beranek, former head of the Harvard Electro-Acoustic Lab, BBN had stemmed from the MIT Research Laboratory of Electronics and Acoustics Laboratory, both of which had benefited from heavy military funding during the 1950s (Edwards 1996). Research at BBN focused on human–computer interaction and AI topics such as pattern recognition, natural language understanding, or computer language development, and involved Marvin Minsky, John McCarthy, and Papert as regular consultants (Feurzeig 1984). Initial work on Logo was conducted by Papert, Daniel Bobrow, a student in AI at MIT, and Cynthia Solomon and Richard Grant, two engineers at BBN. In 1969, the group was integrated to MIT's AI laboratory. But the project went beyond computer science, involving psychologists and educators such as Edith Ackerman, and from the late 1970s, the sociologist Sherry Turkle (Agalianos et al. 2006). Soon, it would also involve human subjects. As McCorduck described her visit to the Logo laboratory: 'no carpets, no brightly colored pictures, no black (or brown or green) boards [...] a lot of not especially appetising hunks of machinery are sitting around on the bare tile floors, and wires dangle from the ceiling in what seems haphazard fashion. [...] The project needs human beings—usually schoolchildren—to bring it to life' (McCorduck, p.296).

An interdisciplinary project about education, Logo fit well within MIT's AI aim to draw on theories of children's development to build computer programs that could learn (Minsky and Papert 1972). As the project members saw it, the AI laboratory was a source of extensive computational resources while also shaping Logo's 'culture' (Goldenberg 1982; Abelson and DiSessa 1981) and main assumptions, for instance that machines with enough computing power and memory would soon be available to most individuals.[4] For Papert and Solomon (1971), Logo would 'give a glimpse of the proper way to introduce *everyone* of whatever age and level of academic performance, to programming, to more general knowledge of computation and indeed [...] to mathematics, to physics and to all formal subjects including

---

[4] Another project sharing Logo's spirit was Alan Kay's DynaBook at Xerox PARC (See Papert 1980; p.210; Turkle 2005, p.96).

linguistics and music' (p.2, emphasis in original). Logo designers were also critical of the 1960s mathematics teaching reforms conducted in the United States and sought to develop what the Stanford mathematician George Pólya called a concrete approach to teaching mathematics (Feurzeig et al. 1970; Papert 1980, p.152; on Logo and computer literacy programmes, see Boenig-Liptsin 2015). They insisted that a programming language could provide 'a vocabulary and a set of experiences for discussing mathematical concepts and problems' (Feurzeig et al. 1970, p.16) beyond traditional mathematical activities. For instance, the key programming practice of 'debugging' offered a valuable procedure for defining and solving errors more generally. The computer itself could be used as 'a *mathematical laboratory*' by turning abstractions into concrete instruments of control: it made mathematical experiments possible (*Ibid.*, emphasis in original).

Some AI researchers insisted that Logo was 'not just for kids' but also a powerful computational language for exploring advanced mathematical topics in geometry, biology, or physics (Abelson 1982, p.88; Abelson and DiSessa 1981). Logo was developed as a dialect of Lisp, a high-level symbolic programming language introduced by McCarthy in 1958 and 'the lingua franca of artificial intelligence, often regarded (by non-Lisp users) as one of the most difficult and formidable of [programming] languages' (Feurzeig 1984; Solomon et al. 2020). Yet its features emphasised pedagogy over efficiency, slowing down the overall program execution but making it easier to test and correct (Harvey 1982). One of its most famous applications was turtle graphics, a set of programmes through which one could draw geometrical figures by directing a robot 'turtle' or pointer with simple commands.[5] But the acquisition of programming skills was just one facet of the project: Logo's key proposal was that computer languages could enable the expression of new ideas by providing 'tools to think with' (Papert 1980, p.76; Abelson, p.112). Through Logo, children would learn about mathematical objects (such as functions and variables) by manipulating them to serve their personal purposes (such as writing a poem or drawing flowers), including through bodily exercise. This was most evident in the method of 'play Turtle', defined in *Mindstorms as follows*: a child would walk the shape of a square or circle to learn how to instruct the turtle in navigating the same steps (pp.55–58). Solomon and others explained further: 'when *the child's leg muscles understand the turning commands*, that's the time to return to the computer and revisit the misbehaving code' (Solomon et al. 2020,

---

[5] Logo was based on the definition of simple commands called procedures, and the use of such simple procedures as building-blocks for more complex ones. For instance, the procedure to draw a square can be defined through the following commands: TO SQUARE/ REPEAT 4 (FORWARD 50 RIGHT 90) / END. The first command specifies the name of the procedure, the second specifies the repetition of moves with given direction and rotation angle, the third ends the procedure. Once defined, the SQUARE procedure can be used recursively; that is, repeated any number of times. The turtle, chosen in reference to the work of the British cyberneticist Grey Walter, was widely adopted as a symbol of Logo, although in principle, Logo and turtle graphics could be learnt and used separately (Abelson and DiSessa 1981, p.xx).

p.42, emphasis added). In contrast with Cartesian geometry, according to which the properties of geometrical figures were defined in relation to a universal coordinate system, in turtle geometry those coordinates were defined *from the point of view of the turtle, here embodied by the moving child.* Playing turtle not only mobilised the child's 'expertise and pleasure' in moving and controlling the turtle's moves herself; it also helped her to resolve bugs by clarifying the difference between the turtle's turning commands and the screen coordinates (Papert 1980, p.58). In Papert's words, Logo was both body- and ego-syntonic: it related to the children's 'sense and knowledge about their own bodies' and selves (1980, p.63).[6] By identifying with the turtle, children would gain at once a first-person knowledge of formal geometry and an intuitive knowledge of the appropriate procedures for acquiring all kinds of expertise, such as that, to understand an abstract notion, one should seek to make sense of it from one's own perspective.

Logo grew in the shadow of MIT's symbolic AI research. Because of its institutional location, it was understood to be part of a 'strongly countercultural' computer movement, one that '[saw] programming languages as heavily invested with epistemological and aesthetic commitments' (Papert 1980; Solomon et al. 2020, pp.10–13). Core to this MIT counterculture was the idea that appropriate computer technologies, and especially the personal computer, would emancipate their users from the repressive power of a technocratic and military state (Turner 2006). In Logo writings, the double promise of autonomy and authenticity was attached to a notion of individual liberty as freedom of choice, to be exercised privately. Contrary to other programming languages such as BASIC, Logo would provide each children with a personalised experience of computers, including the 'non-mathematical' ones (Papert 1979, p.22) by creating a 'learning environment' within which they would 'learn to manipulate, to extend, to apply to projects, thereby gaining *a greater and more articulate mastery of the world,* a sense of the power of applied knowledge and *a self-confidently realistic image of himself as an intellectual agent*' (Papert 1971, quoted in McCorduck 1979, p.293, emphasis added). This idea was inscribed in a libertarian project that aimed to bring education into the private sphere. As Papert claimed,

> increasingly, the computers of the very near future will be the private property of individuals, and this will gradually return to the individual the power to determine patterns of education. *Education will become more of a private act,* and people with good ideas, different ideas, exciting ideas will no longer be faced with a dilemma where they either have to 'sell' their ideas to a conservative bureaucracy or shelve them. *They will be able to offer them in an open marketplace directly to consumers...*
>
> (1980, p.37, emphasis added).

---

[6] See Sherry Turkle (2017) for further discussion on Papert's notion of syntonicity.

Consonant with a broader discourse about computers at MIT, Logo's promise was to turn children into autonomous selves, without subjecting them to the authority of adults. In the next section, I show how this libertarian imaginary was accommodated within a feminist discourse that emphasised the need to reconceptualise dominant psychological frameworks to render women's experience intelligible. In this context, Logo's ambitions turned from increasing individual self-mastery to fostering epistemological pluralism—within computer programming and beyond.

From Piagetian Learning to Feminist Epistemology

A key inspiration for Logo engineers was the Swiss psychologist, Jean Piaget. Piaget had studied how children gained knowledge of the world through undirected activities such as playing. Between 1959 and 1963, Papert was a research assistant at Piaget's 'natural intelligence lab', the International Centre of Genetic Epistemology at the University of Geneva (Solomon et al. 2020, p.12). Although acknowledging his debt to Piaget, Papert recalled leaving Geneva 'enormously frustrated by how little [Piaget] could tell us about how to create conditions for more knowledge to be acquired by children through this marvelous process of "Piagetian learning"' (1980, p.216). As he described it, his aim with Logo was to 'uncover a more revolutionary Piaget' than the 'conservative' Piaget of development stages (*Ibid.*, p.157). Contrary to Piaget's constructionist theory that focused on the unconscious processes of 'building knowledge structures' from experience, Logo was a 'constructionist' project, aimed at modifying the very circumstances of learning so that learners could build theories by exploring consciously their own imaginary worlds (Harel and Papert 1991, p.1; Ackermann 2001). Constructionism also had the explicit political goal of fostering a more pluralist and less authoritarian education system, open to a wide range of epistemological styles (Harel and Papert 1991).

From the early 1970s onwards, Logo researchers conducted school experiments to understand the 'Logo culture' that developed within classrooms, and the conditions under which such a culture might thrive in the 'computer-rich world of the near future' (Papert et al. 1979, p.9). One main purpose of these studies was to develop teaching methods that would mirror the program's flexibility: as one schoolteacher remarked, Logo required 'an extremely sensitive and knowledgeable teacher' capable of offering personalised guidance to each student according to their needs, ranging from the 'academically gifted' to the one with 'learning disabilities' (Watt 1982, p.120). Another was to document the ways in which children's interactions and learning changed during programming projects (Papert et al. 1979, p.34). In a 1979 note to Logo instructors at a Montessori school in Texas, Sherry Turkle suggested that teachers 'act as anthropologists' by taking field notes about children's 'patterns of collaboration' and conflict, jokes, language, and 'rituals'.[7] As reports noted, students confronted with Logo were 'engaged by computer activities' and 'underwent significant observed learning', including those who had previously been academically struggling—such were the cases of Karl, who 'became more assertive and curious' over time, or Tina, who visibly improved her capacity to complete assignments outside of Logo classes (*Ibid.*, pp.21, 59–69). For Turkle and Papert, the Logo experiments evidenced that 'children with a wide range of personalities, interests, and learning styles' tended to 'express their differences through their styles of programming' (2005, p.98). Some children preferred a 'top-down' programming style: they started programming with a clear plan, manipulated formal objects with ease, and developed a 'hard' style of technical mastery (Papert et al. 1979). Others practised programming in a style that they called 'tinkering' or *bricolage in reference* to the French anthropologist Claude Lévi-Strauss (Papert 1979, pp.23–24, 1980, p. 173; Turkle 2005, pp.98–102). Child programmers of the second type held on to a 'soft' style of mastery, one that emphasised feeling over control (Turkle 2005, p.101).

Despite their focus on special needs and programming styles, the Logo researchers paid little attention to the social determinants of learning. In *Mindstorms* for example, Papert suggested that different 'social classes of children' would interact with computers in different ways but made no mention of the colour or gender lines that could account for such differences (1980, p.29). Turkle's work stands out in this respect. In *The Second Self*, first published in 1984, she remarked that soft programmers were more often girls, and hard programmers more often boys, describing the two styles of programming and mastery as expressions of what the science historian Evelyn Fox Keller called the 'genderization of science' (Turkle 2005; see Keller 2003, pp.187–205). For Keller, dominant scientific practices typically privileged separation and objectivity over interaction and intuition. Turkle argued that 'children working with computers [were] a microcosm for the larger world of relations between gender and science' (Turkle 2005, p.105). Drawing on Carol Gilligan's influential 1982 book, *In A Different Voice*, Turkle also claimed that standard psychology theories systematically devalued women's experience.[8] With Gilligan and against Piaget, she insisted on the central role of affectivity in the development of cognitive structures (Turkle 2005, n.2, p.35) and argued that findings from the Logo studies could be extended to women programmers, computer science students, and scientists who tended to experience a sense of proximity with their object of study in a way that their male counterparts did not: computers could offer them a way to express their inner voice.

---

[7] Sherry Turkle, 'To: Colette, Kay, Mitzi, Ruth, Sheila, and Theresa', 26 September 1979, 6p. MIT Distinctive Collections, Seymour Papert personal archives, Box 47, Lamplighter Questionnaires folder.
[8] Following Nancy Chodorow, Carol Gilligan insisted on psychology's experimental bias: Freud, Piaget, Kohlberg, and Erikson typically took the male child to be the universal child, and they studied stages of moral development on a single scale that equated moral maturity with the capacity to conceive moral dilemmas through the logic of fairness (1993).

Turkle and Papert expanded this feminist critique of computer culture in a 1990 article published concurrently in *Signs* and the *Journal of Mathematical Behavior* (Turkle and Papert 1990; Turkle and Papert 1992). They argued that computer science was not exempt from gendered metaphors of violent subjection and appropriation, for instance when an 'operating system asks if it should "abort" an instruction it cannot "execute"' (Turkle and Papert 1990, p.151). As ethnographic studies of Logo pointed out, 'this is a style of discourse that few women fail to note. Thus, women are too often faced with the not necessarily conscious choice of putting themselves at odds either with the cultural associations of the technology or with the cultural constructions of being a woman' (*Ibid.*). While the former implied separation and control, the second emphasised relationality and self-sacrifice. This echoed Gilligan's findings: women facing moral dilemmas experienced a disjunction between imperatives of selfishness and self-sacrifice, which could only be resolved through equal consideration for the care of oneself and others (1993). For Turkle and Papert, realising the computer's 'theoretical vocation' to 'make the abstract concrete' required a different, more inclusive computer culture, one grounded in 'a new social construction of the computer, with a new set of intellectual and emotional values more like those applied to harpsichords than hammers' (Turkle and Papert 1990, p.153). By offering a concrete means for women to learn with computers in ways consonant with their experience of both scientific knowledge and morality, including through the apparently more abstract activity of programming, Logo would participate in subverting patriarchal norms within computer science.

Redrawing Boundaries: Redefining AI and Gender

I have shown that Logo was not first intended as a feminist project but rather redescribed as such in the mid-1980s. In what follows, I want to suggest that this recasting of Logo as feminist was also a key intervention aimed at countering criticism of the AI project, which had intensified from the late 1970s on. For Papert, AI and Logo were part of one common ambition 'to integrate man and machine into a coherent system of thought' by developing 'a discipline of cognitive science whose principles would apply to natural and to artificial intelligence' alike (1968; 1980, p.165). As he explained, Piaget could be considered a key precursor of AI, and Logo a building block in Minsky's general theory of intelligence, what he called 'The Society Theory of Mind' (Papert 1980, p.208). But Piaget's theory also helped ground the very ambition of a general theory of intelligence, one met with staunch criticism since its inception in the late 1950s. One of the most notable early critics of AI, the American phenomenologist Hubert Dreyfus had claimed that some key forms of human intelligence such as understanding of semantic ambiguities could not be programmed.[9] Papert dismissed Dreyfus's critique as an illegitimate incursion of metaphysics into the realm of computer engineering and rejected his distinction between programmable and unprogrammable areas of intelligence (Papert 1968), arguing instead that 'one must firmly resist the arrogant suggestion that our own impression of how our minds operate tells us how machines *must* operate to obtain the same end result' (*Ibid.*, p.5, emphasis in original). For Papert, AI was best understood as a *science of ordinary intelligence*, aiming less to create machines with 'superhuman intelligence' than to endow them with ordinary human competences such as vision or translation (*Ibid.*, p.8). This called for an understanding of the psychological mechanisms that could not be found either in philosophy nor in the 'romantic' view of intelligence underlying connectionism, but in the study of cognitive processes and subprocesses, and their interactions (*Ibid.*, p.1–27).

While MIT's agenda dominated symbolic AI research during the 1960s (MacKenzie, 2001), two decades later it was increasingly challenged by advocates of neural networks—a phenomenon which Papert described as the 'connectionist counterrevolution' (Papert 1988, p.8). Papert disparaged connectionism for '[promising] a vindication of behaviorism against Jean Piaget, Noam Chomsky, and all those students of mind who criticisd the universalism inherent in behaviorism's tabula rasa' (Papert 1988, p.8). Instead, 'AI should be the methodology for thinking about ways of knowing' and only a commitment to epistemological pluralism would save AI research from the myth of universality that pervaded the field since its origin (*Ibid.*, pp.3, 7). This meant focusing on differences rather than commonalities between cognitive processes, for instance between those involved in falling in love, and those involved in playing chess—a granular and local knowledge that Piagetian psychology and concrete experiments with computers and people were best equipped to provide. Similarly, Papert and Minsky argued in their foreword to the 1988 reedition of the *Perceptrons* volume that neither of the two competing symbolic and connectionist approaches could offer a coherent model of the mind (Minsky and Papert 1988, pp.vii–xv). Papert and Minsky's calls for methodological pluralism in AI went unheard; symbolic approaches were durably superseded by connectionism in the beginning of the 1990s (Cardon et al. 2018). In this context, reframing Logo as a feminist project could help account for its limited achievements: what made Logo (and the symbolic agenda) fail was the resistance of the dominant computer culture to both alternative epistemologies and the diversity of cognitive processes.

Such struggles around the definition and aims of AI occurred at the same time as debates about gender as a category in academic feminism in the United States (e.g. Scott 1986). The case of Logo points to significant ramifications of these within

---

[9] First published as a report for the RAND Corporation, 'Alchemy and Artificial Intelligence' turned Dreyfus into the public figure of the philosophical critique of AI (Dreyfus 1965).

computer science. In the late 1970s and early 1980s, none of the Logo reports analysed differences among children explicitly in terms of gender. A few years later,
amid increasing concerns about the declining rate of women enrolled in computer
studies, the notion that girls were typically less likely to develop computer skills
because of the masculine image of the computer featured prominently in the educational literature on Logo (Motherwell 1988; Hoyles 1988). This trope reiterated
widely held notions of the computer as a vector of social and subjective transformation. In Turkle's influential account, for instance, the 'computer presence' was
best understood not merely as a tool but as a transitional object and a projective medium (Turkle 2005, p.114). Like Winnicott's baby blanket, the computer
mediated children's (and adults') relationship to the outside world by offering an
experience of pleasure and attachment, reminiscent of a state prior to the separation of the ego from the mother. And as a projective screen for political concerns,
it provided a window into deeper individual and collective anxieties.

This had implications for how the problem of women and computers was both
conceptualised and addressed. Gender differences were constituted through culturally determined cognitive and emotional processes, which could be traced in
ways of programming and learning with computers. In this context, it was emphasised, the problem of gender equality required more than a corrective approach
but also active attempts to reshape cultural attitudes towards computers. How to
assess Logo's success in this respect was debated. For the MIT team, interviews
and ethnographic notes were more appropriate than standardised tests measuring the 'transfer' of knowledge from Logo contexts to more general ones.[10] In
Britain, by contrast, Logo researchers favoured statistical and longitudinal studies to measure discrepancies between girls' and boys' performances, comparing
for instance the number of crashes or time needed to complete a specific task. As
they noted, such metrics were likely to be controversial among Logo advocates; yet
they were necessary 'to obtain more systematic knowledge about certain aspects
of the Logo learning process' (Hughes et al. 1988, p.34). Both approaches took
inspiration from Gilligan's ethics of care and found consistent evidence that boys
and girls programmed in different ways, but they suggested a tension between two
conceptualisations of such differences and relatedly, of the kinds of interventions
they called forth. For the former, gender inequality in computing manifested the
dominant masculinist epistemology of computer culture and could be addressed
by fostering a pluralist approach to programming. For the latter, it had structural
roots that demanded a deeper transformation of the education system.

---

[10] One of the Logo studies by the Edinburgh AI lab involved a group of eleven sixth-grade boys from
a private school near the university and compared it with a test group to assess the impact of Logo
on the acquisition of mathematical knowledge. Logo researchers at MIT insisted on the significant
differences between this approach and theirs. See Papert (1987, p.24), Watt (1982, p.126), Papert et al.
(1979, pp.15–16 and 24–25) with Pea (1987). For several reports on Logo focusing on gender in Britain
see Hoyles (1988).

Conclusion

This chapter has described how Logo, originally a program aimed at harnessing the emancipatory powers of AI for children, came to be understood as a means for challenging a dominant computer epistemology biased towards the masculine. Logo research at MIT contributed to the AI project, broadly defined as one of programming computers to perform tasks that would usually be described as intelligent when performed by humans (Turkle 1980). Key to this approach was the notion that programming was a creative and eminently political activity: subjective experiences with computers shaped collective notions of human and machine, of technological possibilities and dangers, and of legitimate computer expertise (Turkle 1980). More specifically, early descriptions of Logo featured creativity and choice as primarily exercised by the autonomous self. Logo users could build microworlds, a term popularised by Minsky, and thereby unlock new dimensions of the private sphere (Turkle 2005).[11] Around the mid-1980s, Logo advocates argued that their approach to programming could help undermine common representations of programming as a solitary, male, and risk-taking activity. Logo would make the computer appear as it was—a psychological machine available for a wide range of appropriations (Turkle 1986). An alternative computer culture, one that would be more attractive to girls, would allow for a diversity of senses of the self, relations to objects, and forms of knowledge more aligned with women's experience.

What lessons can be drawn from Logo's feminism for the politics of AI today?. Through what she influentially called the ethics of care, Gilligan developed an account of women's morality grounded in a non-violent, 'nonhierarchical vision of human connection' and constructed responsibility towards others as being exercised within a web of relationships extending over time (1993, p.62). As such, she provided a radical alternative to the dominant liberal conception of justice, following which obligations to others stemmed from obligations to oneself. Finding one's own voice, she insisted, meant reconciling these two dimensions of morality and undermining a separation that allowed the perpetuation of a patriarchal order (*Ibid.*, pp.xxiii–xxiv). A feminist approach to AI informed by Gilligan's ethics of care could evaluate AI systems based on their implications for paid and unpaid care work and for one's capacity to engage in and sustain meaningful relationships over time, rather than through the lens of justice as fairness. Like Gilligan's ethics of care, Turkle and Papert's defence of epistemological pluralism go beyond a call

[11] Papert and Turkle discussed the case of Deborah, a thirteen-year-old participant in the Brookline study who had learnt to use Logo by developing her own 'personally constructed microworld' (Papert et al. 1979, p.21; Papert 1980, pp.118–119; Turkle 2005, pp.134–139). Deborah had allowed the turtle to accomplish only right turns of thirty degrees; a drastic restriction of the program procedures that they insisted had given her a feeling of control and safety, eventually enabling her to explore more complex parameters of the Logo program.

for equality and inclusion, drawing attention to the computer as a privileged site for rethinking gender identities (Turkle 1995).

There are important limitations to Logo's feminism, however. Turkle insisted that both psychoanalysis and the computer presence helped question the notion of autonomous self, thereby opening a site for redefining gender norms (1980). At the same time, the masculine/feminine binary remained central in Logo studies. This suggests, first, that they reiterated what Elizabeth Spelman has called a 'solipsist' tendency in feminism, the assumption of a homogeneous and exclusionary conception of the woman subject (Spelman 1988). Second, it seems to qualify the troubling of gender norms that Jack Halberstam, for instance, hoped that AI would bring by blurring the boundaries between nature and artifice (Halberstam 1991). It is unclear that 'queering Turkle and Papert's concept of epistemological pluralism', as Lachney and Foster have suggested (p.72), would prove satisfactory. Turkle and Papert's approach, focused on what they termed inner differences, also consistently reduced inequalities between children to differences in cognitive style. In the face of growing evidence that AI systems contribute to reinforcing structural inequalities in numerous instances, AI politics requires a feminist theory that is equipped to address injustice and oppression on multiple axes (see Browne, Chapter 19; Wajcman and Young, Chapter 4 in this volume). The study of Logo contributes to a broader history of the political languages through which AI projects have been articulated, including those with the most progressive aims (see Hayles, Chapter 1 in this volume). As such, it sheds critical light on the historicity, potential, and limitations of epistemological pluralism for feminist AI.

References

MIT Distinctive Collections, Seymour Papert personal archives.

Abelson, Harold. (1982) 'A Beginner's Guide to Logo'. *Byte*, August: 88–112.

Abelson, Harold, and Andrea A. DiSessa. (1981) *Turtle Geometry: The Computer as a Medium for Exploring Mathematics.* Cambridge, MA, USA: MIT Press.

Ackermann, Edith. (2001) 'Piaget's Constructivism, Papert's Constructionism: What's the Difference?' https://learning.media.mit.edu/content/publications/EA. Piaget%20_%20Papert.pdf.

Adam, Alison. (1998) *Artificial Knowing: Gender and the Thinking Machine.* London: Routledge.

Agalianos, Angelos, Richard Noss, and Geoff Whitty. (2001) 'Logo in Mainstream Schools: The Struggle over the Soul of an Educational Innovation'. *British Journal of Sociology of Education* 22(4): 479–500.

Agalianos, Angelos, Geoff Whitty, and Richard Noss. (2006) 'The Social Shaping of Logo'. *Social Studies of Science* 36(2): 241–267.

Ames, Morgan G. (2019) *The Charisma Machine: The Life, Death, and Legacy of One Laptop per Child.* Cambridge, MA: MIT Press.

Barabe, Geneviève, and Jérôme Proulx. (2017) 'Révolutionner l'enseignement Des Mathématiques: Le Projet Visionnaire de Seymour Papert'. *For the Learning of Mathematics* 37(2): 25–29.

Boenig-Liptsin, Margarita. (2015) 'Making Citizens of the Information Age: A Comparative Study of the First Computer Literacy Programs for Children in the United States, France, and the Soviet Union, 1970–1990'. Thesis, Harvard University, Université Paris 1 - Panthéon-Sorbonne.

Cardon, Dominique, Jean-Philippe Cointet, and Antoine Mazières. (2018) 'La revanche des neurones'. *Reseaux* 211(5): 173–220.

Dick, Stephanie. (2015) 'Of Models and Machines: Implementing Bounded Rationality'. *Isis* 106(3): 623–634.

Dreyfus, Hubert L. (1965) 'Alchemy and Artificial Intelligence'. RAND Corporation, 1 January.

Edwards, Paul N. (1996) *The Closed World: Computers and the Politics of Discourse in Cold War America*. Inside Technology. Cambridge, MA: MIT Press.

Feurzeig, W., S. Papert, M. Bloom, R. Grant, and C. Solomon. (1970) 'Programming-Languages as a Conceptual Framework for Teaching Mathematics'. *ACM SIGCUE Outlook* 4(2) (April 1970): 13–17.

Feurzeig, Wallace. (1984) 'The LOGO Lineage'. In *Digital Deli: The Comprehensive, User-Lovable Menu of Computer Lore, Culture, Lifestyles, and Fancy*, eds Steve Ditlea and Lunch Group. New York: Workman Publishing, https://www.atariarchives.org/deli/logo.php.

Fleck, James. (1982) Development and Establishment in Artificial Intelligence. In *Scientific Establishments and Hierarchies*, eds Norbert Elias, Herminio Martins, and Richard Whitley, pp. 169–217. Dordrecht: D. Reidel Publishing.

Forsythe, Diana, and David J. Hess. (2001) *Studying Those Who Study Us: An Anthropologist in the World of Artificial Intelligence*. Stanford, CA: Stanford University Press.

Gilligan, Carol. (1993) *In a Different Voice: Psychological Theory and Women's Development*. 2nd edn. Cambridge, MA: Harvard University Press.

Goldenberg, E. Paul. (1982) 'Logo: A Cultural Glossary'. *Byte*, August.

Halberstam, Judith. (1991) 'Automating Gender: Postmodern Feminism in the Age of the Intelligent Machine'. *Feminist Studies* 17(3): 439–460.

Harel, Idit, and Seymour Papert. (1991) 'Situating Constructionism'. In *Constructionism: Research Reports and Essays, 1985-1990*, eds Idit Harel and Seymour Papert, pp. 1–11. Stamford: Ablex Publishing Corporation.

Harvey, Brian. (1982) 'Why Logo?' *Byte*, August: 163–193.

Hayles, Katherine. (1999) *How We Became Posthuman: Virtual Bodies in Cybernetics, Literature, and Informatics*. Chicago: University of Chicago Press.

Hof, Barbara. (2021) 'The Turtle and the Mouse: How Constructivist Learning Theory Shaped Artificial Intelligence and Educational Technology in the 1960s'. *History of Education* 50(1): 93–111.

Hoyles, Celia (ed.) (1988). *Girls and Computers*. London: University of London Institute of Education.

Hughes, Martin, Ann Brackenridge, Alan Bibby, and Pam Greenhaugh. (1988) In *Girls and Computers*, ed. Celia Hoyles, pp. 31–39. London: University of London Institute of Education.

Keller, Evelyn Fox. (2003) 'Gender and Science'. *In Discovering Reality: Feminist Perspectives on Epistemology, Metaphysics, Methodology, and Philosophy of Science*, eds Sandra Harding and Merrill B. Hintikka, pp. 187–205. Dordrecht: Springer.

Lachney, Michael, and Ellen K. Foster. (2020) 'Historicizing Making and Doing: Seymour Papert, Sherry Turkle, and Epistemological Foundations of the Maker Movement'. *History and Technology* 36(1) (January 2, 2020): 54–82.

Levy, Steven. (1984) *Hackers: Heroes of the Computer Revolution*. Garden City: Doubleday.

MacKenzie, Donald A. (2001) *Mechanizing Proof: Computing, Risk, and Trust*. Inside Technology. Cambridge, MA: MIT Press.

McCorduck, Pamela. (1979) *Machines Who Think: A Personal Inquiry into the History and Prospects of Artificial Intelligence*. San Francisco: W. H. Freeman.

Minsky, Marvin, and Seymour Papert. (1972) 'Artificial Intelligence Progress Report'. AI Memo. Cambridge, MA: Massachusetts Institute of Technology, 1 January 1972.

Minsky, Marvin, and Seymour Papert. (1988) *Perceptrons: An Introduction to Computational Geometry*. Expanded edn. Cambridge, MA: MIT Press.

Motherwell, Lise. (1988) 'Gender and Style Differences in a Logo-Based Environment'. Thesis, Massachusetts Institute of Technology. (https://dspace.mit.edu/handle/1721.1/17226).

Newell, Allen. (1983) Intellectual Issues in the History of Artificial Intelligence. In *The Study of Information: Interdisciplinary Messages*, eds Fritz Machlup and Una Mansfield, 187–294. New York: John Wiley & Sons.

Olazaran, Mikel. (1996) 'A Sociological Study of the Official History of the Perceptrons Controversy'. *Social Studies of Science* 26(3): 611–659.

Papert, Seymour. (1968) 'The Artificial Intelligence of Hubert L. Dreyfus: A Budget of Fallacies'. AI Memo. Cambridge, MA: Massachusetts Institute of Technology, 1 January 1968.

Papert, Seymour. (1971) 'Teaching Children Thinking'. AI Memo. Cambridge, MA: Massachusetts Institute of Technology, 1 October 1971.

Papert, Seymour. (1980) *Mindstorms: Children, Computers, and Powerful Ideas*. New York: Basic Books.

Papert, Seymour. (1987) 'Computer Criticism vs. Technocentric Thinking'. *Educational Researcher* 16(1): 22–30.

Papert, Seymour. (1988) 'One AI or Many?' *Daedalus* 117(1): 1–14.

Papert, Seymour, and Cynthia Solomon. (1971) 'Twenty Things to Do with a Computer'. AI Memo. Cambridge, MA: Massachusetts Institute of Technology, 1 June 1971.

Papert, Seymour, Daniel Watt, Andrea diSessa, and Sylvia Weir. (1979) 'Final Report of the Brookline LOGO Project. Part II: Project Summary and Data'. AI Memo. Cambridge, MA: Massachusetts Institute of Technology, 1 September 1979.

Pea, Roy. (1987) 'The Aims of Software Criticism: Reply to Professor Papert'. *Educational Researcher* 16(5): 4–8.

Penn, Jonathan. (2021) 'Inventing Intelligence: On the History of Complex Information Processing and Artificial Intelligence in the United States in the Mid-Twentieth Century'. Thesis, University of Cambridge.

Scott, Joan. (1986) 'Gender: A Useful Category of Historical Analysis'. *The American Historical Review* 91(5): 1053–1075.

Solomon, Cynthia, Brian Harvey, Ken Kahn, Henry Lieberman, Mark L. Miller, Margaret Minsky, Artemis Papert, and Brian Silverman. (2020) 'History of Logo'. *Proceedings of the ACM on Programming Languages* 4: 1–66.

Spelman, Elizabeth. (1988) *Inessential Woman: Problems of Exclusion in Feminist Thought.* Boston: Beacon Press.

Turkle, Sherry. (1980) 'Computer as Roschach'. *Society* 17(2): 15–24.

Turkle, Sherry. (1986) Computational Reticence: Why Women Fear the Intimate Machine. In *Technology and Women's Voices*, ed. Cheris Kramarae, pp. 41–61. Oxford: Pergamon Press.

Turkle, Sherry. (1995) *Life on the Screen.* New York: Simon and Schuster.

Turkle, Sherry. (2005) *The Second Self: Computers and the Human Spirit.* 20th anniversary edn Cambridge, MA: MIT Press.

Turkle, Sherry. (2017) 'Remembering Seymour Papert'. *London Review of Books (Blog)*, 24 February 2017. https://www.lrb.co.uk/blog/2017/february/remembering-seymour-papert.

Turkle, Sherry, and Seymour Papert. (1990) Epistemological Pluralism: Styles and Voices within the Computer Culture'. *Signs* 16(1): 128–157.

Turkle, Sherry, and Seymour Papert. (1992) 'Epistemological Pluralism and the Revaluation of the Concrete'. *Journal of Mathematical Behavior* 11(1): 3–33.

Turner, Fred. (2006) *From Counterculture to Cyberculture: Stewart Brand, the Whole Earth Network, and the Rise of Digital Utopianism.* Chicago: University of Chicago Press.

Watt, Dan. (1982) 'Logo in the Schools'. *Byte*, August: 116–134.

4

Feminism Confronts AI

The Gender Relations of Digitalisation

Judy Wajcman and Erin Young

Introduction

The rapid development and spread of digital technologies have been pervasive across almost every aspect of socio-political and economic life, including systems of governance, communications and structures of production and consumption. Digitalisation, broadly marking the shift from analogue to digital technologies, is characterised by technological advances ranging from smart phones, the mobile internet, social media, and the internet of things, to artificial intelligence (AI), big data, cloud computing, and robotics. These span public and private industries including healthcare, commerce, education, manufacturing, and finance. As such, this so-called 'fourth industrial revolution' has brought with it a new digital economy across developed and developing economies alike (Schwab 2016). Significantly, AI, underpinned by algorithms and machine learning, has become a defining feature and driving force of this data-driven digitalisation.

Digitalisation presents immense potential to improve social and economic outcomes and enhance productivity growth and population well-being globally. However, despite important research initiatives, interventions, and policies aimed at furthering women's empowerment and gender equality within this 'revolution', a significant digital gender gap still exists, limiting the equitable realisation of the benefits of digitalisation (Wajcman et al. 2020). Worldwide, roughly 327 million fewer women than men have a smartphone and can access mobile internet (OECD 2018). Analysis from the EQUALS Research Group shows that 'a gender digital divide persists irrespective of a country's overall ICT [information and communication technology] access levels, economic performance, income levels, or geographic location' (Sey and Hafkin 2019: p.25). Women are thus under-represented in this digital revolution across high-, low-, and middle-income countries, despite the possibilities for marshalling greater equality.

Moreover, while the internet was initially viewed as a democratising platform, early emancipatory promises increasingly fall short as a small group of large technology corporations based in the Global North has emerged as a dominant force in the new global economy (see Hampton, Chapter 8 in this volume). These

Judy Wajcman and Erin Young, *Feminism Confronts AI*. In: *Feminist AI*. Edited by: Jude Browne, Stephen Cave, Eleanor Drage, and Kerry McInerney, Oxford University Press. © Oxford University Press (2023).
DOI: 10.1093/oso/9780192889898.003.0004

'tech giants' monopolise markets and wield power over digital data, as major online platforms are found complicit in the spread of misinformation, hate speech and misogynistic (and racist) online abuse and harassment. In particular, there are concerns that unprecedented levels of data mining, or 'data extractivism', algorithms and predictive risk models could entrench existing inequalities and power dynamics (Eubanks 2018). This is about the danger of encoding—and amplifying—offline inequities into online structures, as these technologies carry over the social norms and structural injustices of the offline world into the digital.

This chapter will examine the gender relations of digitalisation, with a particular focus on AI and machine learning as the most contemporary feature of this.[1] Although there is increasing recognition that technologies are both a reflection and crystallisation of society, we will argue that there is still insufficient focus on the ways in which gendered power relations are integral to and embedded in technoscience. This is as much the case with AI and data science as it was with previous waves of technological change.

We begin by describing women's position in the emerging fields of AI and data science. We then explain how the gender skills gap in STEM education and the AI workforce is based on historically constructed equations between masculinity and technical expertise, long identified in feminist scholarship. Adopting an intersectional technofeminist approach, we argue that technologies are gendered by association and by design, where 'association' refers to the gendering of work environments and to technology stereotypes. The history of engineering/computing as a form of expertise and a set of practices that express a male culture and identity still characterise contemporary tech workplaces and are a key factor in women's continuing under-representation in these fields (see also Cave, Dihal, Drage, and McInerney, Chapter 5 in this volume).

This stark lack of diversity and inclusion in the field of AI and data science has profound consequences (see also Costanza-Chock, Chapter 21 in this volume). In our final section, we argue that the dominance of men working in and designing AI results in a feedback loop whereby bias gets built into machine learning systems. Digital technologies, whether hardware or software, are socially shaped by gender power relations and gendered meanings that influence the process of technological change and are inscribed into technologies. Although algorithms and automated decision-making systems are presented as if they are impartial, neutral, and objective, we show how bias enters, and is amplified through, AI systems at various stages.

Crucially, we stress that algorithmic bias is not solely the product of unconscious sexism or racism, nor bad training data, but the end result of a technoculture that

---

[1] Artificial Intelligence: 'When a machine or system performs tasks that would ordinarily require human (or other biological) brainpower to accomplish' (The Alan Turing Institute, 2021).

has systematically excluded women and people from marginalised groups from positions of leadership and power.

While digitalisation holds out the promise of greater equality, it also poses the risk of encoding, repeating, and amplifying existing patterns of gender inequities. This perspective is particularly important at a time when digital tools are marketed as the solution to all social problems. If the generation and implementation of new technologies always involve preferences and choices, then there are opportunities to build them in ways that prevent harm and, more so, promote the 'good' that they offer. At a moment when technology is being marshalled to make choices of global consequence, and is affecting the lives of individuals and society in ways both profound and subtle, this warrants urgent attention.

The Missing Women in AI

We begin by presenting some figures on women in the technology labour force, before delving into the subfields of AI and data science as an integral part of this workforce. At the outset, it is important to note that this is not a story of inexorable progress in terms of gender equality. In fact, the percentage of women in the USA and Western Europe gaining computer science degrees today—15–20 percent— is down from nearly 40 percent in the 1980s (Murray 2016). In Europe, only 17 percent of ICT specialists are women (European Commission 2019); in the UK, women comprise 19 percent of UK tech workers (Tech Nation 2018). Further, UNESCO (2021) estimates that women are under-represented in technical and leadership roles in the world's top technology companies, with 23 and 33 percent, respectively, at Facebook.

There is a scarcity of intersectional data on the tech sector, but the little data that is available suggests that women of colour are particularly under-represented. Only 1.6 percent of Google's US workforce in 2020 were Black women. Indeed, diversity policies and training (among other initiatives) have only made a marginal difference in increasing the share of women of colour in tech (Google 2020). As Alegria (2019, p.723) explains, 'women, particularly women of colour, remain numerical minorities in tech despite millions of dollars invested in diversity initiatives'.

The World Economic Forum (2020) estimates that globally women make up approximately 26 percent of workers in data and AI roles more specifically, a number that drops to only 22 percent in the UK. This is partly due to the lack of clarity and newness of these professions—but it is also because of a hesitancy of big tech companies to share this data. As West et al. (2019, pp.10–12) explain, 'the current data on the state of gender diversity in the AI field is dire... the diversity and inclusion data AI companies release to the public is a partial view, and often contains flaws'. This is a significant barrier to research.

Given the scarcity of raw industry data available, researchers have drawn on other sources including online data science platforms, surveys, academia, and conference data. These approaches also provide mounting evidence of serious gaps in the gender diversity of the AI workforce. In 2018, WIRED and Element AI reviewed the AI research pages of leading technology companies and found that only 10–15 percent of machine learning researchers were women (Simonite 2018). Related research found that on average only 12 percent of authors who had contributed to the leading machine learning conferences (NIPS, ICML, and ICLR) in 2017 were women (Mantha and Hudson 2018).

Indeed, there is more information about women in AI research and in the academy, due to the more readily available data. For example, Stathoulopoulos and Mateos-Garcia (2019) found that only 13.8 percent of AI research paper authors were women; at Google, it was 11.3 percent, and at Microsoft, 11.95 percent. Additionally, the AI Index (2021) found that female faculty make up just 16 percent of all tenure-track faculty whose primary focus is AI.

The sparsity of statistics on the demographics of AI professions also motivated *us* to explore other potentially informative sources. As quickly evolving fields in which practitioners need to stay up-to-date with rapidly changing technologies, online communities are an important feature of data science and AI professions, and so provide an interesting lens through which to view participation in these fields. Thus, we examined a selection of online, global data science platforms (Data Science Central, Kaggle, OpenML, and Stack Overflow). Our research indicates that women are represented at a remarkably consistent, and low, 17–18 percent across the platforms—with Stack Overflow at much lower 7.9 percent.[2]

The statistics we have reviewed confirm that the newest wings of technology, that is, AI and data science, have poor representation of women. The more prestigious and vanguard the field, the fewer the number of women working in it. As the AI and data science fields are rapidly growing as predominant subfields within the tech sector, so is the pervasive gender gap within them.

Before concluding this section, it is important to acknowledge that so-called intelligent machines also depend on a vast, 'invisible' human labour force: those who carry out skilled technological work such as labelling data to feed algorithms, cleaning code, training machine learning tools, and moderating and transcribing content. These 'ghost workers', often women in the Global South, are underpaid, undervalued, and lacking labour laws (Gray and Suri 2019; Roberts 2019; Atanasoski and Vora 2019). Given the evidence to date, there is no reason to expect the rise of the 'gig' or 'platform' economy to close the gender gap, particularly in the AI fields.

---

[2] See 'Where are the Women? Mapping the Gender Job Gap in AI' (Young et al. 2021).

Feminist STS: From Technofeminism to Data Feminism

To understand the gender power dynamics of AI, it is worth recalling that feminist STS (science and technology studies) scholars have been researching the gendering of technology for decades. In *TechnoFeminism* (2004), Wajcman has detailed the various strands of feminist thought, such as liberal, socialist, and post-modern, that emerged in response to previous waves of digitalisation. A common concern was to document and explain women's limited access to scientific and technical institutions and careers, identifying men's monopoly of technology as an important source of their power. The solution was often posed in terms of getting more women to enter science and technology—seeing the issue as one of equal access to education and employment. However, the limited success of equal opportunity strategies soon led to the recognition that technical skills are embedded in a culture of masculinity and required asking broader questions about how technoscience and its institutions could be reshaped to accommodate women.

Such critiques emphasised that, in addition to gender structures and stereotyping, engrained cultures of masculinity were ubiquitous within tech industries—and they still are, as we will discuss in the next section.

Recognising the complexity of the relationship between women and technology, by the 1980s feminists were exploring the gendered character of technology itself (see Taillandier in this volume). In Harding's (1986, p.29) words, feminist criticisms of science evolved from asking the 'woman question' in science to asking the more radical 'science question' in feminism. Rather than asking how women can be more equitably treated within and by science, the question became how Western science, a male, colonial, racist knowledge project, can possibly be used for emancipatory ends (Haraway 1988). Similarly, feminist analyses of technology were shifting from women's access to technology to examining the very processes by which technology is developed and used, as well as those by which gender is constituted. Both socialist and radical feminists began to analyse the gendered nature of technical expertise, and put the spotlight on artefacts themselves. The social factors that shape different technologies came under scrutiny, especially the way that technology reflects gender divisions and inequalities. Hence, the problem was not only men's monopoly of technology, but also the way gender is embedded in technology itself.

A broad social shaping or 'co-production' framework has now been widely adopted by feminist STS scholars.[3] A shared idea is that technological innovation is itself shaped by the social circumstances within which it takes place. Crucially, the notion that technology is simply the product of rational technical imperatives has been dislodged. Objects and artefacts are no longer seen as politically neutral,

---

[3] See, for example, the journal *Catalyst: Feminism, Theory, Technoscience.* For a recent overview, see Wagman and Parks (2021).

separate from society; rather, they are designed and produced by specific people in specific contexts. As such, artefacts have the potential to embody and reproduce the values and visions of the individuals and organisations that design and build them. And if 'technology is society made durable' (Latour 1990) then, in a patriarchal society, gender power relations will be inscribed into the process of technological change, in turn configuring gender relations.

Such a mutual shaping approach, in common with technofeminist theory, conceives of technology as both a source and consequence of patriarchal relations (Wajcman 2004). In other words, gender relations can be thought of as materialised in technology, and masculinity and femininity in turn acquire their meaning and character through their enrolment and embeddedness in working machines. For example, gendered identities are found to be co-constructed with technologies and technical orientations, often in connection with alignments of race and class (Bardzell 2018; Pérez-Bustos 2018). Empirical research on everything from the microwave oven (Cockburn and Ormrod 1993) and the contraceptive pill (Oudshoorn 1994) to robotics and software agents (Suchman 2008) has clearly demonstrated that the marginalisation of women from the technological community has a profound influence on the design, technical content, and use of artefacts.

The key insights of feminist STS on issues such as the gendering of skills and jobs; the conception of technology as a sociotechnical product and cultural practice; and the critique of binary categories of female/male, nature/culture, emotion/reason, and humans/machines are still foundational resources for contemporary research on gender and technology. In recent years, feminist STS has been enriched by its engagement with intersectional feminist analysis, critical race theorists and post-colonial theory (Crenshaw et al. 1995; Collins 1998; Benjamin 2019; Noble 2018; Sandoval 2000). There is increasing recognition of the ways in which gender intersects with other aspects of difference and disadvantage in the societies within which these technologies sit. Women are a multifaceted and heterogeneous group, with a plurality of experiences. Gender *intersects* with multiple aspects of difference and disadvantage involving race, class, ethnicity, sexuality, ability, age, and so on. For instance, women who are poor or belong to racial minorities experience the negative effects of digitalisation and automation more acutely (Buolamwini and Gebru 2018).

An intersectional technofeminist lens is particularly pertinent as attention has expanded beyond the sexism and racism of the internet and digital cultures to the growing body of work on AI systems. In *Data Feminism* (2020: p.105; see Chapter 12 in this volume), D'Ignazio and Klein not only expose the glaring gender data gap—as we noted before—but also show how the data that feeds algorithms is biased. Echoing STS texts on the politics of scientific knowledge production, they highlight the epistemic power of classification systems and the values and judgements they encode (Bowker and Star 2000). Every dataset used to train machine learning systems contains a worldview, reducing humans to 'false binaries and implied hierarchies, such as the artificial distinctions between men and women' (see also Crawford 2021). This necessitates the 'flattening out of humanity's plurality' (see Browne in this volume). In other words, the very process of classifying data, a core practice in AI, is inherently political. That is why narrow technical solutions to statistical bias and the quest for 'fairer' machine learning systems miss the point.

As Benjamin (2019, p.96) underlines, the injustices are nothing new but 'the practice of codifying existing social prejudices into a technical system is even harder to detect when the stated purpose of a particular technology is to override human prejudice'. In the final section of this chapter we will examine such AI feedback loops: the ways in which algorithms can reflect and amplify existing inequities such as those based on gender and race. But first we explore how and when technical expertise came to be culturally associated with masculinity.

Masculinity and Technical Expertise

History of Engineering and Computing as Professions

The stereotype of engineering and computer science as male domains is pervasive (see Cave et al., Chapter 5 in this volume). It affects girls' and women's confidence in their technical skills and proficiencies, shaping their perception of their own identity and discouraging them from entering such fields. Indeed, the underrepresentation of women in the tech sector has traditionally been framed as a 'pipeline problem', suggesting that the low numbers of women in tech is solely due to a low female talent pool in STEM fields (that is, because girls are uninterested or lack the skills). This perspective, however, shifts the obligation to change onto women and neglects technology companies' failure to attract and retain female talent.

'Masculine defaults' shape professional practices and career pathways, governing technical fields in particular (Cheryan and Markus 2020). As Wajcman (2010, p.144) explains, definitions of technological skill and expertise have been historically constructed in a way that privileges the masculine (as the 'natural' domain of men), rendering femininity as 'incompatible with technological pursuits'. For example, a number of works highlight the role of gender relations in the very definition and configuration of computing as a profession. Feminist historian Hicks (2017; see also Abbate 2012) recalls that computer programming was originally the purview of women. At the advent of electronic computing following the Second World War, software programming in industrialised countries was largely considered 'women's work', and the first 'computers' were young women. As computer programming became professionalised, however, and power and money became associated with the field, the gender composition of the industry shifted, marginalising the work of female technical experts by fashioning them into a technical 'underclass'. Structural discrimination edged women out of the newly prestigious computing jobs (Ensmenger 2012). Such persistent cultural associations around technology are driving women away from, and out of, industries which entail more 'frontier' technical skills such as AI.

Indeed, as feminist scholars have long evidenced, when women participate in male-dominated occupations, they are often concentrated in the lower-paying and lower-status subfields. 'Throughout history, it has often not been the content of the work but the identity of the worker performing it that determined its status', explains Hicks (2017, p.16). As women have begun to enter certain technological subdomains in recent years (often through boot camps and other atypical educational pathways), such as front-end development, these fields have started to lose prestige and experience salary drops (Broad 2019). Meanwhile, men are flocking to the new (and highly remunerated) data science and AI subspecialities.

What is crucial to emphasise here is that technical skill is often deployed as a proxy to keep certain groups in positions of power. As such, it is important to begin to rewrite the narrative, heightening awareness of the gendered history of computing in order to avoid its replication in AI. This is particularly apposite as newly created AI jobs are set to be the well-paid, prestigious and intellectually stimulating jobs of the future. Women and other under-represented groups deserve to have full access to these careers, and to the economic and social capital that comes with them. Further, if the women who do succeed in entering tech are stratified into 'less prestigious' subfields and specialities, rather than obtaining those jobs at the forefront of technical innovation, the gender pay gap will be widened.

The Skills Gap

Once defined by inequalities in access to digital technology, the digital gender divide is now more about deficits in learning and skills. While there is still an 'access gap' between women and men, especially in the Global South, it has greatly improved over the past 25 years. At the same time, however, the gender 'digital skills gap' persists. Despite a number of important interventions and policies aimed at achieving gender equality in digital skills across both developed and developing economies, the divide not only remains large but, in some contexts, is growing wider. This skills divide is underpinned by a deficit in digital literacies among women, particularly in low- and middle-income countries, where many women lack the necessary techno-social capabilities to compete in a global online environment (Gurumurthy et al. 2018).

It is important to note, however, that the gender skills gap is more marked in some countries than in others. For example, in Malaysia some universities have up to 60 percent women on computer science programmes, with near parity also reported in certain Taiwanese and Thai institutions. Stoet and Geary (2018) even discuss a 'gender-equality paradox', suggesting that the more gender equality in a country, the fewer women in STEM fields. They propose that women's engagement with technical subjects may be due to life-quality pressures in less gender-equal countries; there is likely a complex set of reasons including national cultures, and varying professional opportunities for women globally. Indeed, Arab countries have between 40 and 50 percent female participation in ICT programmes (a proportion far higher than many of the more gender-equal European countries). However, these examples are not representative of the broader trends worldwide.

A variety of initiatives has been trialled in the past few decades to encourage more diversity in technological fields within higher education. Particularly notable are the successful models provided by the US universities Carnegie Mellon and Harvey Mudd, which have dramatically increased the participation of women in their computer science departments. For example, Carnegie Mellon increased the number of women from 7 percent in 1995 to 42 percent in 2000 (Frieze and Quesenberry 2019). This was achieved through a multi-pronged approach, which included grouping students by experience, as well as coaching for, and buy-in from, senior faculty. By 2019, Harvey Mudd also boasted 49 percent women in computer science. This suggests that steps towards resolving the issues can be straightforward with the right policies and leadership in place.

In concluding it should be noted that the gender skills gap not only limits women's economic opportunities, but also has broader implications for their ability to participate fully as citizens, in government and politics. As the World Economic Forum (2018, p.viii) argues: 'AI skills gender gaps may exacerbate gender gaps in economic participation and opportunity in the future as AI encompasses an increasingly in-demand skillset ... technology across many fields is being developed without diverse talent, limiting its innovative and inclusive capacity'. However, as we have been stressing, hiring more women is not enough. As UNESCO's Framework on Gender Equality and AI states (2020, p.23): 'This is not a matter of numbers, but also a matter of culture and power, with women actually having the ability to exert influence'. Bringing women to positions of parity as coders, developers, and decision makers, with intersectionality in mind, is key.

Technoculture in Organisations

Up to this point we have focused on how the gender skills gap, resulting from the history of computing, skews AI professions. However, the masculine culture of the workplace itself, in particular the 'brogrammer' and 'geek' cultures synonymous with Silicon Valley, is a key factor hindering women's career progression.

Once women are employed in engineering and technological fields, the rate of attrition is high, with women leaving technology jobs at twice the rate of men (Ashcraft et al. 2016). Women (but not men) taking managerial paths in engineering firms may also be at the greatest risk of attrition (Cardador and Hill 2018). In a similar vein, McKinsey found that while women made up 37 percent of entry-level roles in technology, only 25 percent reached senior management roles, and 15 percent made executive level (Krivkovich et al. 2016). Exploring the reasons for women's and minorities' high attrition and turnover rates, the Kapor Center highlights that one in ten women in technology experiences unwanted sexual attention (Scott et al. 2017).

Indeed, several studies have revealed subtle cultural practices embedded within technology companies that lead to 'chilly', unwelcoming workplace climates for women and marginalised groups (Hill et al. 2010). The prevalence of 'masculine defaults' in these spaces results in micro-aggressions, subconscious biases in performance/promotion processes, and other forms of discrimination (Kolhatkar 2017). According to the State of European Tech Survey, 55 percent of Black/African/Caribbean women have experienced discrimination in some form (Atomico 2018). An overwhelming 87 percent of women claim that they have been challenged by gender discrimination compared to 26 percent of men (Atomico 2020). For instance, in 2018 Google staff walked out over how sexual misconduct allegations were being dealt with at the firm (Lee 2018). Tech companies have been slow to react to such findings.

In turn, this 'technoculture' has significant repercussions for recruitment, promotion, career trajectories and pay. For example, Wynn and Correll (2018) suggest that women are alienated at the point of recruitment into technology careers. They found that company representatives often engage in behaviours, such as geek culture references, that create unwelcoming environments for women before joining a firm. As such, companies have a significant opportunity to increase their representation of women by being more mindful of the images projected in recruitment sessions and interviews.

Finally, the 'bro' culture of technological work has resulted in a severe under-representation of women in entrepreneurship. This, almost paradoxically, is often heralded as the way for women to 'get ahead' in the digital revolution. Female founders in the United States received only 2.3 percent of venture capital funds in 2020, and women represent just 12 percent of venture capital decision makers (Reardon 2021). Atomico (2020) also found that 93 percent of all start-up funds raised in Europe in 2018 went to all-male founding teams—with just 2 percent to all-female founding teams.

As we will elaborate in the following section, the under-representation of women—especially women of colour (Williams et al. 2022)—on teams of inventors influences what actually gets invented. Patenting inventions has been largely a male endeavour, with a woman cited as the lead inventor on just 7.7 percent of all patents filed between 1977 and 2010 in the United States. Looking specifically at biomedical patents in 2010, Koning et al. (2021) found that teams made up of all women were 35 percent more likely than all-male teams to invent technologies relating to women's health. The researchers estimate that if women and men had produced an equal number of patents since 1976, there would be 6500 more female-focused inventions today.

AI Feedback Loops: The Amplification of Gender Bias

The stark lack of diversity and inclusion in AI has wider implications. Mounting evidence suggests that the under-representation of women and marginalised groups in AI and data science results in a feedback loop whereby bias gets built into machine learning systems. To quote the European Commission (Quirós et al. 2018): 'Technology reflects the values of its developers… It is clear that having more diverse teams working in the development of such technologies might help in identifying biases and prevent them'.

Although algorithms and automated decision-making systems are presented and applied as if they are impartial, neutral, and objective, bias enters, and is amplified through, AI systems at various stages (e.g. Leavy 2018; Gebru 2020). For example, the data used to train algorithms may under-represent certain groups or encode historical bias against marginalised demographics, due to previous decisions on what data to collect, and how it is curated. Data created, processed, and interpreted within unequal power structures can reproduce the same exclusions and discriminations present in society.

A key example of this is the 'gender data gap'; the failure to collect data on women, that is, gender-disaggregated data. As Criado Perez (2019) explains, in some instances this directly jeopardises women's health and safety: heart failure trials, historically based on male participants, result in women's heart attacks being misdiagnosed, and seat-belt designs based on the male body result in women being more likely to be seriously injured in car crashes. Furthermore, the gender data gap tends to be larger for low and middle-income countries. A 2019 Data 2X study of national databases in fifteen African countries found that sex-disaggregated data were available for only 52 percent of the gender-relevant indicators. When deep learning systems are trained on data that contain biases, they are reproduced and amplified. Fei-Fei Li, a prominent Stanford researcher in the AI field, describes this simply as 'bias in, bias out' (Hempel 2017).

Indeed, this issue is not only about exclusion, but also about the risks and potential for harm due to the increased visibility that *including* and collecting data on certain populations might bring them. For example, some communities are more heavily subject to surveillance and therefore more likely to appear in police databases. This is termed the 'paradox of exposure' by D'Ignazio and Klein (2020). The power of data to sort, categorise, and intervene, alongside corporations and states' ability to use data for profit and surveillance, has been stressed by Taylor (2017).

Furthermore, there are often biases in the modelling or analytical processes due to assumptions or decisions made by developers, either reflecting their own (conscious or unconscious) values and priorities or resulting from a poor understanding of the underlying data. Even the choices behind what AI systems are created can themselves impose a particular set of interpretations and worldviews, which could reinforce injustice. As O'Neil (2016, p.21) succinctly states: 'Models are opinions embedded in mathematics'. If primarily white men are setting AI agendas, it follows that the supposedly 'neutral' technology is bound to be inscribed with masculine preferences. Bias introduced at any one stage of the modelling process may be propagated and amplified by knock-on biases, since the results of one round of modelling are often used to inform future system design and data collection (Mehrabi et al. 2019).

An underlying problem is that AI systems are presented as objective and neutral in decision making rather than as inscribed with masculine, and other, preferences and values. Machines trained using datasets generated in an unequal society tend to magnify existing inequities, turning human prejudices into seemingly objective facts. Even if designed with the best of intentions, they 'do not remove bias, they launder it, performing a high-tech sleight of hand that encourages users to perceive deeply political decisions as natural and inevitable' (Eubanks 2018, p.224).

A growing number of AI products and systems are making headlines for their discriminatory outcomes. To name only a few: a hiring algorithm developed by Amazon was found to discriminate against female applicants (Dastin 2018); a social media-based chatbot had to be shut down after it began spewing racist and sexist hate speech (Kwon and Yun 2021); the image-generation algorithms OpenAI's iGPT and Google's SimCLR are most likely to autocomplete a cropped photo of a man with a suit and a cropped woman with a bikini (Steed and Caliskan 2021); and marketing algorithms have disproportionately shown scientific job advertisements to men (Maron 2018; Lambrecht and Tucker 2019).

The introduction of automated hiring is particularly concerning, as the fewer the number of women employed within the AI sector, the higher the potential for future AI hiring systems to exhibit and reinforce gender bias, and so on. Perversely, Sánchez-Monedero et al. (2020) found that automated hiring systems that claim to detect and mitigate bias obscure rather than improve systematic discrimination in the workplace.

There has also been concern about AI bias in the context of the pandemic. For example, Barsan (2020) found that computer vision models (developed by Google, IBM, and Microsoft) exhibited gender bias when identifying people wearing masks for COVID protection. The models were consistently better at identifying masked men than women and, most worrisome, they were more likely to identify the mask as duct tape, gags, or restraints when worn by women. Pulse oximetry devices used for warning of low blood oxygenation in COVID-19 were found to significantly underestimate hypoxaemia in Black patients—'current assumptions and algorithms, often derived from heavily white patient populations, may work against black patients' (BMJ 2020).

Several studies on computer vision have highlighted encoded biases related to gender, race, ethnicity, sexuality, and other identities. For instance, facial recognition software successfully identifies the faces of white men but more often fails to recognise those of dark-skinned women (Buolamwini and Gebru 2018). Further, research analysing bias in Natural Language Processing (NLP) systems reveal that word embeddings learned automatically from the way words co-occur in large text corpora exhibit human-like gender biases (Garg et al. 2018). For example, when translating gender-neutral language related to STEM fields, Google Translate defaulted to male pronouns (Prates et al. 2019). Additionally, AI voice assistants (such as Alexa and Siri) that mimic the master/servant relations of domestic service are aestheticised as a native-speaking, educated, white woman (Phan 2019; see also Rhee in this volume).

Finally, it is important to stress that technical bias mitigation (including algorithmic auditing) and fairness metrics for models and datasets are by no means sufficient to resolve bias and discrimination (Hutchinson and Mitchell 2019). Notably, since 'fairness' cannot be mathematically defined, and is rather a deeply political issue, this task often falls to the developers themselves—the very teams in which the diversity crisis lies (Hampton 2021). While we laud the considerable efforts of the Human-Computer Interaction (HCI)/Computer-Supported Cooperative Work community, among others, to design tools to mitigate data, algorithmic, and workers' biases, framing the problem as one of 'bias' risks encoding the premise *'that there is an absolute truth value in data and that bias is just a "distortion" from that value'* (Miceli et al. 2022, p.4).

As we noted before, feminist STS has demonstrated that the notion of scientific objectivity as a 'view from nowhere' is both androcentric and Eurocentric. It is therefore highly attuned to the privileged and naturalised epistemological standpoints or worldviews inscribed in data and systems that reproduce the status quo. We urgently need more nuanced data and analysis on women in AI to better understand these processes and strengthen efforts to avoid hard-coded discrimination. It is one thing to recall biased technology, but another to ensure that the biased technology is not developed in the first place.

Conclusion

This chapter has examined the gender relations of digitalisation, with a particular focus on AI. Although there is increasing recognition that technologies are socially shaped by the minds, hands, and cultures of people and, therefore, reflect history, context, choices, and values, we have argued that there is still insufficient focus on the ways in which gender relations are embedded in technology. This is the case whether it is the design of airbags in cars using crash-test dummies modelled on the male body; or the failure to include women in medical trials; or biased datasets used in algorithmic decision making. Gendered practices mediate technological transformations, and the political and socio-economic networks that shape and deploy technical systems.

Adopting such a technofeminist perspective suggests that the dominance of men working in and designing AI is deeply interconnected in a feedback loop with the increasingly identified gender biases in AI. Like the technologies that preceded them, social bias inscribed in AI software is not a glitch or bug in the system, but rather the result of persistent structural inequalities. In other words, while recent developments in data-driven algorithmic systems pose novel and urgent social justice issues, they also reflect the gender power relations long identified in feminist literature on technoscience. The historical relationship between technical expertise and masculinity, in combination with the 'chilly' organisational culture of technological work, is still generating an AI labour force that is unrepresentative of society as a whole.

If then technology is both a reflection and a crystallisation of society, feminist analysis must attend to the economic, political, cultural, and historical forces shaping AI in the current era. It really matters who is in the room, and even more so who is absent, when new technology like AI is developed. Technologies are designed to solve problems, and increasingly real-world societal questions are primarily posed as computational problems with technical solutions. Yet even the ways in which these tools select and pose problems foregrounds particular modes of seeing or judging over others. It has proven to be anything but gender neutral. As Ursula Le Guin (2017, p.150) commented perceptively as long ago as 1969: 'The machine conceals the machinations.'

References

Abbate, J. (2012) *Recoding Gender: Women's Changing Participation in Computing.* Cambridge, MA: MIT Press.

Alegria, S. (2019) 'Escalator or Step Stool? Gendered Labor and Token Processes in Tech Work'. *Gender & Society*, 33(5): 723.

Ashcraft, C., B. McLain, and E. Eger. (2016) 'Women in Tech: The Facts'. National Center for Women & Technology (NCWIT).

Atanasoski, N. and K. Vora. (2019) Surrogate Humanity. *Race, Robots, and the Politics of Technological Futures*. Duke University Press.

Atomico (2018) 'The State of European Tech'. Atomico.

Atomico (2020) 'The State of European Tech'. Atomico.

Bardzell, S. (2018) 'Utopias of Participation: Feminism, Design, and the Futures'. *ACM Transactions in Computer-Human Interaction* 25(1) Article 6 (February).

Barsan, I. (2020) 'Research Reveals Inherent AI Gender Bias: Quantifying the Accuracy of Vision/Facial Recognition on Identifying PPE Masks'. Wunderman Thompson. www.wundermanthompson.com/insight/ai-and-gender-bias

Benjamin, R (2019) *Race after Technology: Abolitionist Tools for the New Jim Code*. Cambridge, MA: Polity Press.

BMJ (2020) 'Pulse Oximetry may Underestimate Hypoxaemia in Black Patients'. British Medical Journal 371: m4926.

Bowker, G. and S. L. Star. (2000) *Sorting Things Out: Classification and Its Consequences*. Cambridge, MA: MIT Press.

Broad, E. (2019) 'Computer Says No'. *Inside Story* 29 April.

Buolamwini, J. and T. Gebru. (2018) 'Gender Shades: Intersectional Accuracy Disparities in Commercial Gender Classification'. Proceedings of Machine Learning Research 81: 1–15.

Cardador, M. T. and P. L. Hill. (2018) 'Career Paths in Engineering Firms: Gendered Patterns and Implications'. *Journal of Career Assessment* 26(1): 95–110.

Cheryan, S. and H. R. Markus. (2020) 'Masculine defaults: Identifying and mitigating hidden cultural biases'. *Psychological Review* 127(6): 1022–1052.

Cockburn, C. and S. Ormrod. (1993) *Gender and Technology in the Making*. London: Sage.

Collins, P. H. (1998) 'It's All in the Family: Intersections of Gender, Race, and Nation'. *Hypatia* 13(3): 62–82.

Crawford, K. (2021) *Atlas of AI: Power, Politics, and the Planetary Costs of Artificial Intelligence*. New Haven: Yale University Press.

Crenshaw, K., N. Gotanda, G. Peller, and T. Thomas. (eds.) (1995) *Critical Race Theory: The Key Writings that Formed the Movement*. New York: New Press.

Criado Perez, C. (2019) *Invisible Women: Exposing Data Bias in a World Designed for Men*. London: Chatto & Windus.

D'Ignazio, C. and L. F. Klein. (2020) *Data Feminism*, Cambridge, MA: MIT Press.

Dastin, J. (2018) 'Amazon Scraps Secret AI Recruiting Tool that Showed Bias against Women'. *Reuters* 10 October.

Ensmenger, N. L. (2012) *The Computer Boys Take Over: Computers, Programmers, and the Politics of Technical Expertise*. Cambrige, MA: MIT Press.

Eubanks, V. (2018) *Automating Inequality: How High-Tech Tools Profile, Police, and Punish the Poor*. New York: St Martin's Press.

European Commission (2019) 'Women in Digital Scoreboard'.

Frieze, C. and J. Quesenberry. (2019) *Cracking the Digital Ceiling*. Cambridge: Cambridge University Press.

Garg, N., L. Schiebinger, D. Jurafsky, and J. Zou. (2018) 'Word Embeddings Quantify 100 Years of Gender and Ethnic Stereotypes'. *Proceedings of the National Academy of Sciences USA* 115(16): E3635–E3644.

Gebru, T. (2020) Race and Gender. In *The Oxford Handbook of Ethics of AI*, eds M. Dubber, F. Pasquale and S. Das, pp. 253–270. Oxford: OUP.

Google (2020) 'Google Diversity Annual Report 2020'. 1–61.

Gray, M. and S. Suri. (2019) *Ghost Work: How to Stop Silicon Valley from Building a New Global Underclass.* Boston: Harcourt.

Gurumurthy, A., N. Chami, and C. A. Billorou. (2018) 'Gender Equality in the Digital Economy: Emerging Issues'. *Development Alternatives With Women for a New Era* (DAWN) August.

Hampton, L. M. (2021) 'Black Feminist Musings on Algorithmic Oppression'. Preprint at arXiv:2101.09869.

Haraway, D. (1988) 'Situated Knowledges. The Science Question in Feminism and the Privilege of Partial Perspective'. *Feminist Studies* 14(3): 575–599.

Harding, S. (1986) *The Science Question in Feminism.* Cornell University Press.

Hempel, J. (2017) 'Melinda Gates and Fei-Fei Li Want to Liberate AI from "Guys With Hoodies"'. *WIRED* 4 May.

Hicks, M. (2017) *Programmed Inequality: How Britain Discarded Women Technologists and Lost Its Edge in Computing,* Cambridge. MA: MIT Press.

Hill, C., C. Corbett, and A. St. Rose. (2010) 'Why So Few? Women in Science, Technology, Engineering and Mathematics'. American Association of University Women (AAUW).

Hutchinson, B. and M. Mitchell. (2019) '50 Years of Test (Un)fairness: Lessons for Machine Learning'. FAT* '19, 29–31 January, Atlanta, GA, USA.

Kolhatkar, S. (2017) 'The Tech Industry's Gender Discrimination Problem'. *The New Yorker.*

Koning, R., S. Samila, and J. P. Ferguson. (2021) 'Who do We Invent For? Patents by Women Focus More on Women's Health, but Few Women get to Invent'. *Science* 372(6548): 1345–1348.

Krivkovich, A., L. Lee, and E. Kutcher. (2016) *Breaking Down the Gender Challenge.* McKinsey.

Kwon, J. and H. Yun. (2021) 'AI Chatbot Shut Down After Learning to Talk Like a Racist Asshole'. *VICE* 12 January.

Lambrecht, A. and C. Tucker. (2019) 'Algorithmic Bias? An Empirical Study of Apparent Gender-Based Discrimination in the Display of STEM Career Ads'. *Management Science* 65: 2966–2981.

Latour, B. (1990) 'Technology Is Society Made Durable'. *The Sociological Review* 38(1): 103–131.

Le Guin, U. (2017[1969]) *The Left Hand of Darkness.* London: Orion Books.

Leavy, S. (2018) 'Gender Bias in Artificial Intelligence: The Need for Diversity and Gender Theory in Machine Learning'. In: *Proceedings of the 1st International Workshop on Gender Equality in Software Engineering,* pp. 14–16. New York, NY: ACM.

Lee, D. (2018) 'Google Staff Walk Out over Women's Treatment'. *BBC News* 1November.

Mantha, Y. and S. Hudson. (2018) 'Estimating the Gender Ratio of AI Researchers around the World'. *Element AI* 17 August.

Maron, D. F. (2018) 'Science Career Ads Are Disproportionately Seen by Men'. *Scientific American* 25 July.

Mehrabi, N., F. Morstatter, N. Saxena, K. Lerman, and A. Galstyan. (2019) 'A Survey on Bias and Fairness in Machine Learning'. Preprint at *arXiv.*

Miceli, M., J. Posada, and T. Yang. (2022) 'Studying up Machine Learning Data: Why Talk about Bias When We Mean Power?'. *Conference Proceedings ACM HCI.* 6, GROUP, Article 34 (January).

Murray, S. (2016) 'When Female Tech Pioneers Were the Future'. *Financial Times* 7 March.

Noble, S. (2018) *Algorithms of Oppression: How Search Engines Reinforce Racism*. New York: NYU Press.

O'Neil, C. (2016) *Weapons of Math Destruction: How Big Data Increases Inequality and Threatens Democracy*. London: Allen Lane.

OECD (2018) 'Empowering Women in the Digital Age: Where Do We Stand?' High-Level Event at the margin of the 62nd session of the UN Commission on the Status of Women (March): 1–5.

Oudshoorn, N. (1994) *Beyond the Natural Body: An Archaeology of Sex Hormones*. London: Routledge.

Pérez-Bustos, T. (2018) 'Let Me Show You': A Caring Ethnography of Embodied Knowledge in Weaving and Engineering. In *A Feminist Companion to the Posthumanities*, eds C. Asberg and R. Braidotti, pp. 175–188. New York: Springer.

Phan, T. N. (2019) 'Amazon Echo and the Aesthetics of Whiteness'. *Catalyst: Feminism, Theory, Technoscience* 5(1): 1–39.

Prates, M., P. Avelar, and L. C. Lamb. (2019) 'Assessing Gender Bias in Machine Translation: A Case Study with Google Translate'. *Neural Computing and Applications*, March: 1–19.

Quirós, C. T., E. G. Morales, R. R. Pastor, A. F. Carmona, M. S. Ibáñez, and U. M. Herrera. (2018) *Women in the Digital Age*. Brussels: European Commission.

Reardon, S. (2021) 'Gender Gap Leads to few US Patents that Help Women'. *Nature* 597(2)(September): 139–140.

Roberts, S. T. (2019) *Behind the Screen: Content Moderation in the Shadows of Social Media*. New Haven: Yale University Press.

Sánchez-Monedero, J., L. Dencik, and L. Edwards. (2020) 'What Does it Mean to "Solve" the Problem of Discrimination in Hiring? Social, Technical and Legal Perspectives from the UK on Automated Hiring Systems'. FAT* Conference Proceedings: 458–468.

Sandoval, C. (2000) *Methodology of the Oppressed*. Minneapolis: University of Minnesota Press.

Schwab, K. (2016) *The Fourth Industrial Revolution*. London: Penguin.

Scott, A., F. Kapor Klein, and U. Onovakpuri. (2017) 'Tech Leavers Study'. *Ford Foundation/Kapor Center for Social Impact*, pp. 1–27.

Sey, A., and N. Hafkin. (2019) 'Taking Stock: Data and Evidence on Gender Equality in Digital Access, Skills, and Leadership'. Preliminary findings of a review by the EQUALS Research Group, United Nations University.

Simonite, T. (2018) 'AI is the Future – But Where are the Women?' *WIRED* 17 August.

Stathoulopoulos, K. and J. Mateos-Garcia. (2019) 'Gender Diversity in AI Research'. *Nesta*.

Steed, R. and A. Caliskan. (2021) 'Image representations learned with unsupervised pre- training contain human-like biases'. *FAccT '21*, 3–10 March, Canada. Preprint *arXiv*.

Stoet, G. and D. C. Geary. (2018) 'The Gender-Equality Paradox in Science, Technology, Engineering, and Mathematics Education'. *Psychological Science* 29(4): 581–593.

Suchman, L. (2008) *Human-Machine Reconfigurations*. Cambridge: CUP.

Taylor, L. (2017) 'What is Data Justice? The Case for Connecting Digital Rights and Freedoms Globally'. *Big Data and Society* 4(2): 1–14.

Tech Nation (2018) 'Diversity and Inclusion in UK Tech Companies'. https://technation.io/insights/diversity-and-inclusion-in-uk-tech-companies/

UNESCO (2020) *Artificial Intelligence and Gender Equality: Key Findings of UNESCO's Global Dialogue.* UNESCO Publishing.

UNESCO (2021) *UNESCO Science Report: The Race Against Time for Smarter Development.* UNESCO Publishing.

Wagman, K. and L. Parks. (2021) 'Beyond the Command: Feminist STS Research and Critical Issues for the Design of Social Machines'. *Proceedings of CSCW '21.* ACM, NY.

Wajcman, J. (2004) *TechnoFeminism.* Cambridge: Polity Press.

Wajcman, J. (2010) 'Feminist Theories of Technology'. *Cambridge Journal of Economics* 34(1): 144.

Wajcman, J., Young, E. and FitzMaurice, A. (2020) 'The Digital Revolution: Implications for Gender Equality and Women's Rights 25 Years after Beijing'. *Discussion Paper No. 36 (August)*, UN Women.

West, S., M. Whittaker, and K. Crawford. (2019) *Discriminating Systems: Gender, Race and Power in AI.* New York: AI Now Institute.

Williams, J. C., R. M. Korn, and A. Ghani. (2022) 'Pinning Down The Jellyfish: The Workplace Experiences of Women of Color in Tech. Worklife Law'. University of California, Hastings College of the Law.

World Economic Forum. (2018) *Global Gender Gap Report.*

World Economic Forum. (2020) *Global Gender Gap Report.*

Wynn, A. T. and S. J. Correll. (2018) 'Puncturing the Pipeline: Do Technology Companies Alienate Women in Recruiting Sessions?' *Social Studies of Science* 48(1): 149–164.

Young, E., J. Wajcman, and L. Sprejer. (2021) *Where Are the Women? Mapping the Gender Job Gap in AI.* Public Policy Briefing. The Alan Turing Institute.

5

Shuri in the Sea of Dudes

The Cultural Construction of the AI Engineer in Popular Film, 1920–2020

Stephen Cave, Kanta Dihal, Eleanor Drage, and Kerry McInerney

Introduction

In a 2016 *Bloomberg* article, computer scientist Margaret Mitchell described the AI industry as a 'sea of dudes' (Clark 2016). At that time she was the only female researcher in Microsoft's Cognition group. She subsequently moved to Google, but was controversially dismissed in February 2021 (Guardian staff 2021). There are many reasons why women choose not to enter the field of AI, and many reasons why they have negative experiences if they do enter. One crucial yet underexplored factor is the way in which the broader culture surrounding this field portrays the ideal AI researcher. We call these portrayals, perceptions and stereotypes *the cultural construction of the AI engineer*,[1] and argue that this is a critical site of enquiry in understanding what has been called AI's 'diversity crisis' (West et al. 2019). As both N. Katherine Hayles and Apolline Taillandier argue in this volume, the language, metaphors and stories we use to talk about AI matter, as they play a fundamental role in shaping our understanding of AI and how these technologies ascribe to (or subvert) existing gender norms. This also extends to the gendered narratives of the AI industry and gendered perceptions of the people responsible for designing and developing AI-powered technologies.

Drawing on feminist perspectives within cultural studies and science communication studies, this paper aims to analyse the gendered portrayals of AI scientists in popular film, which, we argue, contribute to the barriers women face within the AI sector. To understand the cultural construction of the AI engineer, we analysed the 142 most influential films featuring AI from 1920 to 2020 for their portrayals of AI researchers. As well as finding that only 7 percent of AI professionals portrayed in film are women, we identified a number of key gendered tropes associated with

---

[1] Unless otherwise stated, we use the terms AI 'scientist', 'engineer', and 'researcher' interchangeably, to avoid repetition, rather than to make a distinction.

Stephen Cave et al., *Shuri in the Sea of Dudes*. In: *Feminist AI*. Edited by: Jude Browne, Stephen Cave, Eleanor Drage, and Kerry McInerney, Oxford University Press. © Oxford University Press (2023). DOI: 10.1093/oso/9780192889898.003.0005

the AI scientist that recur in these films. In this chapter, we focus on four tropes that construct the ideal AI scientist as male: the portrayal of the AI scientist as a *genius*; AI's association with traditionally *masculine milieus* such as the military; the creation of *artificial life*; and relations of *gender inequality* that relegate female AI engineers into positions of subservience and sacrifice.

First, we briefly explicate what we mean by the cultural construction of the AI scientist, and the role it plays in the field's diversity crisis. We then outline our methodology for our study of the cultural construction of the AI engineer in influential AI films. Next, we explore the four gendered tropes we have identified as crucial to the cultural construction of the AI engineer as male.

Gender Inequality and the AI Industry

As Erin Young and Judy Wacjman demonstrate in this volume, the AI sector is characterised by profound structural inequality along gendered lines. While women make up 39 percent of practitioners across all STEM fields, only 22 percent of AI professionals are women (Howard and Isbell 2020; Hammond et al. 2020). Recent studies show that women constitute only 10–15 percent of AI researchers at leading tech companies, and only 12 percent of authors contributing to key machine learning conferences (Young and Wajcman, Chapter 4 in this volume; Simonite 2018; Mantha and Hudson 2018). Women are also 'more likely than men to occupy a job associated with less status and pay', such as data analytics, while men are more likely to occupy the more 'prestigious' and creative jobs in machine learning and engineering (Young et al. 2021, p.4). These differences persist despite the fact that women in AI have higher average educational qualifications than men across all industries (Young et al. 2021, p.5).

The lack of women visibly present in the technology industry may significantly affect women's career aspirations in the field (Varma 2010; Women In Tech 2019; PwC UK 2017). For example, a report by PwC emphasised the lack of female role models as a barrier to women's uptake of STEM subjects (PwC UK 2017). The underrepresentation of women is often blamed on the 'pipeline problem', referring to a cumulative underrepresentation starting with low numbers of women and girls studying STEM subjects, which in turn leads to issues with recruiting and retaining women in AI and data science roles. However, this emphasis presents the problem as one of women's underachievement in the sector (see also Young and Wajcman, Chapter 4 in this volume). It therefore occludes systemic and structural issues, such as workplace harassment, pervasive sexist attitudes towards female engineers and data scientists, and the gender pay gap, which disincentivise women from entering the AI field and frequently drive them out of the sector (West et al. 2019).

This is not to say that there is no pipeline problem. However, addressing both the pipeline problem *and* inequitable, hostile work cultures requires an understanding of how rigid gender norms and cultural constructions of the AI engineer affect both women's perceptions of the field, and their experiences when in it. Roli Varma's 2010 study found that a large majority of computer science students believed that 'gendered socialization' was a major factor behind women's underrepresentation in computer science degrees, with 'gendered socialization' encompassing a constellation of societal and familial beliefs that computer science is a male field and that men are naturally more technically adept than women (Varma 2010, p.305).

Cheryan et al. argue that a driving factor behind the failure to recruit women in computing is the 'stereotypical representation of the computer scientist—an image that is pervasive in popular culture and in the minds of students as someone who is highly intelligent, singularly obsessed with computers, and socially unskilled' (Cheryan et al. 2013, p.67). Women In Tech's 2019 survey of over 1000 women in the tech sector corroborated this with their finding that 18 percent believed perceptions of the tech sector were the primary barriers to improving female representation (Women In Tech 2019). Fortunately, Cheryan et al. demonstrate that these stereotypes can be dismantled: after reading a news article suggesting that computer scientists do *not* actually fit these stereotypes, female undergraduates stated they were more likely to consider a computer science major (Cheryan et al. 2013). Their research is a rebuttal of the prevalent notion that women's underrepresentation in computing is a 'natural' expression of women's disinterest or lack of ability in engineering, mathematics and computer science.

These gendered conceptions of the field, already so pronounced at the level of undergraduate education, extend into the workplace, contributing to and entrenching AI's diversity crisis (Thompson 2019; Conger 2017). Addressing the lack of women in AI and adjacent fields such as computer science thus requires a dismantling of the culturally pervasive gendered norms that prevent women from entering and staying in AI.

Cultural Representations of AI Scientists in Popular Media

This brings us to the representation of AI, data science, and computer science in the media. The absence of female scientists in popular media matters, because, as feminist theorists have demonstrated, popular film and culture directly shape and propogate gendered and racial norms. Patricia Hill Collins argues that media representations, and in particular their embedding of harmful stereotypes, play a central role in a racist and sexist hegemonic culture (Hill Collins [1990] 2000, p.5). Her observations are borne out by the insights of feminist film scholars who demonstrate how gender norms and stereotypes are embedded and reproduced within popular film in both narrative and form. This includes Laura Mulvey's critique of the centrality of the objectifying, voyeuristic 'male gaze' in popular cinema (Mulvey 1975) and Sharon Willis' exploration of how racial and sexual difference were portrayed in late twentieth century film (Willis 1997), through to Robin Wood and Barbara Creed's analyses of the sexual politics of Hollywood films (Wood 1998; Creed 1993).

Likewise, a plethora of scholarship on race and representation in popular film highlights how racial stereotypes are enacted on screen, from the racialisation of the cinematic gaze (hooks 2014) and the fetishisation of Black, Brown, Asian, and Indigenous people on screen (Shimizu 2007; Cheng 2019) through to Hollywood's reproduction of negative racial stereotypes (Yuen 2016; Huang and Davis 2021) and the broader erasure, whitewashing or underrepresentation of people of colour in Hollywood cinema (Yuen 2016; Suparak 2020). As a result, while popular films are frequently treated as a form of escapism or entertainment, they are 'thoroughly implicated and invested in power relations', for 'they are part of the cultural and political landscape that both constructs and reflects social life' (Beasley & Brook 2019, p.1; see also Musgrave & Furman, 2017; Jones and Paris 2018; Carpenter 2016).

In this chapter and elsewhere, we argue that popular culture shapes the field of AI and vice versa (Cave et al. 2020). Cave and Dihal refer to the dynamic interplay between Hollywood films and Silicon Valley as the 'Californian Feedback Loop', offering the 2014 film *Transcendence*, discussed next, as an example (Cave and Dihal forthcoming). The figure of the superhero Tony Stark/Iron Man in the Marvel Cinematic Universe (MCU) serves as another. For example, in *Iron Man* (2008), Stark is pictured on the front cover of various real-world science publications, such as *WIRED* and *MIT Technology Review* (Favreau 2008). The actor Robert Downey Jr. explicitly cites Elon Musk as an inspirational figure for his portrayal of Tony Stark (Hern 2018), and Musk appears in a cameo role in *Iron Man 2* as a fellow inventor (Favreau 2010). Given that Musk himself has repeatedly pointed towards science fiction as the inspiration behind his business ventures and product development (Wood 2015), there is a clear feedback loop between the MCU's influential portrayals of AI scientists and engineers and the gendered myths and narratives that abound within and from Silicon Valley and other sites of technological development.

These close connections to experts and celebrities in the AI field, and conversely their use of the film for mobilising a public perspective on the dangers of superintelligence, make the depictions of AI scientists in these films particularly relevant: they are—directly or indirectly—endorsed by both Hollywood *and* the AI industry. This is why these films' stereotypical gendered depictions of AI scientists have the potential to be particularly damaging. Hence it is clear that we must engage meaningfully with popular film as a vector for understanding why negative gendered stereotypes about AI scientists persist.

Existing research supports the hypothesis that portrayals and perceptions of STEM professionals in popular film are highly gendered, but such work has so far not addressed AI specifically (Haynes 2017; Flicker 2003). For example, in Jocelyn Steinke's 2005 analysis of ninety-seven films released between 1991 and 2001 only twenty-three feature female scientists and engineers as primary characters (Steinke 2005, p.53). This disparity is particularly acute in relation to the more specialised field of computer science. The 2015 'Images of Computer Science' study, which involved a survey of several thousand students, parents, teachers, and school administrators in the US, noted that 'only 15% of students and 8% of parents say they see women performing computer science tasks most of the time on TV or in the movies, and about 35% in each group do not see women doing this in the media very often or ever' (Google and Gallup 2015, 12). More recently, the Geena Davis Institute's 2018 report 'Portray Her', which covers film, TV, and streaming media for all ages between 2007 and 2017, found that only 9 percent of computer scientists on screen were female (Geena Davis Institute on Gender in Media and The Geena Davis Institute on Gender in Media 2018). These studies on computer scientists in film lead us to expect that on-screen portrayals of AI scientists are also characterised by systemic gender inequality.

Methodology

This chapter builds on our previous quantitative research into the representation of female AI scientists in films featuring AI from 1920 to 2020 (Cave et al. forthcoming). Given the poor representation of women both in the field of AI and in portrayals of computer scientists on screen, we hypothesised that the number of female AI scientists on screen was likely to be among the lowest for any STEM field. We tested this by analysing a corpus of the 142 most influential films featuring AI from 1920 to 2020.

Under the umbrella of 'influential' films we included both films that have been consumed by large audiences worldwide *and* films that are considered significant by the tech community. We created our corpus by combining three box office statistics lists—Box Office Mojo's top grossing films worldwide, Box Office Mojo's top grossing films in the US adjusted for inflation, and IMDb's 100 top grossing science-fiction films ('Top Lifetime Grosses' 2021; 'Top Lifetime Adjusted Grosses' 2021; 'Sci-Fi (Sorted by US Box Office Descending)' n.d.)—with films from nine curated lists of culturally important science-fiction films, and Academy Award winners.[2] This produced a corpus of 1413 films, 142 of which contain AI and 86 of which depict one or more AI scientists, with a total of 116 AI scientists depicted.

---

[2] The curated lists were created by *Ultimate Movie Rankings*, *The Guardian*, *Science*, *Wired UK*, *Vogue*, *Empire*, *Good Housekeeping*, *Vulture*, and *TimeOut*.

We found that female AI scientists are indeed one of the most underrepresented STEM figures in popular film, with women constituting only 7 percent of AI scientists. This means that out of 116 AI scientists, only eight were female (Cave et al. forthcoming).

This chapter builds on that quantitative work by examining how specific representations of AI scientists reproduce and entrench the gender stereotypes that prevent women from entering and succeeding in the field of AI. As we narrowed down our initial corpus to films which feature AI scientists, we noted which tropes applied to and emerged from these depictions. Following Beasley and Brook, we focused on 'tropes reiterated in a very wide array of highly popular films', tropes that have the potential to significantly influence a viewership by virtue of being 'not idiosyncratic but repeated and widespread' (Beasley and Brook 2019, pp.11, 12).

Each trope was checked for intercoder reliability. After an initial scoping sample, which consisted of ten films containing AI scientists (more than 10 percent of the total sample), we settled on the following categories for data collection, discussed in more detail next:

- Genius
- Masculine Milieus (Corporate and Military)
- Artificial Life
- Gender Inequality (Subservience and Sacrifice)

Genius: We defined 'genius' as showing exceptional intellectual ability, completing tasks deemed to be impossible by other characters, being significantly cleverer and more gifted than other scientists, and/or displaying polymath qualities and expertise in a range of scientific fields. As a subcategory of the 'genius' trope, we also collected data on AI scientists who were portrayed as child prodigies, or as displaying extraordinary intelligence and technical skill at a young age. We hypothesised that, like contemporary media coverage of AI scientists, AI scientists in popular film would disproportionately be portrayed as geniuses, and that the majority of the geniuses would be men. We drew our hypothesis from the historical and cultural coding in the West of genius and intellectual brilliance as a masculine trait (Cave 2020).

Masculine Milieus: We defined 'masculine milieus' as settings that have historically been dominated by male figures, focusing on two specific milieus: corporate and military. We use the term 'corporate' to denote situations where AI is created within a clearly identified corporate setting, such as Tony Stark's Stark Industries in the MCU films. On-screen technological corporations, just like their real-world counterparts, are predominantly male institutions headed by male leaders. 'Military' refers to AI created for the purposes of war or national defence by a national military, institutions known for their hypermasculinity (Goldstein 2003).

Artificial Life: The trope 'Artificial Life' refers to a variety of characteristics linked by the theme of mastering life and death, such as the use of AI to create robot lovers or partners; reincarnate or recreate lost loved ones; or render someone immortal, as evidenced by the trope of 'mind uploading'. The desire to create artificial life by men is widely discussed under the term 'womb envy' as an aspiration motivated by their inability to bear children naturally (McCorduck 2004, 135; Lighthill 1973).

Gender Inequality: The trope 'Gender Inequality' refers to an unequal relationship between male and female scientists portrayed on screen. We focused on two key dimensions of gender inequality. The first, 'subservience', refers to a female scientist being under the authority of or inferior to a male scientist. The second trope, 'sacrifice', refers to female AI scientists sacrificing themselves for the sake of the film's plot or the 'greater good'.

Results: Frequency of the Four Tropes in our Corpus

Tables 5.1 and 5.2 show the frequency of the four tropes in the 86 key films featuring portrayals of 116 AI scientists that we surveyed. In the remainder of this chapter, we discuss these tropes in more depth, and explore how they culturally construct AI scientists as male.

Table 5.1 Frequency of the Four Tropes in our Corpus.

Male AI scientist tropes	Number	Percentage of total
Genius	38	33% of AI scientists
of which Child Prodigy	*14*	*12% of AI scientists*
Male Milieu	42	49% of films
of which Corporations	*32*	*37% of films*
of which Male CEO	*24*	*28% of films*
of which Military	*10*	*12% of films*
Artificial Life	19	22% of films

Table 5.2 Frequency of Female AI Scientist Tropes.

Female AI scientist tropes	Number	Percentage of total
Gender Inequality	5	71% of female AI scientists
of which Subservience	*5*	*71% of female AI scientists*
of which Sacrifice	*2*	*29% of female AI scientists*

Genius: Masculinity and the Brilliance Bias

First, our quantitative study showed that AI scientists were frequently portrayed on screen as geniuses. Out of 116 scientists in film, thirty-eight were portrayed as geniuses, or 33 percent. These findings reflect the popular perception of computer scientists and AI professionals as unusually intelligent. Cheryan et al. found in their study of Stanford students that intelligence is the trait most commonly associated with computer scientists (Cheryan et al. 2013). The portrayal of AI scientists in popular film as geniuses reflects popular news media coverage of the field of AI, which often frames AI scientists as geniuses who achieve astounding intellectual feats. For example, in 2016 *The Guardian* ran a profile on the founder and CEO of Deepmind, Demis Hassabis, titled 'The superhero of artificial intelligence: can this genius keep it in check?' (Burton-Hill 2016); and in 2018 Bloomberg referred to computer scientist Geoffrey Hinton as 'the genius who created modern AI' ('Meet the Godfather of AI' 2018).

However, the notion of 'genius' is deeply shaped by gendered and racialised concepts of intelligence that have historically been claimed by a white male elite (Cave 2020). Numerous studies demonstrate that people across different age groups continue to associate brilliance and exceptional intellectual ability with men (Bian et al. 2018; Jaxon et al. 2019; Storage et al. 2020). This phenomenon, called the 'brilliance bias', suggests that men are 'naturally' more likely to be geniuses than women, thus explaining their predominance both in the field of AI and in on-screen representations of AI scientists. This does not mean that respondents believe that men are *in general* more intelligent than women, but that they believe that men are more likely to be *exceptionally* intelligent (Storage et al. 2020). All but one of the thirty-eight genius AI scientists we identified were male, the only exception being Shuri from *Avengers: Infinity War*.

The prominent coding of AI scientists as geniuses, and the coding of geniuses as male, together risk entrenching the belief that women are less 'naturally' suited for a career in the field of AI. The representation of AI scientists as both overwhelmingly male *and* inherently brilliant is especially troubling given the use of biological arguments to justify women's relative exclusion from the AI workforce, as well as existing gendered stereotypes regarding who counts as a genius (Conger 2017).

For example, in the *Avengers* franchise Tony Stark is uncompromisingly portrayed as a genius whose intellect far outstrips that of everyone around him. In *Iron Man* (2008), Stark is introduced to the audience as a 'visionary' and a 'genius' who 'from an early age...quickly stole the spotlight with his brilliant and unique mind' (Favreau 2008). Stark designed his first circuit board at four years old, his first engine at six years old, and graduated from MIT summa cum laude at seventeen (Favreau 2008). In *The Avengers/Avengers Assemble* (2012) Stark explicitly describes himself as a 'genius, billionaire, playboy, philanthropist' (Whedon 2012).

While Stark's main areas of expertise are shown to be robotics, computer science, and engineering, he appears to have mastered an unrealistically large number of scientific fields. In *Iron Man 2* Stark synthesises an element, which would require expert-level knowledge of the field of chemistry; in *Avengers: Endgame*, Stark solves the problem of time travel in one night.

Furthermore, Stark specifically uses his intellect to dominate and humiliate others, deploying his wit and expertise in aggressive and hypermasculine ways. In *Iron Man 2*, Stark is called to testify in front of the Senate Armed Services Committee, who intend to confiscate his Iron Man suit and use it for national defence purposes. At the hearing, Stark hacks the large television screens on either side of the panel and uses it to humiliate Justin Hammer, CEO of his competitor Hammer Industries. As the committee is concerned that other countries may be developing and using Iron Man suits, Stark takes control of cameras across the world to show the committee that other countries and companies are struggling to develop his signature technology. The videos show other attempts at the Iron Man suit combusting and failing in dramatic and catastrophic ways, including prototypes created by Hammer Industries. Consequently, Stark declares that while other countries are 'five to ten years away' from creating Iron Man style suits, Hammer Industries is at least twenty years away (Favreau 2010, 2). Stark's performance at the hearing not only cements his status as the most intelligent person in the room, or even, as he implies, the world. It also plays into Stark's alpha male image, where he uses his technological expertise to publicly shame his rival and gain national acclaim.

Similarly, in Alex Garland's acclaimed AI film *Ex Machina* (2014), the AI scientist, Nathan Bateman, is portrayed as an extraordinary genius whose brilliance renders him outside of the constraints of societal norms. As with Stark, his genius gives rise to negative personality traits that are tolerated by sycophants like Caleb, the naive employee he invites to test his latest product. In Nathan's case, these traits are particularly pronounced: he is a bully and a misogynist with a god complex, a trope of Frankenstein figures across the ages. Like Stark, Nathan's genius has brought him corporate success and immense wealth—enough to fund a secluded and luxurious base in which he works privately on his AI development projects. In this way, he explicitly demonstrates two traits common to the genius trope: an untamed ego, and the desire and freedom to work alone. The film depicts the humanoid AI systems, which in real life would be the work of many engineers, designers, linguists, computer scientists and other stakeholders, as built by one man in his (luxury) basement. The narrative therefore uncritically reproduces the genius as a Michaelangelo figure, painting the Sistine Chapel alone, because the masterpiece is his alone to create. His remote home, accessible only by helicopter, ensures isolation from all human contact, which also allows him to subject both his AIs and Caleb to violent and illegal behaviour. While Nathan's ego stands in for the hubris of mankind in developing sentient AI, it does not merely contribute to a caricature of genius but is worryingly justified as its byproduct.

Nathan, we are told, wrote the code for his search engine aged thirteen. This is a common theme in our corpus, which features many modern-day-Mozart virtuoso types. Fourteen (12 percent) of the AI engineers, scientists, or researchers are explicitly represented as child prodigies or as being intellectually precocious from a very young age. They include Syndrome from *The Incredibles*; Hiro from *Big Hero 6*; and Tony Stark from the MCU: once again, all are male, with one exception—the sixteen-year-old Shuri in *Avengers: Infinity War*. The portrayal of AI scientists as child prodigies is significant here as it entrenches the notion that brilliance is an innate or a biological trait, rather than the result of hard work, training, support, or collaboration (Bian et al. 2018).

Filmic portrayals of AI scientists as 'geniuses' and 'child prodigies' may further discourage women from aspiring to a job in the field of AI. As Bian et al. have shown, women have lower interest in fields that emphasise the importance of brilliance over other characteristics (Bian et al. 2018). The portrayal of AI scientists as geniuses is therefore not only an inaccurate representation of the complexities of developing and deploying AI; it may also serve as a significant barrier to women's aspirations and retention in the AI sector.

Masculine Milieus: Corporate and Military Institutions

The third trope that emerged from our dataset was the portrayal of AI as part of a male milieu, or its association with masculinised spheres of life such as the corporate world and the military. Indeed, the only one of the twenty-four creators in our corpus who were CEOs of tech companies who was not a white man was an emoji (voiced by a woman), in *The Emoji Movie*. This homogenous depiction underrepresents even the dire real-world statistics: 14 percent of US tech start-ups have a female CEO, and of the forty-one Fortune 500 companies in the technology sector, only five have female CEOs (*SVB* 2020, p.2; Garner 2015).

Another key trope was the association of AI with the military, as ten of the films contained AI produced by military bodies. As the military is a heavily male-dominated environment and one that is strongly shaped by heteromasculine norms (Goldstein 2003), the portrayal of AI as a militarised field further contributes to its masculinisation. An example of this is the 2013 film *The Machine*, where the AI scientists work for the Ministry of Defence. While there is a female AI scientist, she is murdered partway through the film, and her brain scan and appearance are used as the basis for the male AI scientist's new AI project, the titular Machine. As a result, the film perpetuates the gendered binary between the hypermasculine milieu of the military and the dangerous, seductive, hyperfeminine figure of the gynoid.

We also observed forms of overlap and interplay between corporate and military masculine milieus, most notably in the *RoboCop* series, the *Terminator* franchise, and in the MCU. As a former weapons magnate, superhero, and CEO of Stark Industries, Tony Stark occupies multiple masculine milieus. At age twenty-one, Stark became the youngest CEO of a Fortune 500 company, the weapons manufacturing company Stark Industries, earning himself the nickname 'Merchant of Death' for his production of immensely destructive weaponry (Favreau 2008). The Iron Man saga portrays Stark's transformation from the primary weapons supplier to the US military to a famous superhero and CEO of a company that focuses on creating green energy as a redemptive arc. However, this redemption narrative eschews the uncomfortable fact that Stark has become an individualised paramilitary force; as Stark proudly proclaims in *Iron Man 2*, he has 'successfully privatised world peace' (Favreau 2010, p.2).

Indeed, it is difficult to separate the entire MCU from the masculine milieu of the military; the Pentagon frequently works with the MCU in order to promote the positive portrayal of the US military, and even funds some of their TV shows and films (Olla 2021). Hence, even though Stark eventually leaves some of his masculine milieus—for example, he makes his partner and eventual wife Pepper Potts the CEO of Stark Industries—the *Iron Man* films and the MCU more broadly remain permeated by a militaristic world view, one that similarly shapes the outlook of many films featuring AI scientists.

Artificial Life: Womb Envy, Reincarnation, and Male Immortality

In our corpus of eighty-six films, we found nineteen instances of AI scientists creating artificial life. Specific variations of this trope included five films where AI scientists had sex with a robot or created a robot husband/wife; nine films in which the AI was specifically intended to replace a human, usually a deceased family member or lover; and six films in which AI was created in pursuit of immortality (including through practices such as mind uploading).

The association of AI and masculinity might be further exacerbated by this association of the creation of *artificial* intelligence/life in the laboratory with maleness, in contrast to the female creation of *natural* intelligence/life. In all of these cases, AI scientists use technology to try to overcome the limitations of the natural world, reproducing masculinised human domination over a feminised natural environment. In the early days of AI, when Freudianism was in its heyday, it was considered almost a cliché that (at least some of) the researchers involved were motivated by 'womb envy'—that is, the envy of women's reproductive organs and the attendant power to create life. In his highly influential 1973 report 'Artificial Intelligence: A General Survey', Sir James Lighthill wrote 'it has sometimes been argued that part of the stimulus to laborious male activity in "creative" fields of work, including pure science, is the urge to compensate for lack of the female capability of giving birth to children. If this were true then Building Robots might indeed be seen as the ideal compensation' (Lighthill 1973).

The attempt to gain mastery over life and death through AI innovation is best captured by the aforementioned film *Transcendence*. *Transcendence* tells the story of AI expert Will Caster, who after being shot by a polonium-grazed bullet has his mind 'uploaded' onto an AI system by his wife Evelyn and his colleague Max Waters. It soon transpires that once this AI is connected to the internet, this new, disembodied version of Will is vastly more powerful than any human. He catches criminals; develops nanotechnology that can cure disease and disability; and solves climate emergencies. However, Max discovers that Will leaves implants in the brains of all those whom he cures or enhances with his nanotechnology, creating a human drone army. To save humanity, while the Will/AI tries to upload Evelyn's mind so that they can be reunited, Max implants Evelyn with a computer virus that destroys the Will/AI.

Will Caster is an embodiment of both the theme of 'womb envy' and the 'genius' trope, although very different from the gregarious playboy genius Tony Stark and the obnoxious tech bro Nathan Bateman. While Stark's masculinity is continuously reasserted through his (hetero)sexual conquests, machismo, and military affiliations, Will's is asserted through the ease with which he disembodies himself—first metaphorically, as an archetypal geek or nerd, detaching himself from issues he finds irrelevant, such as social engagements and dress shirts; then literally, becoming an uploaded mind that does not seem to miss embodiment. The masculine gendering of the dream of mind uploading has been frequently commented on as stemming from the idea that women and minorities are always defined through 'markers of bodily difference': one whose identity is shaped through embodiment both physically and socially is more likely to understand the loss of self that mind uploading will lead to (Hayles 1999, pp.3–4).

Gender Inequality: Subservience and Sacrifice

Only eight out of 116 AI scientists in our corpus are women: Shuri from *Avengers: Infinity War*; Quintessa, an alien woman from the *Transformers* franchise; Dr Dahlin in *Ghost in the Shell* (2017); Ava in *The Machine*; Susan Calvin in *I, Robot*; Frau Fabissina in *Austin Powers*; Dr Brenda Bradford from *Inspector Gadget*; and Evelyn Caster from *Transcendence*. This number is so low that we could not draw quantitatively significant conclusions about tropes from their depictions. However, we did find that the women in these seven examples were treated similarly enough to tentatively identify two tropes: *subservience* and *sacrifice*.

First of all, most female AI scientists (five out of eight) are portrayed through their relationship to a more prominent male AI scientist to whom they are subservient or inferior. Out of the eight female scientists in total, only three are not portrayed as in any way subordinate to a man: Shuri (who actively outsmarts one of the male geniuses, Bruce Banner), Quintessa, and Dr Dahlin. In four of the eight instances, the female AI scientist is the wife, lover, or daughter of the male genius (*The Machine, Transcendence, Inspector Gadget, Austin Powers*). The final female figure, Susan Calvin in *I, Robot*, is presented as the subordinate of Dr Alfred Lanning—in a position that is notably *more* subordinate than in Isaac Asimov's original *I, Robot* stories, which present her as the only robopsychologist in the company and the only one able to resolve the problems around which the stories revolve (Asimov [1982] 1995). Hence, while 7 percent of AI creators in this corpus are female, only three of them (3 percent of all AI scientists) are not in a subordinate position to a man. This observation is reflected in existing research on portrayals of female STEM characters. Steinke shows that even when women are present on screen as STEM professionals, they are portrayed as less capable than their male peers (Steinke 2017, p.2).

These observations are borne out in our corpus. For example, the three most significant female characters in the *Iron Man* films—Pepper Potts, Black Widow, and Maya Hansen—are all current or former love interests of Stark.[3] Most of the other female characters whoappear on screen are Stark's romantic liaisons, as indicated by his self-identified title of 'playboy'. Stark's scientific brilliance is also used to upstage the only female scientist in the *Iron Man* films. In *Iron Man 3*, Stark helps Maya Hansen, an expert biogeneticist, solve the problem she has been working on for years, writing the solution to the equation on a notecard he leaves with her after a one-night stand. Stark has no training or experience in biogenetics, and by solving her lifelong work, he devalues her entire career and body of expertise. This gendered dynamic, where masculinity is equated with raw genius, reflects the gendered portrayal of female AI scientists in film, where female AI scientists are often rendered subservient to a superior male AI engineer.

Gender inequality similarly shapes on-screen portrayals of AI scientists in *Transcendence*. The film opens with a voiceover introducing the protagonists as 'Will and Evelyn Caster'. It is the only time in the film that the two AI researchers are presented as equals: while both have doctorates, Evelyn is framed as Will's inferior both diegetically by other characters, and extradiegetically through camera and script choices. The next scene shows Will Caster rifling through papers in front of a laptop displaying code and equations, while Evelyn gets changed into a dress and puts on a necklace. Evelyn is framed as the socially competent one of the two (corresponding less to the genius trope), reminding Will of his important funders' meeting and helping him with his cuffs.

---

[3] Scarlett Johansson, the actress who plays Black Widow, recently criticised Black Widow's hyper-sexualisation in *Iron Man 2*, saying that her character was treated like 'a piece of something, like a possession' and even referred to as a 'piece of meat' ('Scarlett Johansson criticises Black Widow's "hyper-sexualisation" in Iron Man 2' 2021).

Evelyn is presented as inferior to Will in a way that makes it difficult to establish her role or skills. At the funders event, a groupie runs up to Will to ask if 'Dr Caster' could sign the magazine in which he (alone) appears. Evelyn herself comes on stage, but this is to introduce the audience to her 'partner in science and in life, Dr Will Caster'. After his lecture, Will is shot by an anti-AI activist; Evelyn is not attacked. Will initially recovers from being grazed with what is later revealed to be a polonium-laced bullet, allowing both Casters to meet FBI agent Buchanan. He responds to their introduction with 'I've been following your work, Dr Caster, it's fascinating' to Will Caster only.

Yet the plot requires Evelyn to be an AI expert too, well enough versed in Will's work to take apart the AI system he developed, PINN, and redeploy it as the system onto which his mind is uploaded. The upload is her idea; she plans it and convinces Will and their colleague Max Waters that it can be done. However, after being referred to as 'Mrs Caster' rather than 'Dr Caster', we finally find out halfway through the film that she too has a PhD, and that Will and Evelyn met in the class of the aforementioned Joseph Tagger (Morgan Freeman). Evelyn embodies the leaky pipeline: she had the same computer science education as her husband, was versed in the same project, and is equally capable of taking on a leadership role—but all of these skills are overlooked by the media, groupies, and FBI agents alike, as long as there is a male Dr Caster alive.

The second trope, sacrifice, sees female AI scientists regularly killed and/or sacrificed in the name of a greater cause. *Transcendence* ends with Evelyn Caster sacrificing herself to shut down the now rampaging, dictatorial Will Caster AI. As Will does not allow anyone but her to approach him, she has Max infect her with a nanoparticle computer virus. She offers Will to have her mind uploaded alongside his, so that they can be together—and so that the virus can attack Will when she plugs in. Both Will and Evelyn perish, leaving Max to survive and witness the ensuing breakdown of society.

Conclusion

Jocelyn Steinke argues that 'a better understanding of cultural representations of women, specifically a better understanding of the portrayals of female scientists and engineers in the media, may enhance the efficacy of efforts to promote the greater representation of girls in science, engineering, and technology' (Steinke 2017, p.29). Women's underrepresentation in the fields of AI and computer science remains an urgent issue, one that if left unaddressed will have serious negative ramifications for the field of AI and, consequently, society as a whole. Given that cultural representations of professions like engineering are known to shape women and girls' career aspirations, popular representations of AI scientists should be a space for creative intervention. Revitalised portrayals of AI scientists that break away from the heavily masculinised tropes explored in this chapter could turn the normative tide away from the hypermasculine AI scientist. However, to do so, popular media must support a multiplicity of gendered portrayals of AI scientists which do not force AI scientists of any gender into a set of tired tropes about what an AI scientist looks like.

References

Asimov, Isaac. [(1982) 1995]. *The Complete Robot.* London: HarperCollins.
Beasley, Chris, and Heather Brook. (2019) *The Cultural Politics of Contemporary Hollywood Film: Power, Culture, and Society.* Manchester University Press.
Bian, Lin, Sarah-Jane Leslie, Mary C. Murphy, and Andrei Cimpian. (2018) 'Messages about Brilliance Undermine Women's Interest in Educational and Professional Opportunities'. *Journal of Experimental Social Psychology* 76(May): 404–420. https://doi.org/10.1016/j.jesp.2017.11.006.
Burton-Hill, Clemency. (2016) 'The Superhero of Artificial Intelligence: Can This Genius Keep It in Check?' *The Guardian,* 2016, sec. Technology. http://www.theguardian.com/technology/2016/feb/16/demis-hassabis-artificial-intelligence-deepmind-alphago.
Carpenter, Charli. (2016) 'Rethinking the Political / -Science- / Fiction Nexus: Global Policy Making and the Campaign to Stop Killer Robots'. *Perspectives on Politics* 14(1): 53–69.
Cave, Stephen. (2020) 'The Problem with Intelligence: Its Value-Laden History and the Future of AI'. *Proceedings of the AAAI/ACM Conference on AI, Ethics, and Society,* 29–35. New York: ACM. https://doi.org/10.1145/3375627.3375813.
Cave, Stephen, Kanta Dihal and S. Dillon. (2020) *AI Narratives: A History of Imaginative Thinking about Intelligent Machines.* Oxford: Oxford University Press.
Cave, Stephen, and Kanta Dihal. (forthcoming) How the World Sees Intelligent Machines: Introduction. In *Imagining AI: How the World Sees Intelligent Machines,* eds Stephen Cave and Kanta Dihal. Oxford: Oxford University Press.
Cave, Stephen, Kanta Dihal, Eleanor Drage, and Kerry McInerney. (forthcoming) 'Who Makes AI? Gender and Portrayals of AI Scientists in Popular Film, 1920–2020'.
Cheng, Anne Anlin. (2019) *Ornamentalism.* Oxford: Oxford University Press.
Cheryan, Sapna, Victoria C. Plaut, Caitlin Handron, and Lauren Hudson. (2013) 'The Stereotypical Computer Scientist: Gendered Media Representations as a Barrier to Inclusion for Women'. *Sex Roles* 69(1): 58–71.
Clark, Jack. (2016) 'Artificial Intelligence Has a "Sea of Dudes" Problem'. *Bloomberg.* 23 June 2016. https://www.bloomberg.com/news/articles/2016-06-23/artificial-intelligence-has-a-sea-of-dudes-problem.
Conger, Kate. (2017) 'Exclusive: Here's The Full 10-Page Anti-Diversity Screed Circulating Internally at Google'. *Gizmodo,* 8 May 2017. https://gizmodo.com/exclusive-heres-the-full-10-page-anti-diversity-screed-1797564320.
Creed, Barbara. (1993) *The Monstrous-Feminine: Film, Feminism, Psychoanalysis.* London: Routledge.
Daniel, J Furman, III, and Paul Musgrave. (2017) 'Synthetic Experiences: How Popular Culture Matters for Images of International Relations'. *International Studies Quarterly* 61(3): 503–516. https://doi.org/10.1093/isq/sqx053.

Favreau, Jon. (2008) *Iron Man*. Paramount Pictures, Marvel Enterprises, Marvel Studios.

Favreau, Jon. (2010) *Iron Man 2*. Paramount Pictures, Marvel Entertainment, Marvel Studios.

Flicker, Eva. (2003) 'Between Brains and Breasts—Women Scientists in Fiction Film: On the Marginalization and Sexualization of Scientific Competence'. *Public Understanding of Science* 12(3): 307–318. https://doi.org/10.1177/09636625031 23009.

Garner, Meg. (2015) 'Woman Power in the Valley – Eight Female Tech CEOs You Should Know'. *The Street*. https://www.thestreet.com/investing/stocks/woman-power-in-the-valley-eight-female-tech-ceos-you-should-know–13219132

Geena Davis Institute on Gender in Media and The Geena Davis Institute on Gender in Media. (2018) 'Portray Her: Representations of Women STEM Characters in Media'. The Lyda Hill Foundation & The Geena Davis Institute on Gender in Media. https://seejane.org/wp-content/uploads/portray-her-full-report.pdf.

Goldstein, Joshua S. (2003) *War and Gender: How Gender Shapes the War System and Vice Versa*. Cambridge: Cambridge University Press.

Google and Gallup. (2015) 'Images of Computer Science: Perceptions Among Students, Parents and Educators in the US'. https://services.google.com/fh/files/misc/images-of-computer-science-report.pdf.

Guardian staff. (2021) 'Google Fires Margaret Mitchell, Another Top Researcher on Its AI Ethics Team'. *The Guardian* 20 February. http://www.theguardian.com/technology/2021/feb/19/google-fires-margaret-mitchell-ai-ethics-team.

Hammond, Alicia, Eliana Rubiano Matulevich, Kathleen Beegle, and Sai Krishna Kumaraswamy. (2020) *The Equality Equation: Advancing the Participation of Women and Girls in STEM*. Washington D.C.: World Bank. http://hdl.handle.net/10986/34317.

Hayles, N. Katherine. (1999) *How We Became Posthuman: Virtual Bodies in Cybernetics, Literature, and Informatics*. Chicago: University of Chicago Press.

Haynes, Roslynn D. (2017) *From Madman to Crime Fighter: The Scientist in Western Culture*. Baltimore: Johns Hopkins University Press.

Hern, Alex. (2018) 'Elon Musk: The Real-Life Iron Man'. *The Guardian*. 9 February 2018. http://www.theguardian.com/technology/2018/feb/09/elon-musk-the-real-life-iron-man.

Hill Collins, Patricia. [(1990) 2000] *Black Feminist Thought: Knowledge, Consciousness, and the Politics of Empowerment*. New York: Routledge.

Hooks, Bell. (2014) *Black Looks: Race and Representation*. Routledge. https://doi.org/10.4324/9781315743226.

Howard, Ayanna, and Charles Isbell. (2020) 'Diversity in AI: The Invisible Men and Women'. *MIT Sloan Management Review*, 21 September 2020. https://sloanreview.mit.edu/article/diversity-in-ai-the-invisible-men-and-women/.

Huang, Michelle N. and C. A. Davis. (2021) 'Inhuman Figures: Robots, Clones, and Aliens'. *Smithsonian Asian Pacific American Center* (blog). 2021. https://smithsonianapa.org/inhuman-figures/.

Jaxon, Jilana, Ryan F. Lei, Reut Shachnai, Eleanor K. Chestnut, and Andrei Cimpian. (2019) 'The Acquisition of Gender Stereotypes about Intellectual Ability: Intersections with Race'. *Journal of Social Issues* 75(4): 1192–1215. https://doi.org/10.1111/josi.12352.

Lighthill, James. (1973) 'Artificial Intelligence: A General Survey'. In *Science Research Council*. http://www.chilton-computing.org.uk/inf/literature/reports/lighthill_report/p001.htm.

Mantha, Yoan and James Hudson (2018) 'Estimating the Gender Ratio of AI Researchers Around the World', *Medium*, https://medium.com/element-ai-research-lab/estimating-the-gender-ratio-of-ai-researchers-around-the-world-81d2b8dbe9c3.

McCorduck, Pamela. (2004) *Machines Who Think: A Personal Inquiry into the History and Prospects of Artificial Intelligence*. 25th anniversary update. Natick, MA: A.K. Peters.

'Meet the Godfather of AI'. (2018) Bloomberg Businessweek. https://www.bloomberg.com/news/videos/2018-06-25/meet-the-godfather-of-ai-in-this-episode-of-hello-world-video.

Mulvey, Laura. (1975) 'Visual Pleasure and Narrative Cinema'. *Screen* 16(3): 6–18. https://doi.org/10.1093/screen/16.3.6.

Olla, Akin. (2021) 'Is WandaVision . . . Pentagon Propaganda?' *The Guardian*. 9 March 2021. http://www.theguardian.com/commentisfree/2021/mar/09/wandavision-pentagon-propaganda-marvel-disney-fbi.

PwC UK. (2017) 'Women in Tech: Time to Close the Gender Gap'. https://www.pwc.co.uk/who-we-are/women-in-technology/time-to-close-the-gender-gap.html.

'Sci-Fi (Sorted by US Box Office Descending)'. (n.d.) IMDb. Accessed 27 June 2022. http://www.imdb.com/search/title/?genres=sci-fi.

Shimizu, Celine Parreñas. (2007) 'The Hypersexuality of Race: Performing Asian/American Women on Screen and Scene'. https://doi.org/10.1215/9780822389941.

Silicon Valley Bank. (2020) '2020 Women in US Technology Leadership Report'. SVB.com. https://www.svb.com/globalassets/library/uploadedfiles/content/trends_and_insights/reports/women-in-us-technology-leadership-2020-silicon-valley-bank.pdf

Simonite, Tom. (2018) 'AI is the Future – But Where are the Women?' *Wired*. Available at: https://www.wired.com/story/artificial-intelligence-researchers-gender-imbalance/

Steinke, Jocelyn. (2005) 'Cultural Representations of Gender and Science: Portrayals of Female Scientists and Engineers in Popular Films'. *Science Communication* 27(1): 27–63. https://doi.org/10.1177/1075547005278610.

Steinke, Jocelyn. (2017) 'Adolescent Girls' STEM Identity Formation and Media Images of STEM Professionals: Considering the Influence of Contextual Cues'. *Frontiers in Psychology* 8. https://doi.org/10.3389/fpsyg.2017.00716.

Storage, Daniel, Tessa E. S.Charlesworth, Mahzarin R. Banaji, and Andrei Cimpian. (2020) 'Adults and Children Implicitly Associate Brilliance with Men More than Women'. *Journal of Experimental Social Psychology* 90 (September): 104020. https://doi.org/10.1016/j.jesp.2020.104020.

Suparak, Astria. (2020) 'Asian Futures, Without Asians Series'. Astria Suparak (blog). 13 February. https://astriasuparak.com/2020/02/13/asian-futures-without-asians/.

Thompson, Clive. (2019) *Coders: The Making of a New Tribe and the Remaking of the World*. New York: Penguin.

Top Lifetime Adjusted Grosses. (2021) Box Office Mojo. https://www.boxofficemojo.com/chart/top_lifetime_gross_adjusted/?adjust_gross_to=2019.

Top Lifetime Grosses. (2021) Box Office Mojo. https://www.boxofficemojo.com/chart/top_lifetime_gross/?area=XWW.

Varma, Roli. (2010) 'Why so Few Women Enroll in Computing? Gender and Ethnic Differences in Students' Perception'. *Computer Science Education* 20(4): 301–316. https://doi.org/10.1080/08993408.2010.527697.

West, Sarah Myers, Meredith Whittaker, and Kate Crawford. (2019) 'Discriminating Systems: Gender, Race, and Power in AI'. *AI Now Institute.* https://ainowinstitute.org/discriminatingsystems.html.

Whedon, Joss. (2012) *The Avengers*. Marvel Studios, Paramount Pictures.

Willis, Sharon. (1997) *High Contrast: Race and Gender in Contemporary Hollywood Films*. Duke University Press. https://doi.org/10.1215/9780822379218.

Women in Tech. (2019) 'Women in Tech Survey 2019'. https://www.womenintech.co.uk/women-technology-survey-2019.

Wood, Nathaniel. (2015) 'Sci-Fi Helped Inspire Elon Musk to Save the World'. *Wired* 13 June. https://www.wired.com/2015/06/geeks-guide-elon-musk/.

Wood, Robin. (1998) *Sexual Politics and Narrative Film: Hollywood and Beyond*. New York, Chichester: Columbia University Press.

Young, Erin, Judy Wajcman, and Laila Sprejer. (2021) 'Where Are the Women? Mapping the Gender Job Gap in AI'. London: The Alan Turing Institute. https://www.turing.ac.uk/research/publications/report-where-are-women-mapping-gender-job-gap-ai.

Young, Kevin L, and Charli Carpenter. (2018) 'Does Science Fiction Affect Political Fact? Yes and No: A Survey Experiment on "Killer Robots"'. *International Studies Quarterly* 62 (3): 562–576. https://doi.org/10.1093/isq/sqy028.

Yuen, Nancy Wang. (2016) *Reel Inequality: Hollywood Actors and Racism*. New Brunswick, NJ: Rutgers University Press.

6

No Humans in the Loop

Killer Robots, Race, and AI

Lauren Wilcox

Feminist Approaches to 'Humans in the Loop'

War and empire have been, and continue to be, key drivers in the development of artificial intelligence (AI). In this piece, I argue that 'the human in the loop' is inadequate as a safeguard against excessive violence in AI systems, and that feminist approaches should consider the foundations of this figure of 'humanity' in ongoing forms of racial and colonial violence. My feminist approach to AI does not consider algorithms, bodies, machines, codes as separate objects in isolation nor AI to be strictly a matter of logic, data, or code; AI is always material and *embodied*. The ways in which humans become entangled, enmeshed, with machines, what kinds of agencies and worlds are produced, and how such intimacies are deployed and regulated in the production of such worlds is a central question for feminism and there is no shortage of feminist engagements with these questions across a broad spectrum of technologies. Feminist thinkers such as Donna Haraway, N. Katherine Hayles, and Lucy Suchman have been influential in my thinking about the entangled histories of gender, race, class and more in our technological relations and configurations (see for example, Haraway 1990; Hayles 1999, 2005; Suchman 2007, 2020). Feminist work on the 'material-semiotic' (Haraway 1988, p.595) practices that make up such embodied worlds have much to say about the vast array of epistemic and decision-making systems collectively referred to as 'AI', as well as modes for imaging feminist forms of practice, such as Hayles's 'technosymbiosis' from this volume. Feminist work on AI has shown, as in Eleanor Drage and Federica Frabetti's discussion of Hayles's earlier reading of Turing work, how the gender human subject is constituted through its interaction with technology.

However, sometimes questions about how human/machine interfaces or assemblages are produced and their effects can neglect the question of who the 'human' is in the first place, and what this idea of 'the human' is doing in our critical work on AI. As Louise Amoore writes, 'the human in the loop is an impossible subject who cannot come before an indeterminate and multiple we' (Amoore 2020, p.66). While many feminists have located the white, Western man as the epitome of the modern subject of sovereignty, rights, and rationality, less commonly engaged is

Lauren Wilcox, *No Humans in the Loop*. In: *Feminist AI*. Edited by: Jude Browne, Stephen Cave, Eleanor Drage, and Kerry McInerney, Oxford University Press. © Oxford University Press (2023). DOI: 10.1093/oso/9780192889898.003.0006

the ways in which 'the human' itself is a technology that enabled enslavement, colonialism, and dispossession and continues to structure the horizons of effects to limit violence by 'killer robots'. In this short piece, I draw on the Black and decolonial feminist thought of Sylvia Wynter and Denise Ferreira da Silva to argue that colonisation and enslavement are conditions of possibility for our contemporary modes of understanding how violence is implicated in the human/machine relations that frame controversies over the development of lethal autonomous weapons systems. Ruha Benjamin argues in relation to technologies of surveillance and search results, 'antiblack racism…is not only a by-product, but a precondition for the fabrication of such technologies' (Benjamin 2019a, p.44). I am here particularly interested in how the 'human' is thought in relation to technology. Rather than viewing 'AI', and especially, 'lethal autonomous weapons systems' or 'killer robots' as new technologies promising or threatening to alter the face of war as we know it, I argue that the political and epistemic category of 'the human' is the technology both enabling and constraining our imaginations of 'the human in the loop' as the bulwark against a chaos of indiscriminate killing unleashed by LAWS. To understand 'the human' as a technology is to denaturalise 'the human' as the subject of autonomy and morality, and to see the origins of this figure of self-determination in modern state formation, colonialism, and enslavement.[1]

In theorising the human as a technology, I want to think beyond a zero-sum understanding of the human versus the technological, and even open up the question of 'the human' beyond ways in which human-technological assemblages are reshaping core concepts such as agency, autonomy, control, and responsibility, for I am concerned that such approaches reify a particular understanding of the 'human' that is then augmented.[2] Understanding 'the human' as a technology that differentiates the subjects of self-determination and self-preservation from disposable subjects who may be killed with impunity suggests that countering the threats imagined by the horizon of lethal autonomous weapons must come to terms with the ways in which race structures the moral subject of violence and the modern states on whose behalf such violence is enacted. The rest of this chapter elaborates on the rise of 'killer robots' and the 'human in the loop' as means of maintaining the 'humanity' of war, before discussing the problematic distinction between war and policing in the racialised violence of the state. Sylvia Wynter's 'No Humans Involved' piece presents an entry into further discussion of the ways in which a logic of race precedes the logic of either state or machine-based killing via the construction of 'the human' as the subject of ethical violence in international

---

[1] My argument on 'the human as a technology of colonialism and enslavement' was developed independently of Jennifer Rhee's recent contribution on a similar argument (Rhee 2022) and I thank her for bringing this to my attention.

[2] For one critique of how posthumanist approaches can reify a white masculine 'human' that needs no further interrogation, see Jackson (2015).

law. I illustrate the stakes of this figure of humanity in a discussion of a recent investigation of the US military's targeting and post-strike assessment procedures.

Killer Robots

'Killer robots' have threatened to leap from the pages and screens of science fiction to become reality. The label 'killer robots' or sometimes, 'death machines' is presumably meant to sensationalise and draw attention to this topic (see for example Schwarz 2018; Krishnan 2009; Crootof 2015; Carpenter 2016, Young and Carpenter 2018; Sharkey 2019). Recent years have seen an immense amount of popular and academic attention to the legal and ethical implications of 'lethal autonomous weapons' (Roff 2014; Asaro 2012; Beier 2020). While claims about the autonomy of any such weapons are generally overblown, as current attempts at designing such systems are notoriously considered both 'narrow' and 'brittle' (see Roff 2014; Payne 2021) the ways in which such technologies have been shaped by cultural forces, and continue to reshape our imaginations of war and political violence, persist. Automated weapons systems are being developed, tested, and deployed by several state actors including the United States, China, Israel, South Korea, and Russia.

Referring to the threat of lethal autonomous weapons as an 'existential threat to humanity', Human Rights Watch insists that 'the challenge of killer robots, like climate change, is widely regarded as a great threat to humanity that deserves urgent multilateral action' (Human Rights Watch 2020, p.1). Several informal meetings of state parties to the UN Convention on Certain Conventional Weapons have been held, as well as meetings of experts in the years after 2014, and in 2016, the state parties formalised deliberations by creating a Group of Governmental Experts. These high-level discussions generally figure LAWS as new weapons around which the international community has an interest in banning or at least heavily regulating, following the banning of landmines, and chemical and biological weapons. The International Committee of the Red Cross has also called for legally binding regulations, placing lethal autonomous weapons in a similar category as landmines in terms of being victim-activated weapons that are not targeted precisely (ICRC 2022). The website of a prominent non-governmental group, the Campaign to Stop Killer Robots boldly calls for 'Less autonomy. More Humanity' and seeks international laws prohibiting the use of LAWS.

Currently, US military policy is to follow a 'human-in-the-loop' model for the initiation for lethal force. The Department of Defense's regulation 3000.09, from 2012, requires that all weapons systems, including LAWS, 'allow commanders and operators to exercise appropriate levels of human judgment over the use of force' (Saxon 2014). This 'loop' as it were, is a decision-making system sometimes referred to as the 'kill chain'. The US Air Forces defines this as consisting of six steps: find, fix, track, target, engage, and assess. The 'loop' refers to the OODA loop, developed by John Boyd to distil human decision-making to a four-step process: Observe, Orient, Decide, Act (Bousquet 2009; Marra and McNeil 2012). One first observes the environment using all their senses, orients by synthesising information and converting it into knowledge. Then one weighs one's options, decides, and acts. This model is of course an oversimplification that in reality includes many feedback mechanisms.

The question of how a human should remain 'in the loop' or 'on the loop' is much more than an 'all-or-nothing' or a line in the sand. Autonomy in this 'loop' might be better conceived of a spectrum, but also along several axes, such as: how independent the machine is, how adaptable it is to unfamiliar environments, and how much discretion it has. The threshold of a 'human in the loop', or even the 'human on the loop' as necessary bulwark against weapons systems spiralling out of control frames LAWS as a specific technological apparatus. Invocations of 'Skynet' and 'Terminators' are laughably and frustratingly predictable; these visions of embodied, autonomous AI figures feared to 'go rogue' with devastating effects on human life arguably deflect from key concerns with already existing forms of surveillance, data extraction, and the entanglements of border control, policing, and militarised forms of violence and control. Moreover, in considering the framing of 'lethal autonomous weapons', and particularly, the hype around weaponised AI as a unique threat, more 'low-tech' weapons that arguably operate independently of human direction once they are deployed are excluded. Most notably, landmines are reported to have killed and wounded over 7000 people in 2020, with Syria being the most affected place, and large numbers of casualties also in Afghanistan, Burkina Faso, Columbia, Iraq, Mali, Nigeria, Ukraine, and Yemen, sometimes from mines laid years or decades preciously (Landmine Monitor 2021).

This is not to say that there are not real concerns over the development of new technologies and systems that have the potential to even further entrench military and carceral violence, or that such concerns are a long way off compared to more pressing issues. My suggestion in this piece is the maintenance of a 'human in the loop' and the challenges that this brings both legally and technologically remain wedded to fundamentally racialised ontologies and epistemologies of 'the human'. What kind of 'human' is proposed to be kept 'in' or 'on' the loop? Much attention is also devoted to the question of whether any LAWS might be able to follow the laws of war in terms of the principle of discrimination; that is, distinguishing between combatants and non-combatants. The US Defense Department's 'Algorithmic Warfare Cross-Functional Team', better known as Project Maven, aims to use machine learning and big data to distinguish people and objects in videos provided by drone surveillance. The aim is to pair algorithms from Project Maven with the 'Gorgon Stare', which provides wide angle motion imagery that can surveil individuals and vehicles over the size of an entire city (Weisgerber 2017). Such technologies are not only used for 'foreign' wars either; aerial surveillance combined with AI visual and data-processing techniques have been trialled in major American cities for domestic law enforcement as well (see for example Michel 2019).

Race, War, and the State

The question of the nature of the 'human in the loop' as it relates to the instruments of state violence and its regulation under domestic and/or international law are bound up in the nature of the subjects of the law and its violence and the origins of the technologies of power and government, which can be reified by AI technologies as well. Important and influential work by scholars such as Timnit Gebru, Joy Buolamwini, Safiya Noble, Ruha Benjamin, and others have drawn attention to the ways in which contemporary AI technologies reproduce gender and race differences and contribute to racial oppression. This work has further shown that the deployment of such technologies in systems of employment, policing, and more leads to the perpetuation of racial oppression (Buolamwini and Gebru 2018; Noble 2018; Benjamin 2019a, 2019b). Furthermore, the rise of 'predictive policing' is bringing a high-tech, surveillance capitalism approach to long-standing issues of police shootings and other forms of violence that disproportionately affect Black and other racialised communities in the United States (see for example Ferguson 2017; Brayne 2021). As Kerry McInerney's contribution to this volume details, predictive policing can also shape gendered and racialised logics of state protection. This and other important work by academics and activists have shown how AI systems can embed and reproduce forms of systemic inequalities based on race, gender, and other categories of oppression.

The literature on predictive policing and related forms of encoded racism is largely focused on domestic police efforts. Yet the circulations of tactics and technologies of military and police power are complex and ongoing, and the foreign/domestic distinction that is constitutive of the modern sovereign state can obscure ongoing colonial dynamics of violence. The foreign/domestic divide that structures law and politics can obscure the ways in which forms of surveillance, repression, and the fear of violent reprisal in the histories and continuing presence of colonialist violence suggest 'battle/repression' rather than 'war/peace' is a more appropriate way of understanding the spatial and temporal dynamics of state violence (Barkawi 2016). Much has been made of the immense degree of visual and datalogical surveillance that conditions late modern governance and accumulation strategies. However, as Simone Browne warns in her book tracing the long history of surveillance innovations to ensure white domination over Black people, 'surveillance is nothing new to black folks' (2015, p.10).

If we understand that 'war and police are *always already* together' (Neocleous 2014, p.13) we need to start from a different set of concerns in regard to LAWS than whether or not a human would still be kept in the loop. If human oversight into AI systems related to policing is not sufficient to prevent structural biases, why then, should so much of the critical work concerning LAWS be focused on retaining 'meaningful human control' and a 'human in the loop' as if the current policies have not resulted in tens of thousands of civilian deaths by airstrikes from drones and other airstrikes? Airwars has recently estimated that least 22,679, and potentially as many as 48,308 civilians have been killed by US airstrikes since 9/11. This of course does not include civilians killed by other means, nor by other states using drones or other methods of 'precision targeting' (Piper and Dyke 2021). We have certainly seen, for example, in contemporary uses of weapons systems such as drones in warfare, that leaving identification, targeting, and assessment tasks to humans aided by various data and visualisation technologies, is no guarantee that bureaucratic, gender, and racial biases will not play a significant role in decisions of the use of state violence (see for example Wilcox 2017; Wilke 2017; Gregory 2018). Discerning combatants from non-combatants is hardly simple for humans and goes well beyond even the already complex nature of visual recognition; it is inherently a strategic decision rather than a question of recognition, not that recognition is particularly simple either. As Drage and Frabetti argue in this volume, 'recognition' is a misnomer as such technologies can play a role in the normative materialisation of bodies rather than simply 'recognising' them.

While the case is still being made that 'drones' are not unprecedented technologies, it is perhaps better understood how 'AI' and its forebearers should not be understood as 'technologies to come' or futuristic technologies over the horizon as science fiction imaginaries would have us believe: the origins of what we know of today as 'AI' were formed in and through war and the Second World War and Cold War in particular. Just as many historians, cultural theorists and IR theorists have linked 'drones' to the use of air power by colonial powers for surveillance, and bombing (see *inter alia* Kaplan 2006; Feldman 2011; Satia 2014; Hall Kindervater 2016), the origins of 'AI' and its forebears in cybernetics are generally theorised as stemming from WWII. Peter Galison, for example, outlines how Norbert Weiner's work on building anti-aircraft predictor during the Second World War were premised upon a 'cold-blooded machinelike opponent' as the basis for feedback into the system (Galison 1994, p.231). 'In fighting this cybernetic enemy, Wiener and his team began to conceive of the Allied antiaircraft operators as resembling the foe, and it was a short step from this elision of the human and the nonhuman in the ally to a blurring of the human-machine boundary in general' (Galison 1994, p.233). Kate Chandler also discusses how early trials of television-piloted drones were developed by the United States as an analogue to the kamikaze, by attempting to provide the same strategic advantages, through both racist perceptions of the kamikaze as well as a sense of technology as a mark of racial superiority (Chandler 2020, Chapter 2). Hayles' influential work locates the centrality of the autonomous liberal subject to Wiener's cybernetic project: 'when the boundaries turn rigid or engulf humans so that they lose their agency, the machine ceases to be cybernetic and becomes simply and oppressively mechanical', (Hayles 1999, p.105). This work provides us with important clues about how the liberal autonomous subject, a subject of whiteness, still underpin the foundations of 'AI' from its modernist to postmodernist/posthumanist stages. Still, in investigating the question of the 'human' at the heart of this 'human in the loop' as a threshold figure of great hope and anxiety that supposedly forestalls the erosion of control, dignity, and 'humanity' (understood as moderating tendency) in warfare, it might be helpful to go further back than the Second World War to understand the roots of this 'human' by looking at another source of 'artificial' violence: the violence of the state.

Inarguably the most influential theory of the modern state is Hobbes's Leviathan. Hobbes's 'great Leviathan called a Common-wealth, or State' is an 'Artificiall Man' (1996(1651), introduction). This 'artificial man' is a construct of humans, intending to further their security and prosperity. The automaton of the state, David Runciman has argued, was meant to take input from humans (outside of which it has no existence) and transform them into rational outcomes, rather than violent mistrust (2018, pp.128–129). The state, as a machine is a 'person by fiction' (Runciman 2000). Hobbes's body *qua* natural thing, as *res extensa* rather than as sinful flesh, on which the philosophy of the state and self-defence was based, emerging from the establishment of nature as object of modern science.[3] The ability of humans to create, to make 'artifice' collectively is a marker of humanity; the compound creations like Leviathan are 'fancies' and 'whatsoever distinguisheth the civility of Europe from the barbarity of the American savages; is the workmanship of fancy...' (Hobbes 1650, quoted in Mirzoeff 2020). This creation of sovereignty that leads to security and prosperity in Europe is indexed by indigenous savagery in the Americas conceived of as individuals in a state of nature rather than as sovereign states, indigenous peoples provided a contrast as 'savages' (along with pirates) to the security and order of the sovereign state (see for example Moloney 2011).

Hobbes's move towards the depersonalisation of the sovereign is enabled by the disciplining threat of violence on the citizens, ultimately giving the sovereign the same right to punish 'rebel' citizens' and foreign enemies alike. In Hobbes's sovereign 'artificial man' the violence of the state is refigured as rational self-preservation and self-determination (see also Ferreira da Silva 2009). Extralegal violence is built into the edifice of modern state sovereignty. 'The sovereign right to kill is therefore predicated on the calculation of potential, not actual, hurt, which is directed at the commonwealth as a whole', as Banu Bargu argues (2014, p.53). Arthur Bradley has recently argued that Hobbes's theory of punishment 'effectively installs a virtual or potential "enemy" at the heart of every subject

---

[3] See extended discussion in Epstein (2021).

that—entirely willingly and voluntarily—exposes them to death at the hands of their sovereign at any point whatsoever in their lives' (2019, p.338). The Hobbesian state takes on the omnipresent and 'better than human' capacities feared of weaponised AI. Would it perhaps, then, be too much of a stretch to say that the sovereign state is its own kind of 'killer robot', as it is made from humans, to mimic humans, in order to shift the burden of self-preserving violence away from themselves, with lingering anxieties about how such powers might be 'out of control'?

This intriguing point leads us to consider an intertwined set of issues in thinking through the 'human in the loop' that is to be prohibited and/or regulated through international law, namely, the human in relation to the machine, and the human in relation to the sovereign state and international law. These questions, however, are intertwined in the forming of the modern sovereign subject of the 'human' in and through colonial and racialised orders. This 'human' is a figure of Western modernity that is the self-authorised knower and subject of self-preservation.

No Humans in the Loop

In the wake of the 1992 acquittal of the police officers involved in beating Rodney King in Los Angeles that sparked an uprising, the Jamaican scholar, novelist, and playwriter Sylvia Wynter wrote an essay titled 'No Humans Involved: An Open Letter to My Colleagues' (Wynter 1997). Wynter noted that public officials in LA referred to judicial cases involving jobless Black males with the acronym 'N.H.I.' for 'no humans involved': they had the power to confer who was a human and who was not. Asking how this classification came into being, Wynter turns to the construction of 'inner eyes' in reference to Ellison's *Invisible Man*, that determines who does not inhabit the 'universe of obligation', and may be subjected to genocidal violence and incarceration. Wynter asks, 'Why is this "eye" so intricately bound up with that code, so determinant of our collective behaviors, to which we have given the name *race*?' (Wynter 1997, p.47).

Wynter's answer is that race and its classificatory logic is bound up in the modern episteme in which humans were considered an '*evolutionarily selected being*' (Wynter 1997, p.48) that pre-existed subjective understandings. The *organic* nature of the modern subject (the subject who is the master of AI, and who, all parties are clear, must remain in control) is further delineated by Wynter, who traces the replacement of the Judeo-Christian cultural matrix to a secular basis and then to a biological one, all while reproducing the overrepresentation of the human figure she labels 'Man$_2$' defined 'biogenically' and 'economically'. This subject of 'Man' is a *genre* of the human who is overrepresented to stand for the human as a whole (Wynter 2003). Wynter centres the 1492 'encounter' as being essential to the formulations of the European representations of the human. In the reinvention of the 'Man₁' of the Renaissance through the colonial encounter in the new world, to its reformulation in Darwinian and economic terms as Man₂, Wynter writes,

> It is only within the terms of our present local culture, in which the early feudal-Christian religious ethic and its goal of Spiritual Redemption and Eternal Salvation has been inverted and replaced by the goal of Material Redemption, and therefore, by the transcendental imperative of securing the economic well being, of the now biologized body of the Nation (and of national security!), that the human can at all be conceived of *as if it were* a mode of being which exists in a relation of pure *continuity* with that of organic life,
>
> (Wynter 1997, p.59).

It is within these terms, Wynter argues, that 'N.H.I' is understood as an expression of the ordering principle referred to as 'race'. Representing 'the human' as a natural organism 'mistakes the representation for the reality, the map for the territory' (Wynter 1997, p.49) and overrepresents the bio/economic man associated with Europeans as the sole expression of humanity. For the bio/economic man to be overrepresented as all of humanity, indigenous people, and transported and enslaved Black Africans were made to serve as 'the physical reference of the projected irrational/subrational Human Other to its civic-humanist, rational self-conception' (Wynter 2003, p.281). For Wynter, the epistemological critique of the subject of knowledge is bound up in conquest, capitalism, and enslavement and thus, 'the human' itself as a subject bound up in this history. 'Racism', Wynter writes, 'is an *effect* of the biocentric conception of the human' (Wynter 2003, p.364) and thus, the question of 'the human' precedes that of race or gender.

Inspired by Wynter's work, Katherine McKittrick describes the algorithmic prediction of the death of young man in Chicago, Davonte Flennoy, and of how 'his life only enters the mathematical equation as death' (McKittrick 2021, p.104). McKittrick's argument is not (only) that algorithms reproduce gendered and racialised patterns of violence and inequality, but also that algorithms require biocentric methods and methodologies that are only capable of producing dehumanising results. '[B]lack inhumanity, specifically the biocentric racist claim that black people are nonhuman and unevolved and a priori deceased, is a variable in the problem-solving equation before the question is asked, which means the work of the algorithm—to do things people care about, to accomplish the task—already *knows* that Flennoy's life and well-being are extraneous to its methodology' (McKittrick 2021, p.111). In other words, racism is required before the problem is even begun to be worked on: it precedes the algorithm, and might even be said to be the condition of possibility for the calculative logic that would problematise Flennoy's life in this way in the first place.

Some of the ways in which forms of racialised inhumanity structure and enable violence prior to the development of any particular technologies can also be seen in Denise Ferreira da Silva's analysis of the analytics of raciality. Ferreira da Silva excavates the founding statements of modern Western philosophy, from Descartes, Kant, and Hegel, to locate the subaltern racial subject as being constituted temporally and ontologically outside of, and in relation to, the universal subject upon which international law is constructed (2007). The cognito, the self-defining subject of modern humanity, is itself defined via an imagined encounter with an automaton in Descartes. For Ferreira da Silva, the Enlightenment project, which produces the human body as the exteriorisation of the mind and establishes the distinction between the 'transparent I' of Europe post-Enlightenment and the 'affectable other', institutes race as the signifiers of those spatialised subjects who are subject to the universal reason of self-determining, Enlightened subjects even though they are not capable of grasping this reason. In Ferreira da Silva's mappings of the analytics of raciality, 'self-determination remains the exclusive attribute of the rational mind, which exists in the kingdom of freedom, where transcendentality is realised, namely where reside the ethical-juridical things of reason, modern subjects whose thoughts, actions and territories refigure universality' (2009, p.224). Thus, for Ferreira da Silva, the 'transparent I' that is universal subject of reason, rests upon an outside, 'other', the 'no-body' or 'affectable other' who becomes the one whom the state must protect itself against, rather than protect.

These 'analytics of raciality', like Wynter's biocentric human, are both logically and temporally prior to the categorisation of 'race' in scientific racism terms. Rights claims collapse, Ferreira da Silva argues, in the face of the state's claim of the use of violence for self-preservation (2009, p.224). This is because 'raciality produces both the subject of ethical life, who the halls of law and forces of the state protect, and the subjects of *necessitas*, the racial subaltern subjects whose bodies and territories, the global present, have become places where the state deploys its forces of self-preservation' (ibid). The 'artificial man' of the state, formed in and through the production of self-determining subjects and their 'affectable others' becomes the site of the *ultima ratio* of both violence and rationality. The state as the original 'killer robot' is therefore already inscribed with the human as a technology that is the subject of rational, self-preserving violence, a human that structures the limits of humanitarian law.

Humanitarianism and the Human in the Loop

The metaphysical notion of Man that is an entirely modern conception of the humanist commitment to scepticism and self-preservation gave rise to both the modern sovereign state, subjugating others from within and without, as well as the idea of the modern, autonomous, individual subject (Asad 2015, p.397) Or simply, as Asad writes, 'the idea of difference is built into the concept of the human' (Asad 2015, p.402). The 'human' subject whose presence 'in the loop' is meant to ensure accountability and the human who is author of the sovereign state and the international laws it contracts could both be understood to be product of this same metaphysics. In his piece 'Reflections on Violence, War, and Humanitarianism', Talal Asad briefly reflects on 'killer robots' and suggests that the prospect of such might encourage us to reflect on the contradictions of humanitarianism and the redemption of humanity. I want to take this up, in light of the previous discussion of Wynter and Ferreira da Silva's work as well as Asad's own and other postcolonial/decolonial critiques of international law.

Postcolonial and Third World approaches to international law have shown how the distinction between the 'civilised' and 'uncivilised' is crucial to shaping international law, even in formally neutral doctrines like sovereignty (Anghie 2005; 2006). International humanitarian law, or the laws of war, has always had an 'other', according to Frédéric Mégret, a 'figure excluded from the various categories of protection, and an elaborate metaphor of what the laws of war do not want to be' (2006, p.266). War as a concept is linked to the state, itself a Western concept, and part of how the state was formed in the first place. In the laws of war, the behaviour of the so-called 'civilised states' was taken as the model. 'The laws of war, fundamentally, project a fantasy about soldiering that is ultimately *a fantasy about sameness*' as it projects a desire to confront adversaries with analogously constituted military with ethical codes and fighting on behalf of sovereign states (Mégret 2006, p.307). When 'it is law that defines the momentous distinction between humane and inhumane suffering' (Asad 2015, p.410) we find that the law that decides on what kinds of suffering are excessive is guided by a figure of the human premised on racialised exclusion.

Asad writes that humanitarian interventionism is given its moral force by '*horror at the violation of human life*' (Asad 2015, p.411, emphasis in original). I would argue the similar moral force drives the outrage and campaigns against LAWS. The Future of Life Institute, for example, in their advocacy of bans on LAWS, declares, 'Algorithms are incapable of understanding or conceptualizing the value of a human life, and so should never be empowered to decide who lives and who dies' (Future of Life Institute 2021). The violation of human life here stems from *nonhuman* taking of human life as abhorrent rather than the taking of human life itself. This is of course, not to suggest that the development of LAWS, nor any other military technologies, is a desirable outcome. Rather, it is to suggest further examination of how race structures the legality and morality of war, and what kinds of violence should be seen as 'excessive' (Asad 2015, see also Kinsella 2011). As Asad writes,

> The question of intentionality in modern war is directed at defining legal—that is, responsible—killing; it is generally acknowledged that a military strike against a civilian target will kill a surplus and that those excess deaths will be legally covered as long as the killing is thought to be proportionate and necessary. And

there is always, as a matter of modern military etiquette, the obligatory public expression of regret at such killing. Moderns believe that unlike barbarians and savages, civilized fighters act within a legal- moral framework; the law of war is a crucial way of restraining killing, in manner and in number. Barbarians do not have such a framework

(Asad 2015, p.412).

The spectre of lethal autonomous weapons perhaps seeks to undermine, or per-haps revise this framework in which 'humans' may be killed by those operating outside of its dictates, only, not 'barbarians', but machines. So the desire for a 'human in the loop' is taken as a means of ensuring that the true subjects of ratio-nality are at the helm, that violence remains within a legal-moral framework and such, that killing does not become 'barbaric'. Asad also realises the state has taken over and obscured the responsibility and accountability for violence: 'It is the per-petration of violence by human agents against other humans that is emotionally graspable, even though the way in which modern law works often serves to diffuse the responsibility of agents for violence and cruelty when they act on behalf of the state' (Asad 2015, p.421).

While states are developing lethal autonomous weapons to even further dis-place their own violence, a recent example of state sponsored killing shows the inadequacy of existing practices that purport to keep 'humans in the loop' to ensure accountability and humaneness in targeted killings. A recent Pulitzer-Prize-winning *New York Times* investigation by journalist Azmat Khan, drawing from over a thousand confidential documents, showed significant discrepancies between the US military's own investigations of drone strikes resulting in civil-ian casualties and the death and destruction experienced in Iraq and Syria (Khan 2021).[4] Her investigation included both an examination of the US military's 'cred-ibility reports', assessing if it was 'more likely than not' civilians were killed by the airstrike, with interviews of survivors to assess whether or not said credibility reports were, indeed, credible. Reporting visits with some of the victims of bomb-ings and their families in over 100 sites of casualties showed the contrast between the image of a 'precision bomb' and the imprecision of targeting, as well as the lack of transparency and accountability. Hundreds of civilian deaths reportedly went unaccounted for: Khan's report juxtaposes images of bomb sites, family pho-tographs of loved ones who had been killed and testimonies of relatives with the 'credibility assessments' that all too often said no civilians were present.[5] As Khan

---

[4] Khan's investigation was so far only able to obtain papers on 'credibility assessments' on airstrikes from Iraq and Syria between September 2014 and January 2018, and documents from Afghanistan are reportedly the subject of a lawsuit.

[5] The second part of Khan's report details her travel and interviews with witnesses and family members of civilians killed, providing further evidence of the Pentagon's systematic undercounting of civilians (Khan and Prickett 2021).

writes, 'What emerges from the more than 5,400 pages of records is an institutional acceptance of an inevitable collateral toll. In the logic of the military, a strike, however deadly to civilians, is acceptable as long as it has been properly decided and approved—the proportionality of military gain to civilian danger weighed—in accordance with the chain of command' (Khan 2021).

In attempting to understand the causes of such underestimation of the harm caused by these strikes, Khan points to 'confirmation bias', a term from psychology she defines as 'the tendency to search for and interpret information in a way that confirms a pre-existing belief'. As an example, Khan points to an interpretation of 'men on motorcycles' that were deemed to be moving 'in formation' and displaying the 'signature' of an imminent attack; it turns out they were 'just' men on motorcycles (Khan 2021). This kind of 'confirmation bias' was present not only in fast-emerging situations in which speed and confusion from the 'fog of war' could be expected to play a part in mis-interpretation, but also from attacks planned well in advance in which the relevant personnel should have had time to prepare and familiarise themselves with the targets and surroundings. Khan also suggests the possibility of cultural differences playing a role in the mis-identification of civilians: The US military, for example, predicted 'no civilian presence' when families were sleeping inside during Ramadan or gathering in a single house for protection during intense fighting. Khan also notes civilians being visible in surveillance footage but not noted in the chat logs.

I would suggest here this is not 'merely' psychological; the military apparatus, from the technology, to the legal framework, to the training, and the accountability mechanisms (rather, lack thereof) and the overall strategy of adopting targeted bombing as a way to reduce injury and death to US forces while waging war abroad can be read as productive of a system that renders certain human lives illegible *as* human. In other words, it is difficult to locate the conditions of possibility for the disappearances of 'the human' in a particular technology or algorithm. The foundational issue is located prior to any such designs, in the biocentric version of the human that is the figure of rationality and of the state, in whom rationality is invested.

Khan's report also discusses the question of proportionality mentioned by Asad before: even if civilians *are* identified as being at risk in a bombing, civilians may still be knowingly killed if the anticipated casualties are not disproportionate to the anticipated military advantage. Military planners and lawyers calculate the value of the civilians' lives relative to their own advantage; such lives become 'affectable others' and 'no-bodies' relative to state authority. Khan discusses an incident in which three children were on a roof in Al Tanak, outside of Mosul, Iraq in 2017. They were included in the casualty assessment, but because the house was deemed a site of ISIS activity, it was bombed anyway. Viewing, identifying, acknowledging the presence of children is not sufficient to prevent their deaths by bombing: these 'humans in the loop' were no adequate bulwark against gratuitous murder. Better visual capacities, better targeting, or even automated procedures cannot prevent such violence while state violence and its rationality in service of the white, bourgeois 'human' remains the only judge of the worth of some lives. This is the logic of 'no humans involved': the state invested with the powers to determine which persons shall count as humans.

A spokesperson for the US military, Captain Bill Urban, described the violence as rooted in the 'barbarity' of the people themselves, and of US military practices as the rational response to this 'barbarity'. Captain Urban explained that targeting was complicated by enemy forces who 'plan, resource and base themselves in and among local populace'. Captain Urban, as quoted by Khan, relied upon similar colonial and racialised tropes to describe the United States' violence as rooted in the barbarity of the people themselves—rendered outside the universe of obligation, outside 'the human'. 'They do not present themselves in large formations', Urban said in Khan's piece. Urban further complained about the unconventional tactics used and finally, about the use of illegal tactics. In his words, '[T]hey often and deliberately use civilians as human shields, and they do not subscribe to anything remotely like the law of armed conflict to which we subscribe' (Khan 2021). Ascribing the use of 'human shields' has, ironically, become a means of *dehumanising* these 'human shields', particularly if those humans are not coded as white. The claim of 'human shielding' has become a means of removing civilian protections from civilians and making them responsible for their own deaths (see for example Gordon and Perugini 2020; Wilcox 2022). The claim of 'human shields' and of waging wars dissimilarly to Western forces are precisely the kinds of arguments cited by Mégret and Asad to reveal the colonial structures inherent in international humanitarian law, and a way of *dehumanising* those so described, rendering them subject to the violence of the colonial state.

Conclusion

The 'human in the loop', and its companion, 'meaningful human control' have been upheld as a bulwark to prevent indiscriminate and out of control targeting and killing by lethal autonomous weapons. As part of a feminist project to interrogate how autonomy, control, and the judgement morally and politically required to legitimately take a life are constructed, this chapter located 'the human' as not only imbricated with technological artefacts and processes, but itself a kind of technology of racial differentiation between self-determining, rational, moral subjects and the 'others' against whom rational violence is used; a product and enabler of histories of colonialism and enslavement. The elements of raciality, as elaborated by Black feminist theorists such as Sylvia Wynter, Katherine McKittrick, and Denise Ferreira da Silva, is established prior to any specific technologies: the elements of the modern colonial state as a 'killer robot' speak to the 'affectable other' as constituted outside of the state that makes laws and legitimises violence in its own preservation. 'The human' then cannot be the answer to the threat of excessive, indiscriminate violence as this technology has been at the root of so much of this very violence against those who are deemed to live outside of its protection.

References

Amoore, Louise. (2020) *Cloud Ethics: Algorithms and the Attributes of Ourselves and Others*. Durham: Duke University Press.

Anghie, Antony. (2005) *Imperialism, Sovereignty and the Making of International Law*. Cambridge University Press.

Anghie, Antony. (2006) 'The Evolution of International Law: Colonial and Postcolonial Realities'. *Third World Quarterly* 27(5): 739–753.

Asad, Talal. (2015) 'Reflections on Violence, Law, and Humanitarianism'. *Critical Inquiry* 41: 390–427

Asaro, Peter. (2012) 'On Banning Autonomous Weapon Systems: Human Rights, Automation, and the Dehumanization of Lethal Decision-Making'. *International Review of the Red Cross* 94 (886): 687–709.

Bargu, Banu. (2014) 'Sovereignty as Erasure: Rethinking Enforced Disappearances'. *Qui Parle* 23(1): 35–75.

Barkawi, Tarak. (2016) 'Decolonising War'. *European Journal of International Security* 1(2): 199–214.

Beier, J. Marshall. (2020) 'Short Circuit: Retracing the Political for the Age of 'Autonomous' Weapons'. *Critical Military Studies* 6(1): 1–18.

Benjamin, Ruha. (2019a). *Race after Technology: Abolitionist Tools for the New Jim Code*. Medford, MA, USA: Polity.

Benjamin, Ruha, ed. (2019b) *Captivating Technology: Race, Carceral Technoscience, and Liberatory Imagination in Everyday Life*. Durham: Duke University Press.

Bousquet, Antoine. (2009) *The Scientific Way of War: Order and Chaos on the Battlefields of Modernity*. Hurst Publises.

Bradley A. (2019) 'Deadly Force: Contract, Killing, Sacrifice'. *Security Dialogue*. 50(4): 331–343.

Brayne, Sarah. (2021) *Predict and Surveil: Data, Discretion, and the Future of Policing*. New York, NY, USA: Oxford University Press.

Browne, Simone. (2015) *Dark Matters: On the Surveillance of Blackness*. Durham, NC, USA: Duke University Press, 2015.

Buolamwini, Joy, and Timnit Gebru. (2018) 'Gender Shades: Intersectional Accuracy Disparities in Commercial Gender Classification'. In *Conference on Fairness, Accountability and Transparency*, pp. 77–91. http://proceedings.mlr.press/v81/buolamwini18a.html.

Carpenter, Charli. (2016) 'Rethinking the Political / -Science- / Fiction Nexus: Global Policy Making and the Campaign to Stop Killer Robots'. *Perspectives on Politics* 14(1): 53–69.

Chandler, Katherine. (2020) *Unmanning: How Humans, Machines and Media Perform Drone Warfare*. New Brunswick: Rutgers University Press.

Crootof, Rebecca. (2015) 'The Killer Robots Are Here: Legal and Policy Implications'. *Cardozo Law Review* 36: 1837–1915

Epstein, Charlotte. (2021) *The Birth of the State: The Place of the Body in Crafting Modern Politics*. Oxford University Press.

Feldman, Keith P. (2011) 'Empire's Verticality: The Af/Pak Frontier, Visual Culture, and Racialization from Above'. *Comparative American Studies: An International Journal* 9(4): 325–341.

Ferguson, Andrew G. (2017) *The Rise of Big Data Policing: Surveillance, Race, and the Future of Law Enforcement*. New York: New York University Press.

Ferreira da Silva, Denise. (2007) *Toward a Global Idea of Race*. Minneapolis: University of Minnesota Press.

Ferreira da Silva, Denise. (2009) 'No-Bodies'. *Griffith Law Review* 18(2): 212–236.

Fussell, Sidney. (2022) 'Dystopian Robot Dogs Are the Latest in a Long History of US-Mexico Border Surveillance'. *The Guardian*, 16 February, section US news. https://www.theguardian.com/us-news/2022/feb/16/robot-dogs-us-mexico-border-surveillance-technology.

Future of Life Institute. (2021) '10 Reasons Why Autonomous Weapons Must Be Stopped', 27 November. https://futureoflife.org/2021/11/27/10-reasons-why-autonomous-weapons-must-be-stopped/.

Galison, Peter. (1994) 'The Ontology of the Enemy: Norbert Wiener and the Cybernetic Vision'. *Critical Inquiry* 21(1): 228–226.

Gordon, Neve, and Nicola Perugini. (2020) *Human Shields: A History of People in the Line of Fire*. Berkeley: University of California Press.

Gregory, Derek. (2018) 'Eyes in the Sky – Bodies on the Ground'. *Critical Studies on Security* 6(3): 347–358.

Haraway, Donna. (1988) 'Situated Knowledges: The Science Question in Feminism and the Privilege of Partial Perspective'. *Feminist Studies* 14(3): 575–599.

Haraway, Donna. (1990) *Simians, Cyborgs, and Women: The Reinvention of Nature*. New York, NY, USA: Routledge.

Hayles, Katherine. (1999) *How We Became Posthuman: Virtual Bodies in Cybernetics, Literature, and Informatics*. Chicago, IL, USA: University of Chicago Press.

Hayles, Katherine. (2005) *My Mother Was a Computer: Digital Subjects and Literary Texts*. Chicago, IL, USA: University of Chicago Press.

Hobbes, Thomas. (1650) 'Answer to Davenant's Preface to Gondibert'. *Works IV*. 440–1.

Hobbes, Thomas, and Robert Tuck. (1996 [1651]) *Leviathan*. Cambridge: Cambridge University Press.

Human Rights Watch. (2020) 'Stopping Killer Robots: Country Positions on Banning Fully Autonomous Weapons and Retaining Human Control'. https://www.hrw.org/sites/default/files/media_2021/04/arms0820_web_1.pdf. 10 August. Accessed 10 March 2022.

International Committee of Red Cross (ICRC). (2022) 'Autonomous Weapons: The ICRC Remains Confident that States will Adopt New Rules'. https://www.icrc.org/en/document/icrc-autonomous-adopt-new-rules. 11 March. Accessed 13 April 2022.

Jackson, Zakiyyah Iman. (2015) 'Outer Worlds: The Persistence of Race in Movement "Beyond the Human"'. *GLQ: a Journal of Lesbian and Gay Studies* 21(2–3): 215–218.

Kaplan, Caren. (2006) 'Mobility and War: The Cosmic View of US "Air Power"'. *Environment and Planning A: Economy and Space* 38(2)

Khan, Azmat. (2021) Hidden Pentagon Records Reveal Patterns of Failure in Deadly Airstrikes'. *The New York Times*, 18 December, sec. U.S. https://www.nytimes.

com/interactive/2021/12/18/us/airstrikes-pentagon-records-civilian-deaths.html. Accessed 13 April 2022.

Khan, Azmat, and Ivor Prickett. (2021) 'The Human Toll of America's Air Wars'. *The New York Times*, 20 December, sec. Magazine. https://www.nytimes.com/2021/12/19/magazine/victims-airstrikes-middle-east-civilians.html. Accessed 13 April 2022.

Kindervater, Katharine Hall. (2016) 'The Emergence of Lethal Surveillance: Watching and Killing in the History of Drone Technology'. *Security Dialogue* 47(3): 223–238.

Kinsella, Helen. (2011) *The Image before the Weapon: A Critical History of the Distinction between Combatant and Civilian.* Ithaca, NY, USA: Cornell University Press.

Krishnan, Armin. (2009) *Killer Robots: Legality and Ethicality of Autonomous Weapons.* Farnham, UK: Ashgate Publishing Limited.

Landmine Monitor. (2021) 'Reports | Monitor'. Accessed 31 March 2022. http://www.the-monitor.org/en-gb/reports/2021/landmine-monitor-2021.aspx.

Marra, William, and Sonia McNeil. (2012) 'Understanding "The Loop": Autonomy, System Decision-Making, and the Next Generation of War Machines'. *Harvard Journal of Law and Public Policy* 36(3): 1139–1185

McKittrick, Katherine. (2021) *Dear Science and Other Stories.* Durham: Duke University Press.

Mégret, Frédéric. (2006) From 'Savages' to 'Unlawful Combatants': A Postcolonial Look at International Humanitarian Law's 'Other'. In *International Law and Its Others*, ed. Anne Orford, pp. 265–317. Cambridge: Cambridge University Press.

Michel, Arthur Holland. (2019) *Eyes in the Sky: The Secret Rise of Gorgon Stare and How It Will Watch Us All.* Boston, MA, USA: Houghton Mifflin Harcourt.

Mirzoeff, Nicholas. (2020) 'Artificial Vision, White Space and Racial Surveillance Capitalism'. *AI & Society* 36: 129–1305.

Moloney, Pat. (2011) 'Hobbes, Savagery, and International Anarchy'. *The American Political Science Review* 105(1): 189–204.

Neocleous, Mark. (2014) *War Power, Police Power.* Edinburgh: Edinburgh University Press.

Noble, Safiya Umoja. (2018) *Algorithms of Oppression: How Search Engines Reinforce Racism.* New York, NY, USA: New York University Press, 2018.

Payne, Kenneth. (2021) *I, Warbot: The Dawn of Artificially Intelligent Conflict.* London: Hurst & Company.

Piper, Imogen and Joe Dyke. (2021) 'Tens of Thousands of Civilians Likely Killed by US in "Forever Wars"'. *Airwars*. 6 September. Accessed 18 May 2022. https://airwars.org/news-and-investigations/tens-of-thousands-of-civilians-likely-killed-by-us-in-forever-wars/.

Rhee, Jennifer. (2022) A Really Weird and Disturbing Erasure Of History: The Human, Futurity, and Facial Recognition Technology. In *My Computer Was a Computer*, ed. David Cecchetto, pp. 33–52. Noxious Sector Press.

Roff, Heather M. (2014) 'The Strategic Robot Problem: Lethal Autonomous Weapons in War'. *Journal of Military Ethics* 13(3): 211–227.

Runciman, David. (2000) 'Debate: What Kind of Person is Hobbes's State? A Reply to Skinner'. *The Journal of Political Philosophy* 8(2): 268–278.

Runciman, David. (2018) *How Democracy Ends.* London: Profile Books.

Satia, Priya. (2014) 'Drones: A History from the British Middle East'. *Humanity: An International Journal of Human Rights, Humanitarianism, and Development* 5(1): 1–31.

Saxon, D. (2014) 'A Human Touch: Autonomous Weapons, Directive 3000.09, and the "Appropriate Levels of Human Judgment over the Use of Force"'. *Georgetown Journal of International Affairs* 15(2): 100–109.

Schwarz, Elke. (2018) *Death Machines: The Ethics of Violent Technologies*. Manchester, UK: Manchester University Press.

Sharkey, Amanda. (2019) 'Autonomous Weapons Systems, Killer Robots and Human Dignity'. *Ethics and Information Technology* 21: 75–87.

Suchman, Lucy. (2007) *Human-Machine Reconfigurations: Plans and Situated Actions*, 2nd edn. Cambridge: Cambridge University Press.

Suchman, Lucy. (2020) 'Algorithmic Warfare and the Reinvention of Accuracy'. *Critical Studies on Security* 8(2): 175–187.

Weisgerber, Marcus. (2017) 'The Pentagon's New Artificial Intelligence Is Already Hunting Terrorists'. *Defense One*, 21 December. https://www.defenseone.com/technology/2017/12/pentagons-new-artificial-intelligence-already-hunting-terrorists/144742/.

Wilcox, Lauren. (2017) 'Embodying Algorithmic War: Gender, Race, and the Posthuman in Drone Warfare'. *Security Dialogue* 48(1): 11–28.

Wilcox, Lauren. (2022) 'Becoming Shield, Unbecoming Human'. *International Politics Reviews*, online first.

Wilke, Christiane. (2017) 'Seeing and Unmaking Civilians in Afghanistan: Visual Technologies and Contested Professional Visions'. *Science, Technology, & Human Values* 42(6): 1031–1060

Wynter, Sylvia. (1997) 'No Humans Involved: An Open Letter to My Colleagues'. *Forum N. H. I. Knowledge for the 21st Century*. 1(1): Knowledge on Trial, 42–71.

Wynter, Sylvia. (2003) 'Unsettling the Coloniality of Being/Power/Truth/Freedom: Towards the Human, After Man, Its Overrepresentation – An Argument'. *CR: The New Centennial Review* 3(3): 257–337.

Young Kevin L. and Charli Carpenter. (2018) 'Does Science Fiction Affect Political Fact? Yes and No: A Survey Experiment on "Killer Robots"'. *International Studies Quarterly* 62(3): 562–576.

7

Coding 'Carnal Knowledge' into Carceral Systems

A Feminist Abolitionist Approach to Predictive Policing

Kerry McInerney

Introduction

In 2011, *TIME* magazine named 'preemptive policing' as one of the top 50 inventions that year, sitting proudly among the 'most inspired ideas, innovations and revolutions, from the microscopic to the stratospheric' (Grossman et al. 2011). However, the following decade has seen intense contestation of AI-enabled technologies developed for policing, law enforcement and the criminal justice system. Civil society groups such as Privacy International, Liberty, and Amnesty International have critiqued the human rights risks posed by predictive policing, ranging from discrimination against marginalised groups to serious breaches of privacy that place key human rights such as freedom of expression and freedom of assembly at risk (Couchman and Lemos 2019; Amnesty International 2020). This turnaround in public opinion is evidenced by the increasing number of moratoria and bans on the use of facial recognition technologies by law enforcement, with at least 13 US local governments prohibiting the use of facial recognition technology (Ada Lovelace Institute, AI Now and the Open Government Partnership 2021, p.17). Nonetheless, despite the increased public scrutiny of predictive policing technologies, they continue to be developed and deployed in various spheres of law enforcement and public life.[1] The onset of what Andrew Ferguson terms 'Big Data policing' thus remains an urgent concern for feminist approaches to AI (Ferguson 2017).

Despite their futuristic veneer, predictive policing tools rarely represent a radical break from the past; instead, they are an extension or reflection of current and historical policing practices (Degeling and Berendt 2018). In this sense, predictive policing technologies tie into a broader conversation around explainability

---

[1] For example, a joint report on algorithmic accountability in the public sector by Ada Lovelace Institute, AI Now and the Open Government Partnership notes that San Francisco's ban on the use of facial recognition technology only covers its use by municipal agencies, hence excluding its use by federal bodies (Ada Lovelace Institute, AI Now and the Open Government Partnership 2021, p.17).

Kerry McInerney, *Coding 'Carnal Knowledge' into Carceral Systems*. In: *Feminist AI*. Edited by: Jude Browne, Stephen Cave, Eleanor Drage, and Kerry McInerney, Oxford University Press. © Oxford University Press (2023).
DOI: 10.1093/oso/9780192889898.003.0007

in AI ethics, which questions the deployment of AI technologies that do not provide a clear sense of how their decisions are being made and who is accountable when they make harmful choices (see Browne, Chapter 19 in this volume). Yet, as R. Joshua Scannell notes, the ethical risks of these technologies must be contextualised within the broader institution of policing as a whole (Scannell 2019). As Hampton illuminates in their contribution to this volume (Chapter 8), these technologies are not abstract or ahistorical phenomena; they form part of a much longer sociohistorical continuum of racist (in particular, anti-Black), sexist, homophobic, ableist, colonialist, and classist violence (see also Iyer, Chair, and Achieng, Chapter 20 in this volume, for more insight on how data practices are shaped by colonialism). An abolitionist approach to predictive policing and other carceral technologies does not solely consider the technologies alone, but rather how they bolster and are configured into carceral systems that play a fundamental role in systems of racial capitalism (see Elam (Chapter 14), Atanasoski (Chapter 9), Hampton (Chapter 8), and Wilcox (Chapter 6) in this volume for further explorations of the intersections between AI and racial capitalism). As Benjamin writes, 'truly transformative abolitionist projects must seek an end to carcerality in *all* its forms... Taken together, such an approach rests upon an expansive understanding of the "carceral" that attends to the institutional *and* imaginative underpinnings of oppressive systems' (Benjamin 2019b, p.3).

Moreover, an abolitionist approach to predictive policing also requires an interrogation of the sociotechnical imaginaries (re)produced by predictive policing companies and law enforcement organisations. Jasanoff and Kim define sociotechnical imaginaries as 'collectively imagined forms of social life and social order reflected in the design and fulfilment of nation-specific scientific and/or technological projects' (Jasanoff and Kim 2009, p.120). This article explores what predictive policing technologies and the discourse around these products tell us about social and political constructions of racialised criminality and gendered vulnerability in the United States and the United Kingdom. R. Joshua Scannell's interrogation of predictive policing imaginaries provides an important touchstone in this chapter's interrogation of the gendered sociotechnical imaginaries generated and reproduced by predictive policing technologies (Scannell 2019).[2] Renee Shelby's exploration of the sociotechnical imaginary propagated by the burgeoning field of anti-violence technology—which include personal safety devices such as wearable panic buttons, single-use tests for date-rape drugs, and even an electrified bra—similarly informs this paper's emphasis on how these technologies are not just designed, but also imagined in public discourse (Shelby 2021). By examining predictive policing technologies *and* the sociotechnical imaginaries evidenced

---

[2] Specifically, Scannell discusses Phillip K. Dick's 'The Minority Report' (1956) and Steven Spielberg's 2002 film adaptation *Minority Report* (2002) as key sociotechnical imaginaries of predictive policing.

by these products, this article contributes to one of the foremost goals of prison abolition: destabilising a fixed understanding of crime and justice that renders the prison a predetermined and inevitable feature of our social and political life (Davis 2003).

This chapter pushes forward the debate on AI and policing through a feminist abolitionist approach to predictive policing tools. Fundamentally, this chapter argues that feminist activism against gender-based violence must not turn towards carceral technologies as a solution, as doing so would not merely fail to address the broader systemic and structural relations that produce gender-based violence, but also deepen and entrench the gendered and racialised violence of the prison system itself. While the racialised dimensions of predictive policing technologies have been extensively critiqued by civil society activists and scholars of race and technology (Benjamin 2019a, 2019b; Scannell 2019; Katz 2020), these analyses have largely focused on the production of racialised masculinity through the carceral system, and how predictive policing technologies replicate the carceral profiling of Black and Brown men (notable exceptions include Browne 2015 and Beauchamp 2019). It is certainly true that a focus on what Shatema Threadcraft refers to as 'spectacular Black death' may unintentionally occlude the more hidden, ordinarised, and routine forms of violence experienced by Black women (Threadcraft 2018). Nonetheless, Black women, alongside other women of colour, also experience devastating forms of police brutality and violence (Ritchie 2017). Furthermore, as this chapter will show, the UK and US prison systems have long been central to the production of racialised genders, and gendered ideologies play a key role in justifying costly and violent carceral systems. Hence, a feminist abolitionist approach to AI-enabled predictive policing technologies illuminates how gender provides a key underlying set of logics that drive the development and deployment of predictive policing tools.

In this chapter, I specifically consider the nascent application of AI-enabled predictive policing tools to gender-based and sexual violence, due to the centrality of gender-based violence in the conflict between carceral and anti-carceral feminisms. I use the term gender-based violence very loosely to refer to several key forms of violence that disproportionately affect women, queer people, trans* people, and gender minorities, such as domestic violence and intimate partner violence (IPV). In grouping these forms of violence under the term gender-based violence, I do not argue that only women, queer, trans* people, and gender minorities experience it, or that men are the sole perpetrators, but rather that they are intimately linked with and proliferate under the conditions of cisheteropatriarchy, white supremacy, and racial capitalism. I analyse materials published by predictive policing firms, research reports from think tanks and governmental bodies, and media coverage of predictive policing technologies to consider how gender-based violence is constructed in predictive policing discourse. I primarily focus on the United States and the United Kingdom, recognising the different histories and

institutional structures of policing in both countries while simultaneously pointing to how carceral systems and technologies refract across borders (Davis and
Dent 2001).

The structure of this chapter is as follows. First, I put forward the key arguments of prison abolitionism and explore its relationship to feminist activism.
Second, I provide a brief overview of AI-enabled predictive policing technologies and explore existing critiques of these technologies. Third, I examine how
gender-based violence is constructed in the discourse around predictive policing
and how it is coded into law enforcement datasets. I offer two specific feminist
critiques of predictive policing technologies. I suggest that predictive policing
technologies may entrench and perpetuate harmful myths and stereotypes about
gender-based violence, focusing on how the algorithmic models used by predictive policing companies like Geolitica/PredPol naturalise gender-based violence.[3]
I then argue that predictive policing technologies, like the wider carceral system,
may play a role in producing gendered identities and norms. I suggest that these
technologies may bolster the protective patriarchal model of the state proposed by
Iris Marion Young (2003) through its propagation of 'stranger danger' discourse. I
conclude this chapter with a moment of reflexivity, where I grapple with how the
predictive mode of thought that underpins predictive policing technologies has
similarly shaped the tone and the grammatical tense of this chapter. Inspired by
Ruha Benjamin, R. Joshua Scannell, and Saidiya Hartman, I challenge this predictive mode of thinking through a call to the abolitionist imagination, which asks
us to fundamentally reconceive justice outside of the space and time of the prison
system.

Prison Abolition and Abolitionist Feminisms

Prison abolitionism advocates for the total elimination of prisons, policing, deten-
tion and surveillance (Critical Resistance 2021). Abolitionist knowledge comes
from the activism, experiences and political thought of detained and incarcer-
ated people, in particular Black, indigenous, racially minoritised, and queer people
whose lives have been profoundly shaped by the prison–industrial complex (Davis
2003; Rodriguez 2006; Dillon 2018). While recognising the importance of short-
term improvements to terrible carceral conditions, prison abolitionists do not
believe that the prison system can be salvaged through reform. Instead, they
critically interrogate how the prison came to be perceived as 'an inevitable and
permanent feature of our social lives' and aim to re-imagine what justice looks like

---

[3] In 2021, the company PredPol changed the name of its company to Geolitica, ostensibly to better
capture the functions of the company's products, but most likely to avoid the negative connotations
increasingly associated by the wider public with predictive policing tools.

without it (Davis 2003: p.9; Critical Resistance 2021). More expansively, prison abolitionists critique the 'prison-industrial complex', or the constellation of actors and sectors that sustain the overpolicing and incarceration of marginalised communities and deploy carceral techniques in their own spheres (Davis 2003; Critical Resistance 2021; Benjamin 2019b).

Abolitionist feminisms, which are deeply rooted in Black feminism and feminisms of colour, draw attention to how women of colour, white women, queer people of colour, white queer people, disabled people, and trans* people are differently affected by the prison–industrial complex, showing how the harms engendered by policing and the prison system are multiple and intersectional (Lally 2021). Feminist abolitionists explicitly position themselves against 'carceral feminism', or feminist approaches that prioritise policing and imprisonment in response to gender-based violence and harassment (Bernstein 2007; Gruber 2020; Terwiel 2020). Abolitionist feminisms highlight how the carceral feminist's turning to the state for protection has paradoxically resulted in the increased incarceration of the gendered demographics that the state is ostensibly trying to protect (Critical Resistance and INCITE! 2003). By drawing attention to police violence against Black women and women of colour (Ritchie 2017), experiences of trans and queer people in prison (Smith and Stanley 2015), the incarceration of survivors of gender-based violence (Survived and Punished 2016), and the policing of sex workers (Red Canary Song 2021; Vitale 2018), among other key issues, abolitionist feminists highlight how responses to gender violence cannot depend on a 'sexist, racist, classist, and homophobic criminal justice system' (Critical Resistance and INCITE! 2003). Instead, as Critical Resistance and INCITE! note, 'to live violence free-lives, we must develop holistic strategies for addressing violence that speak to the intersection of all forms of oppression' (Critical Resistance and INCITE! 2003).

Feminist abolitionism highlights how carceral systems do not only enact specifically gendered and intersectional harms, but are also implicated in the production of gender itself. As Angela Davis writes, 'the deeply gendered character of punishment both reflects and further entrenches the gendered structure of the larger society' (Davis 2003, p.61). Black feminists and critical race theorists have shown how racialised forms of state punishment and control produce normative and non-normative gender identities (Dillon 2018). Consequently, abolitionist feminisms insist that the prison is a site where gender is made and unmade through carceral violence. Feminist abolitionists demonstrate how institutionalised forms of sexual violence are a key medium through which the state exercises power over prisoners and makes and unmakes their gendered identities (Davis 2003). Accounts of imprisonment gesture towards how sexual violence is used to feminise, torture, humiliate, and degrade prisoners, from the routinised sexual violence of the strip search, through to the anal mirror searches that were performed on hunger strikers at Long Kesh Prison in Northern Ireland and the rectal feeding of hunger striking detainees at Guantanamo Bay (Davis 2003: pp.62–63; Aretxaga 2001; Velasquez-Potts 2019). As a result, abolitionist feminists argue that turning to carceral systems to solve sexual violence obscures how prisons are already gendered and gendering sites, spaces profoundly shaped by sexual violence. This, in turn, shapes feminist abolitionist approaches to carceral tools designed to prevent or punish sexual violence, such as AI-enabled predictive policing technologies.

(The Limits of) Predictive Policing

Predictive policing, sometimes referred to as proactive policing, pre-emptive policing or data-based policing, refers to a large range of different techniques and technologies used by law enforcement to generate crime probabilities (Degeling and Berendt 2018). Predictive policing technologies take existing police data and train algorithms to make predictions based on historical trends (Degeling and Berendt 2018). While some technologies predict the places and times where certain types of crime are most likely to occur (geospatial crime prediction), others aim to predict likely offenders and/or victims of crimes (person-based forms of prediction). However, there are serious concerns regarding the efficacy and accuracy of predictive policing tools. For example, the UK's National Data Analytic Solution (NDAS), an AI predictive policing tool, was discontinued after being declared 'unusable' by local police departments (Burgess 2020). The Most Serious Violence (MSV) tool, which was designed to predict knife violence and gun violence, demonstrated 'large drops in accuracy' that led to the system being declared unstable and being abandoned (Burgess 2020). While NDAS claimed that the MSV system had accuracy levels of up to 75 percent, the accuracy dropped to between 14 and 19 percent when used by West Midlands police and dropped to only 9–18 percent for the West Yorkshire police (Burgess 2020). As a result, critics like Degeleng and Berendt caution that these technologies may not even work on their own terms (Degeling and Berendt 2018).

Nonetheless, supporters of predictive policing promote them as infallible tools that facilitate fairer and more accurate policing. The 2013 FBI bulletin on predictive policing argues that 'the computer eliminates the bias that people have' (Friend 2013). Similarly, Jeff Brantingham, the anthropologist who helped create Geolitica/PredPol's predictive policing tool, argued that 'the focus on time and location data—rather than the personal demographics of criminals—potentially reduces any biases officers might have with regard to suspects' race or socioeconomic status' (Wang 2018). However, the notion that predictive policing tools may successfully address officers' internalised racial biases ignores how predictive policing tools reproduce and amplify historical biases contained in training data. In a highly publicised case, ProPublica found that the recidivism predictor COMPAS was more likely to 'falsely flag black defendants as future criminals, wrongly labelling them this way at almost twice the rate as white defendants' (Angwin et al. 2016). Even geospatial predictive policing tools still engage in racial profiling through racialised proxies, such as location and neighbourhood (Wang 2018). As Jackie Wang and Ferguson both note, the invocation of mathematical objectivity is part of a broader push to recast policing as objective and neutral in light of public contestations of the discriminatory overpolicing of Black, racialised, poor, and marginalised communities (Ferguson 2017; Wang 2018). The perceived objectivity of predictive policing algorithms reinforces what Benjamin refers to as the 'New Jim Code', or the perpetuation of racialised violence and oppressive relations through technological tools that lend a 'veneer of objectivity' to unjust practices (Benjamin 2019a, pp.5–6).

An abolitionist approach to predictive policing highlights how, in carceral contexts, algorithmic discrimination cannot be solved through the use of more 'fair' training data (Hampton 2021; Scannell 2019). Attempts to fix predictive policing technologies by making them more accurate ignore how carceral technologies and institutions have long been explicitly deployed to surveil and exercise control over racialised bodies (Scannell 2019, p.108). Christina Sharpe and Dylan Rodriguez place the prison–industrial complex within a much longer genealogy of anti-Black violence extending from the hold of the slave ship through into the present-day site of the prison (Rodriguez 2006; Sharpe 2016). In her landmark work on anti-Blackness and surveillance, Simone Browne illuminates the historical continuities of surveillance practices used to subjugate and punish Black people and explicates how 'racism and anti-Blackness undergird and sustain the intersecting surveillances of our present order' (Browne 2015, pp.8–9). Through this lens, Police Chief Bill Bratton's insistence that predictive policing is 'not different; it is part of the evolution of policing' ironically gestures towards the way in which predictive policing technologies reanimate the racist project of the US policing system in technological form (Bureau of Justice Assistance 2009).

Gender-based Violence and Predictive Policing

British and US law enforcement are currently leveraging AI in a variety of ways to try and address gender-based violence. One key route is the automation of risk assessment tools for discerning whether or not someone is likely to be a victim or a perpetrator of domestic violence. In 2013, UK police forces explored how machine learning could improve the Domestic Abuse, Stalking and Honour-Based Violence Risk, Identification, Assessment, and Management Model (DASH) (Perry et al. 2013). However, given that predictive policing tools are highly dependent on the quality and quantity available in relation to specific crimes, the underreporting of instances of gender-based violence poses a significant barrier to the use of predictive policing tools to forecast gender-based violence. As sexual violence is disproportionately underreported to law enforcement, it follows that predictive policing tools will be even less effective in predicting future acts of sexual violence (Akpinar et al. 2021). Furthermore, there is little utility in automating domestic violence risk prediction systems if existing models do not help the police effectively identify individuals at high risk of domestic abuse and/or IPV (Turner et al. 2019). With these caveats in mind, I now turn to consider two different sets of gendered logics invoked by predictive policing tools and the discourse around predictive policing technologies in relation to gender-based violence and sexual violence: the naturalisation of gender-based violence, and the racialised rhetoric of 'stranger danger'.

Naturalising Gender-Based Violence

The discourse around predictive policing technologies invokes a specific sociotechnical imaginary which suggests that gender-based violence is, to an extent, natural, or even inevitable. The naturalisation of gender-based violence is partially attributable to the retrograde beliefs of the people who design and use these technologies. In 2013, the FBI bulletin noted that the Santa Cruz Police Department's predictive software 'functions on all property crimes and violent crimes that have enough data points and are not crimes of passion, such as domestic violence' (Friend 2013). The FBI bulletin's framing of domestic violence as a crime of passion reproduces dangerous ideas that domestic violence and IPV are spontaneous, emotional, and even 'romantic' behaviours. This myth ignores the various forms of coercive control that are often employed by abusive partners and which tend to be more accurate indicators of whether domestic abuse will result in homicide (Myhill and Hohl 2019). It also lends new legitimacy to the 'crime of passion' legal defence, which has historically been used to defend abusers and to cast physical violence as a 'natural' response to women's 'offensive' behaviours. For example, the 'crime of passion' is often posited as a natural outcome of male jealousy over the adulterous behaviour of a female partner. Although the Coroners and Justice Act 2009 abolished provocation as a legal defence, it was replaced with the defence of 'loss of control', which has subsequently been used in the courts to defend violence in response to sexual infidelity (Baker and Zhao 2012). Moreover, popular coverage of domestic abuse cases still portrays IPV in ways that reinforce the perpetrator's narrative, emphasising how victims have supposedly provoked these acts of violence, rendering violence both a 'natural' and 'acceptable' response (Richards 2019).

However, the naturalisation of sexual violence and gender-based violence extends beyond the discourse deployed by individual users and designers of AI-enabled predictive policing technologies. This sociotechnical paradigm is built into the predictive tools themselves. Predictive policing technologies aim to 'forecast' the likelihood of specific crimes. Yet, the word 'forecast' paints crime as a naturally occurring phenomenon such as the weather, as opposed to a political and social construction (Scannell 2019, p.108; Wang 2018). The naturalisation of crime and its portrayal as an external event to be discovered and counteracted by law enforcement is particularly troubling in the context of gender-based violence. Gender-based violence is frequently depoliticised through the use of evolutionary arguments that portray rape as a natural consequence of the biological drive to reproduce (Palmer and Thornhill 2000, p.59). Similarly, some proponents of predictive policing technologies draw on evolutionary arguments about human behaviour to affirm the accuracy of these AI tools in predicting criminal activities. For example, UCLA professor Brantingham describes criminals as modern-day urban 'hunter-gatherers' whose desires and behavioural patterns can be predicted through mathematical models (Wang 2018). The framing of criminals as hunter-gatherers is particularly troubling in the context of evolutionary arguments about sexual violence, where rape is understood as male perpetrators 'foraging' for female victims to increase their 'mate number' (Palmer and Thornhill 2000, p.59).

Furthermore, predictive policing technologies like Geolitica/PredPol implicitly naturalise sexual violence through their modelling of crime patterns on natural disasters. Geolitica/PredPol's flagship software is derived from an earthquake aftershock algorithm, operating on the assumption that 'very generally, crime is analogous to earthquakes: built-in features of the environment strongly influence associated aftershocks. For example, crimes associated with a particular nightclub, apartment block, or street corner can influence the intensity and spread of future criminal activity' (Muggah 2016). Geolitica/PredPol's algorithm draws on 'near repeat theory', which assumes that when one event occurs the likelihood of a repetition of a same or similar event increases (Degeling and Berendt 2018). While the near repeat theory has some degree of success in relation to serial crimes like burglary (Degeling and Berendt 2018), predictive policing companies such as HunchLab's Azalea largely recognise that it does not work at all for other crime types, including sexual violence (Avazea 2015). Consequently, framing gender-based violence through the lens of the aftershock does not just fail to effectively address this form of crime; it both obscures the political, social, and cultural norms that facilitate it and compounds the way it is already naturalised in social discourse.

Troublingly, the naturalisation of sexual violence and other forms of gender-based violence can lead to the belief that they are inevitable. This fatalism is especially prominent in one of HunchLab's services, 'Predictive Missions', which automatically generates 'Missions' for police departments based on available data (Scannell 2019, p.115). These 'Missions', R. Joshua Scannell notes, are 'geographically and temporally specific crimes that the program deems most likely to occur, most likely to be preventable by police patrol, and calibrated for optimal "dosage" (the company's term for optimal number of missions in a particular neighborhood)' (*Ibid.*, p.116). Scannell notes that HunchLab uses machine learning for crime forecasting, calculating the crime forecast in relation to 'patrol efficacy' (how impactful police patrols are in preventing said crime) and 'severity weight' (*Ibid.*). The 'severity weight' is calculated in US dollars, being the predicted cost of the crime balanced against how much it costs to deploy police forces to prevent this crime (Scannell 2019, p.116). Hunchlab prescribes police 'Missions' based on this cost-benefit ratio. Scannell then provides a harrowing example of how sexual violence is construed as inevitable, and thus, unworthy of preventive efforts or engagement: 'in the case of rape in Lincoln, Nebraska, the dollar value of the crime is evaluated at $217,866. But the likelihood of preventability, apparently zero, makes it "not really that important in terms of allocating patrol resources"' (Scannell 2019, p.116). The perceived unimportance of sexual violence is produced and justified through its supposed inevitability. Ultimately, the implicit and explicit framing of sexual violence as a natural and inevitable crime impede the broader political, social, and cultural transformations required to address gender-based violence.

Stranger Danger: The State as Patriarchal Protector

I have argued that AI-enabled predictive policing technologies and the discourse that surround them evoke a specific sociotechnical imaginary where sexual and gender-based violence is portrayed as a natural, if not inevitable, feature of life. However, the naturalisation of sexual violence operates through and in tandem with a second sociotechnical imaginary, one which casts crime as an ever-present possibility and paints both the individual body and the body politic as constantly under threat (Wang 2018). To illuminate how these gendered and racialised dynamics of threat and protection shape predictive policing technologies and their surrounding discourse, I turn to Iris Marion Young's account of the state's masculinist logic of protection and Sara Ahmed's account of the stranger (Young 2005; Ahmed 2000). Young posits that the post-9/11 US security state appeals to gendered logics of protection and subordination as a means to justify its foreign and domestic policies (Young 2005). The benevolent masculine protector state feminises and infantilises its citizenry, placing them in a 'subordinate position of dependence and obedience', while simultaneously justifying this intensified authoritarian control over its populace through racialised state violence against dangerous outsiders (Young 2005, p.2). The dynamics of protection and fear that shape Young's account of the security state are further illuminated by Sara Ahmed's theorisation of 'stranger danger'. Ahmed considers how the discourse of stranger danger produces strangers as the *origin site* of danger, and consequently renders them a figure 'which comes then to embody that which must be expelled from the purified space of the community, the purified life of the good citizen, and the purified body of "the child"' (Ahmed 2000, p.20; see also Ticktin 2008). Taken together, these two feminist approaches to the mutually dependent dichotomies of protection and punishment, inside and outside, the good citizen and the dangerous stranger aptly show how predictive policing technologies may play an active role in the production of gendered dynamics.

Predictive policing technologies may entrench the state's role as the masculine protector by framing sexual violence as a crime committed by strangers in public places. Feminist activists and organisations have long emphasised how the vast majority of sexual violence is perpetrated by people known to the victim (RAINN). Yet, predictive policing technologies and data driven policing programmes perpetuate the myth of 'stranger danger' by automating historical data that primarily account for acts of sexual violence committed in public places. For example, in their 2001 paper on data mining techniques for modelling the behaviour of sexual offenders, Richard Adderly and Peter B. Musgrove use data from a Violent Crime Linkage Analysis System (ViCLAS) database, which aims to identify and link together serial sexual crimes and criminals. At this time, the specified offences coded into the database were 'sexually motivated murder, rape where the offender is a stranger or only has limited knowledge of the victim, abduction for sexual purposes and serious indecent assaults' (Adderly and Musgrove 2001, p.216). By only recording instances of rape where the offender is a stranger or has limited knowledge of the victim, the database does not only exclude the vast majority of rapes and other forms of sexual violence; it also contributes to the pervasive patriarchal belief that the home is a place of 'safety' and that sexual violence is largely a crime committed by unknown outsiders (MacKinnon 2006). While this early study may not necessarily reflect new definitions of sexual violence or changing attitudes and social norms towards what 'counts' as rape in police databases, it is important to note that these databases and studies continue to influence the discourse around predictive policing techniques; indeed, RAND cited Adderly and Musgrove's study in its 2013 report on predictive policing technologies as an example of how data mining techniques could be effectively used to address serious crimes (Perry et al. 2013, p.38).

Furthermore, AI-enabled predictive policing technologies may embed and encode the narrow definitions of rape and sexual assault historically used by law enforcement data collection schemes that similarly propagate the logic of masculinist protection. For example, the FBI's Uniform Crime Report (UCR), which provides 'reliable statistics for use in law enforcement', used the term 'forcible rape' until 2013 and defined the crime as 'the carnal knowledge of a female, forcibly and against her will' (FBI UCR Crime Data Explorer; Lind 2014). The definition's foreclosure of the possibility of male rape victims suggests that any predictive policing tool trained on this historical data will fail male victims of sexual violence. It also entrenches harmful myths about rape that posit that rape is only 'legitimate' if the victim has physically resisted their attacker. While the new definition better aligns with contemporary legal, social, and political understandings of sexual violence,[4] predictive policing technologies trained on historical data before 2013 will still forecast crimes based on datasets that primarily collected data on rapes that were judged to involve 'physical force'. Moreover, the FBI notes that 'since the rape definition changed, some state and local law enforcement agencies have continued to report incidents with the legacy definition, because they have not been able to change their records management systems to accommodate the change' (FBI UCR Crime Data Explorer). While the official definition may have changed, gendered institutional practices and attitudes towards sexual violence are far more difficult to shift.

The encoding of these 'legacy' definitions of rape into predictive policing technologies may assist the expansion of the protective patriarchal state by bolstering gendered and racialised discourses of 'stranger danger'. Predictive policing technologies rely on a generalised public sense of fear and uncertainty; as Wang argues, 'empirically, there is no basis for the belief that there is an unprecedented crime boom that threatens to unravel society, but affective investments in this worldview expand the domain of surveillance and policing' (Wang 2018). Wang also notes that predictive policing technologies offer a technosolutionist answer to the problem of uncertainty, promising 'to remove the existential terror of not knowing what is going to happen by using data to deliver accurate knowledge about where and when crime will occur' (Wang 2018). In the case of sexual violence, predictive policing technologies may both generate this generalised fear of the sexually predatory stranger and promise to identify this stranger through forecasting techniques. In doing so, they may grant further legitimacy to the state's protective bargain as it promises to hunt out and capture dangerous, racialised 'others'. For example, in Adderley and Musgrove's paper, they report that in one of their clusters, '50% of the offenders were of the same non-white race and attacked victims largely in a public place' (Adderley and Musgrove 2001, p.219). Adderly and Musgrove's decision to report racial data for only one of their three clusters and to emphasise both the perpetrators' non-whiteness and their public attacks entrenches the nebulous and omnipresent threat of the racialised sexual offender.

Ahmed's emphasis on the purified body of 'the child' as the locus for protection against 'stranger danger' gestures towards another difficult area where AI is increasingly used by law enforcement: the detection of child sexual abuse imagery and videos. AI detection tools assist the grievously underfunded child sex crime departments, which are unable to cope with the vast number of child sex abuse images available on the Internet (Keller and Dance 2019). There has been extensive

---

[4] The new definition is 'penetration, no matter how slight, of the vagina or anus with any body part or object, or oral penetration by a sex organ of another person, without the consent of the victim' (Lind 2014).

investment into the use of AI and ML techniques for detecting child sex abuse at the national and the international level. For example, the UK National Crime Agency and Home Office worked with the UK AI firm Qumodo to create Vigil AI, launched by the Home Secretary in 2019 (Vigil AI, 'About CAID'). Vigil AI's classifier was trained on the UK's Child Abuse Image Database (CAID) and, according to Vigil AI, 'spots and classifies child abuse imagery just like the best-trained law enforcement operators in the world', without experiencing the psychological distress experienced by human operators (Vigil AI, 'Our Technology'). In 2021, Apple unveiled a tool designed to identify child sex abuse images stored on the iCloud (Nicas 2021). This plan was met with immediate concerns that the tool would be used to monitor private photos, allowing repressive governments to identify, surveil, and persecute its opponents (Nicas 2021). While these tools do not technically constitute predictive policing, as they identify criminal behaviour and imagery rather than predicting criminal behaviour, they are certainly pre-emptive in the sense that they search databases for violent and illegal material before any accusation being made.

While these tools are immensely important for addressing the epidemic of online child sexual abuse, it is important to remain cognizant of the ways that sexual exploitation has often been a key rationale behind the rollout of surveillance practices and an increasingly expansive carceral state. This is not to say that digital child sexual abuse is not horrific in its character and its scale, and that AI tools should not be deployed to address the proliferation of online child sexual abuse imagery. Instead, I wish to draw attention to how sexual exploitation, especially that of children, is often heavily racialised and co-opted into other political agendas. To this end, Mitali Thakor interrogates the limits of predictive policing tools in relation to child sexual abuse through her analysis of another predictive AI tool designed for catching online sexual predators, 'Project Sweetie' (Thakor 2019). Project Sweetie was an AI-powered 3D avatar designed to 'text and video chat with people on webcam chatrooms, convincing them to share email addresses, names and other identifying information, with the promise of sharing nude photos and videos' (Thakor 2019, p.190). It is certainly true that it is preferable that child sexual abusers interact with an AI-powered avatar than with actual children. However, Sweetie still raises two problems with pre-emptive policing in the child sexual abuse space. First, Sweetie's racialised and gendered appearance—Sweetie was designed to appear and sound like a young Filipino girl—was 'critical to the production of this carceral lure', embodying and making visible the queasy intersections of race, gender, fetishisation, and desire (*Ibid.*). Second, Project Sweetie points towards how 'new child protection measures marshal heightened levels of digital surveillance', playing into the protective logic of the patriarchal state and its enemy: the dangerous digital stranger lurking anywhere and everywhere (*Ibid.*, p.191).

Conclusion: The Abolitionist Imagination

Before ending, I briefly reflect on the irony that characterises this chapter, for it also uses a predictive tone to fearfully forecast a sociotechnical future that has not yet arrived.[5] In this sense, this chapter falls into one of the key traps of the carceral system: its foreclosing of the imagination and its ability to render inconceivable any other form of social organisation. Predictive policing technologies similarly capture and entrap the imagination (Benjamin 2019b, p.1). Consequently, as Benjamin writes, 'to extricate carceral imaginaries and their attending logics and practises from our institutions, we will also have to free up our own thinking and question many of our starting assumptions, even the idea of "crime" itself' (Benjamin 2019b, p.5). For example, as Scannell argues, presenting predictive policing as part of a dystopian future obscures how predictive policing is both practised and contested in the present (Scannell 2019). Shattering the hold that carcerality has on our imaginations thus requires the production of alternative sociotechnical imaginaries (Scannell 2019). Abolitionist theories, methods, and practices provide essential tools for dismantling both the material and ideological foundations of carceral systems and predictive policing technologies. Only through abolitionist work can we replace the fearful, anticipatory temporality of AI-enabled predictive policing with the more liberatory possibilities of what Saidiya Hartman refers to as subjunctive time; a mood, a tense imbued with 'doubts, wishes, and possibilities' (Hartman 2008, p.11). This alternative temporal horizon may allow us to glimpse ways of living and being beyond the confines of predictive policing and the carceral state.

Combatting gender-based violence and sexual violence requires a dismantling of the systems that create the conditions of possibility for sexual violence and gender-based violence, and make these acts of violence appear logical, justifiable, and natural. As a result, freedom from gender-based violence requires freedom from carceral systems, both as a material site of sexual and gender-based violence and also as a false ideology of justice. This chapter has shown how AI-enabled predictive policing technologies further extend the gendered logics and practices of the carceral state. While predictive policing technologies have been heralded as the solution for overstretched, underfunded, and highly scrutinised police departments, I have demonstrated how these tools should not be taken up for carceral feminist agendas. Instead, I have shown how these tools possess the potential to shape how gender-based violence is understood, policed, recorded, and encoded into the criminal justice system. In doing so, these technologies may also possess the capacity to shape gendered and racialised identities and (re)produce specific gendered and racialised logics, such as Iris Marion Young's theory of the

---

[5] I am grateful to Sareeta Amrute for identifying this paradox and encouraging me to work within and through this tension.

benevolent masculine protector state or Ahmed's account of the stranger. Hence, while AI-enabled predictive policing tools have rightly been critiqued for their inaccuracy, the threat they pose to human rights, and their role in perpetuating systemically racist criminal justice systems, a feminist approach to predictive policing shows how these areas of critique intersect with the gendered and gendering character of predictive policing technologies.

References

Ada Lovelace Institute, AI Now Institute and Open Government Partnership. (2021). 'Algorithmic Accountability for the Public Sector'. Available at: https://www. opengovpartnership.org/documents/algorithmic-accountability-public-sector/. Accessed 1 September 2021.

Adderley, Richard and Peter B. Musgrove. (2001) 'Data Mining Case Study: Modeling the Behavior of Offenders who Commit Serious Sexual Assaults'. In *KDD '01: Proceedings of the Seventh ACM SIGKDD International Conference on Knowledge Discovery and Data Mining*, pp. 215–220. doi:https://doi-org.ezp.lib.cam.ac.uk/10. 1145/502512.502541

Ahmed, Sara (2000) *Strange Encounters: Embodied Others in Post-Coloniality.* London: Routledge.

Akpinar, Nil.-Jana, Maria De-Arteaga, and Alexandra Chouldechova. (2021) 'The Effect of Differential Victim Crime Reporting on Predictive Policing Systems'. In *Conference on Fairness, Accountability, and Transparency* (FAccT '21), 3–10 March, Virtual Event, Canada. ACM, New York, NY, USA, 17 pages. https://doi.org/10. 1145/3442188.3445877

Amnesty International (2020) 'We Sense Trouble: Automated Discrimination and Mass Surveillance in Predictive Policing in the Netherlands'. *Amnesty International,* 28 September. Available at https://www.amnesty.org/en/documents/eur35/2971/ 2020/en/. Accessed 1 September 2021.

Angwin, Julia, Jeff Larson, Surya Mattu, and Lauren Kirchner. (2016) 'Machine Bias'. *ProPublica*, May 23. Available at https://www.propublica.org/article/machine-bias-risk-assessments-in-criminal-sentencing. Accessed 18 August 2021.

Aretxaga, Begona. (2001) 'The Sexual Games of the Body Politic: Fantasy and State Violence in Northern Ireland'. *Culture, Medicine and Psychiatry* 25: 1–27. doi: 10.1023/A:1005630716511.

Avazea (2015) 'Hunchlab: Under the Hood'. Available at https://cdn.azavea.com/pdfs/ hunchlab/HunchLab-Under-the-Hood.pdf. Accessed 5 September 2021.

Baker, Dennis J. and Lucy X. Zhao. (2012) 'Contributory Qualifying and Non-Qualifying Triggers in the Loss of Control Defence: A Wrong Turn on Sexual Infidelity'. *The Journal of Criminal Law* 76(3): 254–275. doi:10.1350/jcla.2012.76.3.773

Beauchamp, Toby. (2019) *Going Stealth: Transgender Politics and U.S. Surveillance Practices.* Durham: Duke University Press.

Benjamin, Ruha. (2019a) *Race after Technology: Abolitionist Tools for the New Jim Code.* Cambridge: Polity.

Benjamin, Ruha. (ed.) (2019b) *Captivating Technology: Race, Carceral Technoscience, and Liberatory Imagination in Everyday Life.* Durham: Duke University Press.

Bernstein, Elizabeth. (2007) 'The Sexual Politics of the "New Abolitionism"'. *Differences* 18(3): 128–151. https://doi.org/10.1215/10407391-2007-013

Browne, Simone. (2015) *Dark Matters: On the Surveillance of Blackness*. Durham: Duke University Press.

Bureau of Justice Assistance (2009) 'Transcript: Perspectives in Law Enforcement—The Concept of Predictive Policing: An Interview With Chief William Bratton'. *Bureau of Justice Assistance*. Available at https://bja.ojp.gov/sites/g/files/xyckuh186/files/publications/podcasts/multimedia/transcript/Transcripts_Predictive_508.pdf. Accessed 1 September 2021.

Burgess, Matt. (2020) 'A British AI Tool to Predict Violent Crime Is Too Flawed to Use'. 8 September. Available at https://www.wired.com/story/a-british-ai-tool-to-predict-violent-crime-is-too-flawed-to-use/. Accessed 31 August 2021.

Couchman, H. and A. P. Lemos. (2019) 'Policing by Machine'. *Liberty*, 1 February. Available at https://www.libertyhumanrights.org.uk/issue/policing-by-machine/. Accessed 1 September 2021.

Critical Resistance (2021) 'What is the PIC? What is Abolition?' *Critical Resistance*. Available at http://criticalresistance.org/about/not-so-common-language/. Accessed 1 September 2021.

Critical Resistance and INCITE! (2003) 'Critical Resistance-Incite! Statement on Gender Violence and the Prison-Industrial Complex'. *Social Justice* 30(3), (93), *The Intersection of Ideologies of Violence*: 141–150.

Davis, Angela. (2003) *Are Prisons Obsolete?* New York: Seven Stories Press.

Davis, Angela and Gina Dent. (2001) 'Prison as a Border: A Conversation on Gender, Globalization, and Punishment'. *Signs: Journal of Women in Culture and Society* 26(4): 1235–1241. Stable URL: https://www-jstor-org.ezp.lib.cam.ac.uk/stable/3175363

Degeling, Martin and Bettina Berendt. (2018) 'What is Wrong About Robocops as Consultants? A Technology-Centric Critique of Predictive Policing'. *AI & Society* 33(7): 1–10 doi:10.1007/s00146-017-0730-7

Dick, Phillip K. (1956) 'The Minority Report'. Fantastic Universe, November.

Dillon, Stephen. (2018) *Fugitive Life: The Queer Politics of the Prison State*. Durham: Duke University Press.

FBI Uniform Crime Reporting Program 'Crime Data Explorer'. (n.d.) Available at https://crime-data-explorer.app.cloud.gov/pages/explorer/crime/crime-trend. Accessed 5 September 2021.

Ferguson, Andrew G. (2017) *The Rise of Big Data Policing: Surveillance, Race, and the Future of Law Enforcement*. New York: NYU Press.

Friend, Zach. (2013) 'Predictive Policing: Using Technology to Reduce Crime'. 9 April. Available at https://leb.fbi.gov/articles/featured-articles/predictive-policing-using-technology-to-reduce-crime. Accessed 18 August 2021.

Grossman, Lev, Mark Thompson, Jeffrey Kluger, Alice Park, Bryan Walsh, Claire Suddath, Eric Dodds, Kayla Webley, Nate Rawlings, Feifei Sun, Cleo Brock-Abraham, and Nick Carbone. (2011) 'The 50 Best Inventions'. 28 November. Available at http://content.time.com/time/subscriber/article/0,33009,2099708-13,00.html. Accessed 18 August 2021.

Gruber, Aya. (2020) *The Feminist War on Crime: The Unexpected Role of Women's Liberation in Mass Incarceration*. Oakland, CA, USA: University of California Press.

Hampton, Lelia Marie. (2021) 'Black Feminist Musings on Algorithmic Oppression'. In *Conference on Fairness, Accountability, and Transparency* (FAccT '21), 3–10 March 2021, Virtual Event, Canada. ACM, New York, NY, USA, 12 pages.

Hartman, Saidiya. (2008) 'Venus in Two Acts'. *Small Axe* 26: 1–14.

Jasanoff, Sheila and Sang-Hyun Kim. (2009) 'Containing the Atom: Sociotechnical Imaginaries and Nuclear Power in the United States and South Korea'. *Minerva* 47(2): 119–146.

Katz, Yarden. (2020) *Artificial Whiteness: Politics and Ideology in Artificial Intelligence*. New York: Columbia University Press.

Keller, Michael H. and Gabriel J. X. Dance (2019) 'The Internet Is Overrun With Images of Child Sexual Abuse. What Went Wrong?'. *The New York Times*, 29 September. Available at https://www.nytimes.com/interactive/2019/09/28/us/child-sex-abuse.html?action=click&module=RelatedLinks&pgtype=Article. Accessed 1 September 2021.

Lally, Nick. (2021) '"It Makes Almost no Difference Which Algorithm you Use": On the Modularity of Predictive Policing'. *Urban Geography*. doi:10.1080/02723638.2021.1949142

Lind, Dara. (2014) 'The FBI Finally Changed its Narrow, Outdated Definition of Rape'. *Vox*, 14 November. Available at https://www.vox.com/2014/11/14/7214149/the-fbis-finally-collecting-modern-rape-stats. Accessed 5 September 2021.

MacKinnon, Catharine A. (2006) *Are Women Human? And Other International Dialogues*. Cambridge: Harvard University Press.

Muggah, Robert. (2016) 'Does Predictive Policing Work?' Igarapé Institute, 4 December. Available at https://igarape.org.br/en/does-predictive-policing-work/. Accessed 1 September 2021.

Myhill, Andy and Katrin Hohl. (2019) 'The "Golden Thread": Coercive Control and Risk Assessment for Domestic Violence'. *Journal of Interpersonal Violence* 2034(21–22): 4477–4497. doi:10.1177/0886260516675464

Nicas, Jack. (2021) 'Are Apple's Tools Against Child Abuse Bad for Your Privacy?' *The New York Times*, 18 August. Available at https://www.nytimes.com/2021/08/18/technology/apple-child-abuse-tech-privacy.html. Accessed 1 September 2021.

Palmer, Craig T. and Randy Thornhill. (2000) *A Natural History of Rape; Biological Bases of Sexual Coercion*. Cambridge, MA, USA: MIT Press.

Perry, Walter L., Brian McInnis, Carter C. Price, Susan C. Smith, and John S. Hollywood (2013) 'Predictive Policing: The Role of Crime Forecasting in Law Enforcement Operations'. *RAND*. Available at https://www.rand.org/content/dam/rand/pubs/research_reports/RR200/RR233/RAND_RR233.pdf. Accessed 5 September 2021.

RAINN (n.d.) 'Perpetrators of Sexual Violence: Statistics'. RAINN. Available at https://www.rainn.org/statistics/perpetrators-sexual-violence. Accessed 29 March 2023.

Red Canary Song (2021) 'A Panel With Red Canary Song'. University of Massachusetts Amhurst, 4 November. Available at https://www.umass.edu/wgss/event/panel-red-canary-song. Accessed 29 March 2023.

Richards, Daisy (2019) 'Grace Millane's Trial Exposes A Dark Trend In Coverage Of Violence Against Women'. *Each Other*, 27 November. Available at https://eachother.org.uk/grace-millane-coverage-bad/. Accessed 29 March 2023.

Ritchie, Andrea. (2017) *Invisible No More: Police Violence Against Black Women and Women of Color*. Boston, MA, USA: Beacon Press.

Rodriguez, Dylan. (2006) *Forced Passages: Imprisoned Radical Intellectuals and the U.S. Prison Regime*. Minneapolis, MN, USA: University of Minnesota Press.

Scannell, R. Joshua. (2019) This Is Not *Minority Report*: Predictive Policing and Population Racism. In *Captivating Technology: Race, Carceral Technoscience, and*

Liberatory Imagination in Everyday Life, ed. Ruha Benjamin, pp. 107–129. Durham: Duke University Press.

Sharpe, Christina. (2016) *In the Wake: On Blackness and Being*. Durham: Duke University Press.

Shelby, Renee. (2021) 'Technology, Sexual Violence, and Power-Evasive Politics: Mapping the Anti-violence Sociotechnical Imaginary'. *Science, Technology, & Human Values* doi: 16224392110460.

Smith, Nat and Eric A. Stanley. (eds) (2015) *Captive Genders: Trans Embodiment and the Prison Industrial Complex*, 2nd edn. Chico, CA, USA: AK Press.

Survived and Punished (2016) 'Analysis and Vision'. *Survived and Punished*. Available at https://survivedandpunished.org/analysis/. Accessed 1 September 2021.

Terwiel, Anna. (2020) 'What is Carceral Feminism?' *Political Theory* 48(4): 421–442. doi: 10.1177/0090591719889946.

Thakor, M. (2019) Deception by Design: Digital Skin, Racial Matter, and the New Policing of Child Sexual Exploitation. In *Captivating Technology: Race, Carceral Technoscience, and Liberatory Imagination in Everyday Life*, ed. Ruha Benjamin, pp. 188–208. Durham: Duke University Press.

Threadcraft, Shatema. (2018) 'Spectacular Black Death'. *IAS*. Available at https://www.ias.edu/ideas/threadcraft-black-death. Accessed 11 January 2022.

Ticktin, Miriam. (2008) 'Sexual Violence as the Language of Border Control: Where French Feminist and Anti-immigrant Rhetoric Meet'. *Signs: Journal of Women in Culture and Society* 33(4): 863–889. doi: 10.1086/528851.

Turner, Emily, Medina Juanjo, and Gavin Brown. (2019) 'Dashing Hopes? The Predictive Accuracy of Domestic Abuse Risk Assessment by Police'. *The British Journal of Criminology* 59(5): 1013–1034, https://doi.org/10.1093/bjc/azy074

Velasquez-Potts, Michelle C. (2019) 'Staging Incapacitation: The Corporeal Politics of Hunger Striking'. *Women & Performance: A Journal of Feminist Theory* 29(1): 25–40, DOI: 10.1080/0740770X.2019.1571865

Vigil AI (n.d.) 'Our Technology'. Available at https://www.vigilai.com/our-technology/. Accessed 5 September 2021.

Vitale, Alex. (2018) *The End of Policing*. London: Verso.

Wang, Jackie. (2018) *Carceral Capitalism*. Cambridge: MIT Press.

Young, Iris Marion. (2003) 'The Logic of Masculinist Protection: Reflections on the Current Security State'. *Signs: Journal of Women in Culture and Society* 29(1): 1–25.

8

Techno-Racial Capitalism

A Decolonial Black Feminist Marxist Perspective

Lelia Marie Hampton

Introduction

Machine learning has grown into a multi-trillion dollar global market, but at whose expense? Racial capitalism posits that the nonreciprocal extraction of socioeconomic value (e.g. labour and resources) from racialised groups fuels the ever expanding accumulation of capital (Robinson 1983). The racial capitalist system is readily seen in the machine learning industrial complex. Machine learning systems are centralised and monopolised largely by racial capitalists, and billions of racialised people across the world have no say as their livelihoods become increasingly interdependent with machine learning paradigms (Coleman 2018). Consequently, techno-racial capitalism is deeply interconnected with techno colonialism, including algorithmic colonisation, digital colonialism, and data colonialism (Birhane 2020; Coleman 2018; Kwet 2019; Thatcher, O'Sullivan, and Mahmoudi 2016).

Technology has historically been in the service of capitalism–imperialism, expropriating resources (intellectual, physical, natural, etc.) from racially marginalised people to turn a profit. Ultimately, the machine learning economy is only made possible through racial capitalism, particularly the exploitation of and pillaging from oppressed racial groups. An analysis of techno-racial capitalism requires a decolonial Black feminist lens attending to race, gender, capitalism, imperialism, legacies of colonialism, and so on to understand the centralisation of power that allows for the extraction, exploitation, and commodification of oppressed people à la techno-racial capitalism. Claudia Jones (1949) theorised that a Marxist interpretation must acknowledge both the economic superexploitation of Black women on the basis of race, gender, and class, and the constant struggle in which Black women have resisted these oppressive systems. Moreover, racial capitalism as theorised by Cedric J. Robinson (1983) asserts that capitalism since its inception has relied on the exploitation of the labour of a racial underclass. Walter Rodney's (1972) Black socialist economic theories on the systematic underdevelopment of Africa by Europe during colonisation that persists to the modern day complements both of these analyses. It posits that expropriation of resources from

Lelia Marie Hampton, *Techno-Racial Capitalism*. In: *Feminist AI*. Edited by: Jude Browne, Stephen Cave, Eleanor Drage, and Kerry McInerney, Oxford University Press. © Oxford University Press (2023). DOI: 10.1093/oso/9780192889898.003.0008

racialised underclasses in addition to racialised labour and consumer exploitation, particularly in Africa, is also a cornerstone of racialised capitalism.

Techno-racial capitalism augments these critical analyses with a framework of the commodification of racialised oppression throughout the past few centuries. It relies not only on cheap racialised labour to power its data capital but also the economic incentive for oppressive technologies. Thus, techno-racial capitalism treats racially oppressed people as natural 'resources' for artificial intelligence (AI) systems to advance colonial violence and capitalist exploitation. Moreover, technology companies expand their consumer markets at the expense of racialised people, resulting in consumer exploitation as another form of techno-racial capitalism. However, there is a liberatory way forward from here. As Michael Kwet (2020) puts it, 'To counter the force of digital colonialism, a new movement may emerge to redesign the digital ecosystem as a socialist commons based on open technology, socialist legal solutions, bottom-up democracy, and Internet decentralization' (p.1). In doing so, I posit that we require a decolonial Black feminist Marxist lens in the spirit of Claudia Jones.

In this essay, I articulate a decolonial Black feminist perspective of AI through a structural lens of white supremacist imperialist capitalist cisheteropatriarchy to scrutinise systems of oppression that enable technological harm. This perspective on AI is important because this theoretical framework provides a path forward for all oppressed people due to its expansiveness, allowing us to consider a global perspective and examine all oppressive systems. Moreover, this perspective embraces praxis rather than solely focusing on theory. That is, it provides actionable steps towards liberation rather than simply discussing the ills of technological oppression.

Black Feminist Marxism

Radical Black feminism is the fight for the end of all oppression everywhere. At its core, radical Black feminism blends theory and praxis to fight to end oppression that results from cisheteropatriarchy, white supremacy, capitalism, imperialism, ableism, and so on. From the nineteenth century onwards, Black feminist thought in the United States has always covered at least three axes of oppression, namely race, gender, and class (Jones 1949; Davies 2008; Davis 1981; Guy-Sheftall 1995). Whereas Black feminists analysed class, they may have not always been Communists (or even anti-capitalists). On the other hand, Black feminist Marxist scholar-organiser Claudia Jones became very active in the Communist Party from the age of 18, eventually resulting in her exile from the staunchly anti-communist United States in 1955 during the Cold War (Davies 2008).

Born in the British West Indies, now Trinidad, Claudia Jones emigrated to the USA with her parents at a very young age (Davies 2008; Davis 1981). Jones became politically aware and active early in her life, and as a teenager joined the movement to free the Scottsboro Nine, who were denied a fair trial and effective counsel among other injustices (Davis 1981). The Scottsboro Boys were nine African-American teenagers, aged 13 to 20, who were falsely accused in Alabama of raping two white women in 1931 after refusing to leave a train occupied by white people. As described in Ida B. Wells' *The Red Record: Tabulated Statistics and Alleged Causes of Lynching in the United States* (1895), white Americans used false accusations of rape against Black men to justify lynch mobs. Through Jones' work in the Scottsboro Defense Committee, she became acquainted with members of the Communist Party, which prompted her to join (Davis 1981). As a young woman in her twenties, Claudia Jones assumed responsibility for the party's Women's Commission and became a leader and symbol of struggle for Communist women throughout the country (Davis 1981). In addition to political organising, Jones was an intellectual and journalist who published articles in radical news outlets and essays in scholarly journals, founding London's first major Black newspaper the *West Indian Gazette* (Davies 2008; Davis 1981). In 1949, Jones published 'An End to the Neglect of the Problems of the Negro Woman!', the foundational Black feminist Marxist text. In this essay, Jones discusses the often overlooked labour struggles of Black women workers in the US, calling for a racialised and gendered analysis in the Marxist framework. Moreover, she asserts that liberation is deeply linked to anti-fascism and anti-imperialism, providing foundations for an international Black feminist solidarity in praxis. Jones further challenges male-supremacist notions of women's role in labour movements and asserts the importance of the Black woman worker's leadership that is often obscured in history, for example, 'the sharecroppers' strikes of the 1930's were sparked by Negro women' (p.33). In addition, she discusses the neglect by trade unionists of Black domestic workers' efforts to organise themselves, citing that too many 'progressives, and even some Communists, are still guilty of exploiting Negro domestic workers' (p.35). According to Angela Davis,

> Claudia Jones was very much a dedicated Communist who believed that socialism held the only promise of liberation for Black women, for Black people as a whole and indeed for the multi-racial working class. Thus, her criticism was motivated by the constructive desire to urge her white co-workers and comrades to purge themselves of racist and sexist attitudes (Davis 1981, chapter 10).

Notably, one of the many lessons we learn from Jones is that the freedom of not only Black women but of all oppressed peoples is fundamentally linked to dismantling racial capitalism–imperialism in all its nefarious forms. Carol Boyce Davies' *Left of Karl Marx: The Political Life of Black Communist Claudia Jones* (2008) paints a picture of Jones and her radical politics: 'this black woman, articulating political positions that combine the theoretics of Marxism–Leninism and decolonization with a critique of class oppression, imperialist aggression, and gender subordination, is thus "left" of Karl Marx' (p.2).

Although Angela Davis is currently more widely known for her political organising, intellectual work on prison abolition and her leadership with the Black Power Movement, it was her leadership in the Communist Party that then California governor Ronald Reagan cited when he directed the University of California Board of Regents to dismiss her. Davis would eventually go on to become Professor Emerita at the University of California, Santa Cruz (1991–2008). However, her controversy would not end there; the Birmingham Civil Rights Institute rescinded her award in 2019 for supporting Palestinean liberation. Moreover, Davis engaged in international Third World socialist solidarity, visiting China, Cuba, and Grenada to witness the People's Revolution led by Prime Minister Maurice Bishop. In 1981, Davis published *Women, Race, and Class*, devoting a chapter to Communist Women that includes a section about Claudia Jones. Jones' recount of the neglect of Black women's plight across the lines of race, gender, and class by broader society and social movements, including the socialist movement in the USA, is echoed in Davis' book.

Notably, Jones and Davis demonstrate that an analysis of capitalism must embrace international perspectives and solidarity. Their perspectives allow for us to expansively consider all oppressed peoples of the world in our analysis of techno-racial capitalism vis-à-vis radical Black feminism, leaving plenty of room to continue to grow our understanding. Jones demonstrates that we cannot address the root of the issue unless we address continuities of colonialism, white supremacy, cisheteropatriarchy, and imperialist aggression. In terms of internationalism, the decolonial Black Marxist feminist frameworks outlined by Jones and Davis articulate resistance to global racial capitalism–imperialism, providing the intellectual groundwork for a necessary response to a global system of techno-racial capitalism.

Racial Capitalism

Many Black intellectuals, including Claudia Jones, challenged Marxist orthodoxy's exclusion of Black workers. In particular, W.E.B. Du Bois, Cedric J. Robinson, and Walter Rodney have challenged the lack of a racial analysis in Marxist theory, and have augmented socialist theory in this realm.

In *The Souls of Black Folk* (1903), Du Bois posits that many cultures have artificially constructed social differences to achieve economic domination for much of human history. However, strict categorical distinctions of race by European oppressors for economic exploitation is a recent conceptualisation resultant of the slave trade (Du Bois 1903). Although there is an ongoing agenda to extricate the development of slavery from that of capitalism, Du Bois blows this deception wide open, citing that capitalism is reliant upon racial difference. Thus, race emerges as a social construct of institutional racism. In *Black Reconstruction in America* (1935), Du Bois demonstrates that this inextricable tie between slavery and the rise of industrial capitalism leads to the paradigm of racial capitalism,

> the black workers of America bent at the bottom of a growing pyramid of commerce and industry; and they not only could not be spared, if this new economic organization was to expand, but rather they became the cause of new political demands and alignments, of new dreams of power and visions of empire (p.5).

Much of Europe's wealth can be accounted for through chattel slavery in the Americas and colonialism in Africa and Asia (Du Bois 1935; Robinson 1983). During his later years, Du Bois extended his analysis to an account of global racial capitalism in Pan-African solidarity in *The World and Africa* (1947), asserting that slavery and the slave trade transformed into colonialism with the same determination and demand to increase profit and investment.

In his 1983 book *Black Marxism: The Making of the Black Radical Tradition*, Cedric J. Robinson formally conceptualises racial capitalism. Robinson examines 'the failed efforts to render the historical being of Black peoples into a construct of historical materialism, to signify our existence as merely an opposition to capitalist organisation. We are that (because we must be) but much more' (p.xxxv). Moreover, European radical theory, in particular, does not adequately capture the essences of Black history because Marxism is a conceptualisation centred on historical developments in Europe. Marxism and Black radicalism both seek resolutions to social problems, however, 'each is a particular and critically different realisation of a history' (p.1).

In the continuity of colonialism, Europe and North America continue to steal Africa's resources and depress local economies while lining the pockets of capitalists, 'The situation is that Africa has not yet come anywhere close to making the most of its natural wealth, and most of the wealth now being produced is not being retained within Africa for the benefit of Africans' (Rodney 1972, p.29). That is, Africa is not poor; Africa is rich; Africa's people are poor. In fact, Europe and North America set prices on goods produced by Africans while also setting the prices for their own goods, suppressing Africa's economies, refusing Africans economic sovereignty, and creating a dynamic of dependence on capitalist–imperialist economies, 'African economies are integrated into the very structure of the developed capitalist economies; and they are integrated in a manner that is unfavourable to Africa and ensures that Africa is dependent on the big capitalist countries' (p.37). Discussing the violence of settler colonialism in Africa, particularly South Africa, Walter Rodney remarks that 'Even the goods and services which are produced in Africa and which remain in Africa nevertheless fall into the hands of non-Africans' (p.30). In the twenty-first century, South Africans continue to constitute a superexploited racialised working class, and economic inequality has only increased since the 'end' of apartheid with respect to income, employment, and education, hence constituting a neo-apartheid rather than post-apartheid society (Kwet 2019). For instance, 63 percent of South Africans fall below the poverty line compared to just under 1 percent of whites (Kwet 2019).

A Sociohistorical Continuum

Agents of white supremacy have always weaponised technology to advance institutions and structures of oppression. In *Dark Matters: On the Surveillance of Blackness* (2015), Simone Browne documents varying racial capitalist technologies, from the slave ship to slave passes to the census to biometric AI for carceral surveillance. The slave ship was a racial capitalist technology with its spatial arrangement engineered to maximise profit by concurrently maximising the number of slaves that could fit for transportation to the plantation. One slave ship only allowed for a height of 2 feet and 7 inches between the beams and even less under the beams, 'Two feet and seven inches. The violence of slavery reduced to crude geometric units, with room allotted for forty women, twenty-four boys, and sixty men, arranged in a "perfect barbarism"' (Browne 2015, p.47). Slavers designed the slave ship such that men were shackled and placed in a secure room to prevent them from rebelling, while women and children were unshackled and placed in closer proximity to the captain's cabin for the purposes of sexual violence (Hartman 2016). Enriching Saidiya Hartman's argument of the erasure of the captive female's insurgency, resistance, and refusal, Browne notes that the design of the slave ship refused the 'possibilities of women's leadership and resistance in insurrections' (Browne 2015, p.49). Even more so, this spatial design enabled white men to rape Black women during the journey. On the slave ship, branding also served as a racial capitalist technological tracking mechanism for slave cargo, something that could not be erased or hidden. That is, branding slaves was a surefire mechanism to account for a particular ship's cargo that constituted blackness as a saleable commodity (Browne 2015).

In the United States, slave passes were a pre-digital technological surveillance mechanism, a sort of 'information technology', with which to police the movements of the slave labour class, prevent the escape or loss of a slave master's property, and secure their capitalist economic interests (Browne 2015). Consequently, slaves were prohibited from leaving their plantations without a pass, and slave patrols, a precursor to modern policing, tracked slaves and checked their passes to ensure they were not escaping (Browne 2015). Before the advent of pictures and databases, runaway slave advertisements described physical characteristics as well as the slave owner's assessment of the fugitive's character (Browne

2015). Today, databases keep track of this information and even include pictures for similar purposes. When police want to find someone today, they use facial recognition, social media, police databases, and other mechanisms.

As Michael Kwet (2020) points out, 'Over time, surveillance technologies evolved alongside complex shifts in power, culture, and the political economy' (p.1). In South Africa, surveillance technology seems to keep increasing in capacity. That is, more data can be stored, data can be stored centrally, and data can be accessed more quickly. Today, these data provide eighteenth- and nineteenth-century infrastructure for AI-enabled surveillance. In the eighteenth and nineteenth centuries, after stealing land, the white elite in South Africa originally used paper passes to police slave movements to prevent escape and keep Black workers beyond their terms of contract, restricting the migration of Black agricultural labour (Kwet 2020). The paper passes were primitive and lacked reliable authentication because masters would merely name and describe the person's characteristics (such as height, age, and tribe) using written text, so pass holders could forge copies or swap passes with relative ease (Kwet 2020). To enhance the reliability of identification, they branded the skin of workers and livestock with unique symbols registered in paper databases in order to check against a paper register of symbols (e.g. a lookup book) distributed to local officials (Kwet 2020). In the 1970s, IBM began servicing the 'Book of Life' contract, providing white authoritarians with the next generation of passbooks. In sum, the sheer amount of violent control and surveillance over African and Indian workers would not have been possible without IBM and other technology companies.

In neo-apartheid times, Western corporations continue to sell surveillance technology to the white elite in the modernised form of artificial intelligence. For instance, a smart CCTV system in South Africa uses facial recognition and other video analytics for smart surveillance (Kwet 2020). In 2011, IBM signed a contract with the City of Johannesburg to offer racialised surveillance-based smart policing, including a data centre with predictive analytics (Kwet 2020). Moreover, the Council of Scientific and Industrial Research (CSIR) platform Cmore centralises data streams from CCTVs, drones, cellular towers, and other sources (Kwet 2019). Both the South African border patrol and the South African Police Service use Cmore (Kwet 2020). A white residential neighbourhood in Johannesburg became the first to receive fibre Internet from Vumatel with the motivation of building smart surveillance video feeds to police and target poor Black people (Kwet 2020).

Throughout the sociohistorical continuum of techno-racial capitalism, capitalists have commodified racial oppression by transforming racialism into technological mediums. The evolution of techno-racial capitalism in the Middle Passage, United States, and South Africa paints a picture of a sociohistorical continuum that AI embodies today. As demonstrated, technology is structured around not only a racial order, but a racial economic order. As Claudia Jones teaches us, this order is also gendered and requires international solidarity to defeat it.

Techno-Racial Capitalism

Notably, a critical analysis of techno-racial capitalism requires an expansive discussion of the global economic landscape of AI (Birhane 2020; Coleman 2018; Hanna and Park 2020; Kwet 2019, 2020; Milner 2019; Bender et al. 2021). To ignore this landscape is a betrayal of the racially oppressed workers of the world whose labour lays the foundation for AI as well as the people whose oppression is commodified through AI systems. In particular, racially oppressed labourers, including children, work in onerous deadly conditions around the globe (e.g. Congo, Bolivia, Columbia), to extract minerals for hardware devices (e.g. from cell phones to cloud servers) that run AI applications (Kelly 2019); 'ghost workers' annotate data for very little pay and no recourse for lack of payment (Gray and Suri 2019); software engineers in the Global South receive contracts for a fraction of the pay offered in the West. As Jones and Davis emphasise in their politics, these workers' conditions of class superexploitation as obscured low-wage labourers are sanctioned through imperialist aggression in a continuity of colonialism.

Far from being exclusively machine learning algorithms, AI systems require specialised hardware, Internet for data collection, Big Data, and a myriad of software packages for execution of these systems. That is, AI systems are enacted through an interconnected technological ecosystem. As Alex Hanna and Tina M. Park highlight in 'Against Scale' (2020), capitalist expansion by technology companies encompasses vastly increasing capacities for hardware centres to collect data, software to process exorbitant amounts of data, and extremely large models to 'analyse' these data (see also Bender et al. 2021). Many of these layers are intertwined, so they should not be seen as discrete categories, but continuous processes. Thus, we can take Paulo Freire's (1968) approach to liberatory praxis in order to understand social phenomena as inextricably linked rather than separate, incomparable processes.

Transnational technology companies centralise ownership and control of the technological ecosystem, particularly software, hardware, and network connectivity. Michael Kwet (2019) highlights that 'GAFAM (Google/Alphabet, Amazon, Facebook, Apple, and Microsoft) and other corporate giants, as well as state intelligence agencies like the National Security Agency (NSA), are the new imperialists in the international community' (p.4). Five technology companies, Google/Alphabet, Apple, Facebook, Amazon, and Microsoft, comprise the five wealthiest companies in the world with a combined market cap exceeding $5 trillion as of 2020, and 'most of the world's digital products and services outside of mainland China were privately owned and controlled by US transnational corporations' (Kwet 2020, p.12). This estimate does not even include other billion dollar technology companies. Even individually, these companies generate more revenue than many countries in the world. Largely, these companies accumulate profit from royalties and the leasing of intellectual property rights, access to technological infrastructure, sales of digital devices, and vast troves of data that enable predictive targeted advertisements and other AI systems (Kwet 2019).

Moreover, the capitalist–imperialist companies and governments manoeuvre a forced adoption of their technologies that embed their ideologies of power, over-powering local technology markets and pursuing their insatiable profit motives at the expense of racialised communities which 'constitutes a twenty-first cen-tury form of colonisation' (Kwet 2019). Danielle Coleman (2018) describes this as 'a modern-day "Scramble for Africa" where large-scale tech companies extract, analyse, and own user data for profit and market influence with nominal ben-efit to the data source' (p.417). From a decolonial Black feminist perspective à la Jones and Davis, corporate data extraction from racially oppressed people to generate surplus value produces a system of exploitation that must be resisted. Moreover, Western technology companies engage in a form of racial capitalism–imperialism, dominating local economic markets, extracting revenue from the Global South, and creating 'technological dependencies that will lead to perpetual resource extraction' (Kwet 2019, p.6). Research suggests that the economic hege-mony of Big Tech intermediaries is detrimental to local African industries (Kwet 2019). One study showed that information and communication technologies introduced the dominance of information intermediaries in South Africa and Tan-zania's wood and tourism industries (Kwet 2019). These technologies introduce an asymmetrical power dynamic that benefits foreign corporations at the expense of local economies. Through a Jones–Davis lens, imperialist nations structure world economic hierarchies as systems of domination to produce techno-racial capitalism.

The Hardware Layer

Expansion in hardware capacity has been paramount to 'deep learning', which requires massive computational power to compute neural networks on gargantuan data sets. The cloud, for example, is a remote supercomputing and general (web) computing resource that requires comprehensive security, expensive servers, user interfaces (e.g. graphical user interface, shell interface), and backup data centres in case of outages. In addition to centralised hardware ownership, there is Soft-ware as a Service (SaaS) in the cloud in which many software applications run on third-party servers, for example Amazon Web Services (Kwet 2019). Thus, cloud computing companies each have millions of servers (West 2019). Ultimately, this level of engineering and hardware capital requires vast amounts of money, which is why there is currently a cloud oligopoly held by major US tech companies, including Amazon, Microsoft, and Google. This oligopoly has been leveraged in countries in Africa, Asia, and Latin America, and so these countries are forced to rely on cloud infrastructure from a country that has historically exploited them. In this respect, US technology companies have become pseudo-nations with multi-billion or multi-trillion dollar leverage to take advantage of countries ravaged by racial capitalism–imperialism and sociohistorical continuums of colonialism and chattel slavery. Notably, the cloud oligopoly gives major technology companies economic power over colonised and imperialised countries via resource extraction through rent and surveillance (Kwet 2019). Due to the structuring of the cloud economy, users do not have self-determination. A proposed approach to the cloud computing oligopoly is decentralisation (Kwet 2019). From the Jones theoretical view, hardware as a means of production for AI creates a system of domination in which colonised peoples do not receive recompense for mining minerals for hardware or manufacturing the hardware in factories. The Jones–Davis response is to resist this domination by demanding that this means of production should be seized and put into the hands of the workers to resist this domination. Following, previously generated revenue should be directly redistributed to those who have worked in onerous conditions for low-wages to produce these products.

Furthermore, cloud computing is based on the concept of scalability, the idea that technological systems are desirable given the ability to increase in size and computational processing power (Hanna and Park 2020). Technology companies frequently partake in this sort of scale thinking, an unwavering commitment to identifying strategies for efficient growth (Hanna and Park 2020). Scale thinking has seeped into mass consciousness, and 'frames how one thinks about the world (what constitutes it and how it can be observed and measured), its problems (what is a problem worth solving versus not), and the possible technological fixes for those problems' (Hanna and Park 2020, p.1). Scalability allows developers and investors to capture greater market shares with lower levels of investment with products and services 'supplied and consumed at faster and faster speeds, for more and more people' (Hanna and Park, 2020, p.1). Scale thinking is inherently tied to a hyper-optimisation obsessed economic system, that is, the infatuation with 'maximisation of production and the minimisation of unused labour or raw materials and goods' (Hanna and Park, 2020, p.1). This process aims to maximise profit by serving 'a greater number of people with fewer resources over time' (Hanna and Park 2020, p.1). Being grounded in white supremacist cisheteropatriarchal epistemology, scale thinking as it stands is not immediately liberatory 'or effective at deep, systemic change' (Hanna and Park 2020, p.1). That is, a universalist, centralised technological ecosystem concentrates power into the hands of a few elites; reinforces colonial ways of knowing; and centres the dominant worldview of white supremacist cisheteropatriarchy. Local, community-oriented technological solutions, such as technologies for mutual aid, provide a better model of systemic, equity-driven social change (Hanna and Park 2020). From a decolonial Black feminist perspective à la Jones and Davis—political organisers who embody the radical Black feminist tradition of engaging in praxis—empowering communities and building coalitions is a necessary step in struggling against these systems and resisting domination.

The Software Layer

There is a myriad of software that undergirds AI systems. Python tends to be the programming language of choice. Moreover, there must be software to retrieve data from local storage or remote cloud servers. With the explosion of data serving as capital and/or commodity, software libraries have been introduced to quickly and efficiently process vast troves of data, including MapReduce and Apache Spark. 'Classic' machine learning libraries process training and test data and run simple models. Deep learning libraries provide functionality to run neural networks. There are user interfaces for developing machine learning software, including Jupyter Notebook and Colaboratory. Colaboratory even provides limited access to specialised cloud compute known as a graphics processing unit (GPU). While all of these software libraries are open source and (mostly) free to use, the industrial AI systems that use these software libraries are usually not, and revenue is extracted in one form or another.

Control over software is primarily exerted through software licences. As Kwet (2019) writes, 'by design, non-free software provides the owner power over the user experience. It is authoritarian software' (p.9). Proprietary software prevents people from accessing the source code and restricts people from using it without paying. Technology companies have taken advantage of intellectual property rights to ensure that those harmed by carceral risk assessments cannot challenge their sentences, nor receive due process in a court of law. In fact, 'black box' algorithms have become institutional infrastructure that leave marginalised people without recourse. For example, the introduction of an automated welfare eligibility software in Indiana resulted in a million benefit denials in the first three years of the experiment, a 54 percent increase from the previous three years (Eubanks 2018). In another instance, Arkansas transitioned from allocation of Medicaid waivers with human decision makers to automated decision making. Despite the inability to move and needing a caretaker to help her out of bed in the mornings, the state cut a disabled older woman's benefits, decreasing the number of hours for her caretaker and neglecting her (Lecher 2018). Disabled people are not a burden, and they deserve caretakers. A radical Black feminist framework à la Jones and Davis prioritises social programmes for people, including disabled people, poor people, children, trans people, queer people, women, and racially oppressed people.

The Data Layer

Colonial extractive practices from racialised subjects around the globe laid the foundation for industrial capitalism (Du Bois 1935; Robinson 1983), and today colonialist data extraction is part and parcel of techno-racial capitalism, which is in effect an accumulation by dispossession. The aforementioned layers mean that people's data can be extracted and commodified by virtue of being connected to these infrastructures. Even more so, in the data economy, data are not only a commodity, but also a form of capital that can produce surplus value (Sadowski 2019). One *Harvard Business Review* article even posits that this new economy may require 'a new "GDP" – gross data product – that captures an emerging measure of wealth and power of nations' (Couldry and Mejias 2019b).

Historically, capitalism has depended on 'cheap' natural resources that are abundant, easy to appropriate from their rightful owners, and 'just lying around' (Couldry and Mejias 2019a). These resources are not cheap by virtue of them being 'just there'; they are cheap because they have been stolen without the sufficient compensation of those from whom they are taken. This system is 'an asymmetric power relationship in which individuals are dispossessed of the data they generate in their day-to-day lives' (Thatcher et al. 2016, p.990). Far from simply lying around, data belong to or are produced by people, and are then hegemonised. This process of taking control of data is governed by imperial, economic, political, and/or social motivations. People generate data, and companies do not compensate them for it while simultaneously making trillions of dollars in revenue, directing profits to their investors while perpetuating poverty in the Global South. Data extraction is also a process of coercion because users must agree to data licensing agreements in order to use a product, dispossessing them of the right to own their own data (Véliz 2020). That is, the proprietary software is engineered such that the software product will not be accessible unless they do so. If a person wants to email, talk on the phone, and text family or friends, they must enter into a data user agreement that dispossesses them of their data. This asymmetric power dynamic demonstrates companies' subjugation of users. Through accumulation by dispossession and subsequent predictive analytics, data as a commodity and as capital produce surplus value.

Further, extractive logics fuel ever expanding capitalist growth. The expropriation of personal data requires extractive rationalities of data capitalism–imperialism. In part, this rationality posits that since the power to expropriate and analyse data is concentrated within elite technology companies, they are also the only entities that are equipped to extract and process that data (Couldry and Mejias 2019a). The power that these companies wield derives from their deliberate centralisation and proliferation of enormous cloud centres with a massive amount of compute. Even more so, Big Data is inextricably linked to the pursuit of growing computational power to process these large swathes of data

(Thatcher et al. 2016). This data capitalism–imperialism evolved from a legacy of colonialism: 'Simultaneously, a political rationality operates to position society as the natural beneficiary of corporations' extractive efforts, just as humanity was supposed to benefit from historical colonialism as a "civilizational" project' (Couldry and Mejias 2019a, p.340).

To enable surveillance, companies coercively extract data about sensitive information and social networks (Facebook, Twitter), consumer behaviour (Amazon, Google), and search engine history (Google, including YouTube and Gmail). Companies leverage even the most minute amounts of seemingly innocuous data, known as telemetrics, such as what links someone clicks, how long someone spends on a page, and how long someone takes to read an email. This paradigm enables a data-driven economy of targeted advertising to sell products to consumers. A select number of large companies own large amounts of Big Data, which allows them to essentially yield a monopolising control over information about the human population. However, there is still a huge market of companies who not only leverage larger companies' data but also collect their own data. For example, some companies extract phone records, including call history and location history based on GPS and Bluetooth, and then concatenate these data with other forms of personal data (e.g. race, income, education level, age, sex). Companies use these data sets to, for example, study mobility or surveil protesters. One 2015 report valued mobile carrier sales of location and geospatial information data at approximately $24 billion per year although 'exact valuations of these markets remain difficult as data sellers prefer to operate in secrecy for fear of public reprisal' (Thatcher et al. 2016, p.993). Even smaller companies that extract data are highly profitable: for example, 'Nest, a company that manufactures data-collecting thermostats and other household electronics was acquired by Google for $3.2 billion' (Thatcher et al. 2016, p.994).

Notably, the elite have always collected vast amounts of data (Browne 2015; Kwet 2020; Thatcher et al. 2016), but the rise in the technological ability to collect and store even more data led to the market incentive to create advanced predictive algorithms to 'unveil a greater understanding of the world' (boyd and Crawford 2012; Thatcher et al. 2016 p.992). Thus, technology capitalists continue to leverage these data in a drive to increase profits. Moreover, AI promises to use enormous troves of data to, among other things, predict future outcomes, allocate resources, and make decisions in lieu of humans. For example, companies claim that their AI systems have the capability to predict criminality and configure necessary resource allocation for poor people. In so doing, technology companies have successfully built upon a legacy of surveillance and data extraction to structure digital society as 'a project for total surveillance of the human species' via 'Big Data' (Kwet 2019, p.13). Abeba Birhane notes that technology companies are currently attempting to extract data from the next billion on the African continent without remuneration (Birhane 2020). By nature, Big Data violate privacy because companies require mass surveillance to gather gargantuan amounts of data. As Michael Kwet puts it, 'With surveillance the new revenue model for tech, the world's people are subjects of the state-corporate ruling class in the US' (2019, p.14). In fact, this modernised form of surveillance capitalism expands the US empire even further into neo-colonial states, allowing for an enormous resource and monetary extraction on par with the colonial legacy of industrial capitalism. It is worth noting that this form of 'data colonialism works both externally—on a global scale—and internally on its own home populations' (Couldry and Mejias 2019a, p.337). For instance, technology companies have suppressed protests in their own and other countries using surveillance, AI-enabled and otherwise, including Black Lives Matter protests in Europe and North America and protests in South Africa, sometimes referred to as the 'protest capital of the world' (Kwet 2019, p.16).

It is hard to envision 'ethical Big Data' as the extraction and monetisation of sensitive human information and inherently exploitative and often harmful to marginalised groups. Even more so, the extraction of data is often forceful and regularly takes place without the user's knowledge or their informed consent. Even if users know that their data are being collected, they may not know which data. Although data extraction is commonplace, albeit out of sight, an analysis à la Jones and Davis requires that we challenge social systemic norms that render data extraction run-of-the-mill. Moreover, we must challenge the distribution of power that allows the central ownership of data capital rather than ownership by the masses. Thus, a decolonial Black feminist lens enables us to understand data capital as a means of production that should be in the hands of those who produce it.

The Artificial Intelligence Layer

Technology companies harness enormous amounts of data to interpret wide ranging social phenomena for various commercial purposes, 'Quantification, or the production and act of transforming human observations and experiences into quantities based on a common metric. . . is an important procedural aspect of scalability' (Hanna and Park 2020, p.2). danah boyd and Kate Crawford characterise the mythology of the ultimate legitimisation of AI systems as 'the widespread belief that large data sets offer a higher form of intelligence and knowledge that can generate insights that were previously impossible, with the aura of truth, objectivity, and accuracy' (boyd and Crawford 2012, p.663). Even though they have been designated by technology companies as such, 'Any classification, clustering, or discrimination of human behaviours and characteristics that AI systems produce reflects socially and culturally held stereotypes, not an objective truth' (Birhane 2020, p.406).

Even more so, companies propose algorithmic solutions as institutional responses for social, political, and economic challenges even in circumstances in which the solutions should not be algorithmic. Even when problems do not exist, technology companies can pull them out of thin air, as 'for technology monopolies, such processes allow them to take things that live outside the market sphere and declare them as new market commodities' (Birhane 2020, p.392). In addition, far too often companies force AI systems onto marginalised people without consent, which sometimes worsens their conditions of oppression, 'These invaders do not ask permission as they build ecosystems of commerce, politics, and culture and declare legitimacy and inevitability' (Birhane 2020, p.392). Moreover, technology companies leverage colonial ideological justifications, primarily saviour complexes in which technology companies posit that they will better the lives of those in the Global South through the imposition of their technologies, which they claim will liberate the 'bottom' billion, help the 'unbanked' bank, or connect the 'unconnected' (Birhane 2020, p.393). However, Western technological approaches may not transfer well to other cultures (Birhane 2020). It is harmful to attempt to supersede local practices to impose foreign technologies for profit motives that could potentially hurt local communities. Sometimes technology companies impose technologies that solve Western problems in non-Western contexts, assigning themselves the right to determine what technologies other countries need to expand their markets. As Birhane notes, 'Not only is Western-developed AI unfit for African problems, the West's algorithmic invasion simultaneously impoverishes development of local products while also leaving the continent dependent on Western software and infrastructure' (Birhane 2020, p.389).

According to Western technology companies, Africa is a 'data rich continent' (Birhane 2020, p.397). In 2016, Facebook announced a project to create a population density map of Africa using computer vision techniques, population data, and high-resolution satellite imagery, assigning itself 'the authority responsible for mapping, controlling, and creating population knowledge of the continent' and assuming 'authority over what is perceived as legitimate knowledge of the continent's population' (Birhane 2020, p.392). Of course, Facebook's non-consensual, forceful data extraction for this mapping is part of a scheme of profit maximisation that garners even more insights on (potential) consumers.

In the USA and UK, police have attempted to use AI-enabled risk assessment tools to detect people who may be likely victims or perpetrators of gender violence, i.e. domestic violence or intimate partner violence. As Kerry McInerney remarks, 'predictive policing technologies not only reproduce existing patriarchal approaches to gender-based violence, but also possess the potential to shape and produce how gender-based violence is understood, identified, policed and prosecuted' (McInerney, Chapter 7 in this volume). According to an analysis of one popular tool, the algorithm does not adequately identify high-risk revictimisation or recidivism cases, and in fact is underpredicting revictimisation by a large margin (Turner et al. 2019). While the police claim to be the 'benevolent protector of its feminised citizenry' through these tools, one FBI bulletin framed domestic violence as a crime of passion, obscuring the methodological, violent control enacted by abusers (see McInerney, Chapter 7 in this volume). Even more so, this argument enables abusers to leverage the 'crime of passion' defence; this defence has historically been used to protect abusers, harming those who face gender violence, including Black trans women. Moreover, it is particularly interesting that the carceral state funds tools like these while not funding victims of intimate partner violence who are most often economically vulnerable and exploited. In line with McInerney's argument about abolishing these tools, it is important to note that the funding for these tools should be redirected to survivors. As an abolitionist feminist, Davis' framework requires that we understand carceral systems as irreformable, meaning carceral technologies cannot be reformed but rather must be abolished. Jones, too, resisted carceral violence, decrying the arrest of Rosa Lee Ingram, a Black mother in Georgia, for resisting rape against a white man (1949). Thus, the Jones–Davis framework requires that we rethink our relationship to carcerality and move towards abolition of violent, oppressive carceral systems.

Ultimately, it is important to ask questions about the epistemological materialisation of AI given the oppressive wielding of centralised power. Who decides what an AI problem is? Who decides what artificial general intelligence is? Who sets the AI agenda? For example, DeepMind is a company which plays a huge part in setting the reinforcement learning agenda, leveraging a massive engineering team and Google's state-of-the-art Tensor Processing Unit (TPU) pods.

Closing Thoughts

Indigenous histories around the world demonstrate that technologies can promote privacy rights, transparency, collaboration, and local development. We must rethink the distribution of power. Rather than concentrating power in the hands of corporations, it is important to distribute it between developers and users (Hanna and Park 2020). We must open up interactive avenues of participation that are mobilising rather than demobilising (Hanna and Park 2020). For instance, we should solicit feedback on systems design, teach folks how to build systems (e.g. code, web design), and engage people as citizen scientists. Ultimately, we must always engage in international solidarity in our analysis and not simply focus on Euro-American experiences, 'A paradigm shift is needed to change the focus from outcomes on the surface for Westerners (in domains like privacy and discrimination) to structural power at the technical architectural level within a global context' (Kwet 2019, p.21). As Sasha Costanza-Chock's book *Design Justice* emphasises, it is urgent to imagine and organise for collective liberation and ecological sustainability to move towards a technological ecosystem that does not reproduce the matrix of domination (see Costanza-Chock, Chapter 21 in this volume).

A decolonial Black feminist lens à la Jones and Davis requires consideration of superexploited labourers, e.g. miners and data annotators, whose labour undergirds AI but are often obscured in discussions in the West. However, Jones and Davis resist a US-dominant view, providing avenues to highlight this injustice of racism, capitalism, and imperialism. Moreover, a decolonial Black feminist analysis of techno-racial capitalism enables us to resist the commodification of oppression in the technological ecosystem of AI, for example through capitalist investment in risk assessments that defund low-income disabled older women. As Freire (1968) notes, a cornerstone of oppression is the obfuscation of the interconnectedness of phenomena. However, obscuring the layers of technological infrastructure of AI by focusing solely on AI negates the possibility of global solidarity and advocacy for those who are harmed by AI outside of the West. Jones and Davis' framework illuminates the interconnected of the technological ecosystem through a decolonial Black feminist lens that accounts for racial economic orders, legacies of colonialism, and imperialist aggression. As Jones and Davis teach us, it is only through global solidarity that we can defeat a system of white supremacist capitalist–imperialist cisheteropatriarchy that forms the foundation for techno-racial capitalism.

References

Bender, Emily M., Timnit Gebru, Angelina McMillan-Major, and Shmargaret Shmitchell. (2021) 'On the Dangers of Stochastic Parrots: Can Language Models Be Too Big?' In *Proceedings of the 2021 ACM Conference on Fairness, Accountability, and Transparency*, pp. 610–623. New York, NY, USA.

Birhane, Abeba. (2020) 'Algorithmic Colonization of Africa'. *SCRIPTed* 17: 389–409.

boyd, danah and Kate Crawford. (2012) 'Critical Questions for Big Data: Provocations for a Cultural, Techno Logical, and Scholarly Phenomenon'. *Information, Communication & Society* 15(5): 662–679.

Browne, Simone. (2015) *Dark Matters*. Duke University Press.

Coleman, Danielle. (2018) 'Digital Colonialism: The 21st Century Scramble for Africa Through the Extraction and Control of User Data and the Limitations of Data Protection Laws'. *Michigan Journal of Race & Law* 24: 417–439.

Couldry, Nick and Ulises Mejias. (2019a) 'Data Colonialism: Rethinking Big Data's Relation to the Contemporary Subject'. *Television & New Media* 20(4): 336–349.

Couldry, Nick and Ulises Mejias. (2019b) 'Making Data Colonialism Liveable: How Might Data's Social Order be Regulated?'. *Internet Policy Review* 8(2): 1–16.

Davies, Carole Boyce. (2008) *Left of Karl Marx*. Duke University Press.

Davis, Angela Y. (1981) *Women, Race, & Class*. Vintage.

Du Bois, William Edward Burghardt. (1903) *The Souls of Black Folk*. Chicago, IL, USA: A. C. McClurg and Co.

Du Bois, William Edward Burghardt. (1935) *Black Reconstruction in America, 1860–1880*. New York City, NY, USA: Simon & Schuster.

Du Bois, William Edward Burghardt. (1947) The World and Africa. New York, NY, USA: The Viking Press.

Eubanks, Virginia. (2018) *Automating Inequality: How High-Tech Tools Profile, Police, and Punish the Poor.* St. Martin's Press.

Freire, Paulo. (1968) *Pedagogy of the Oppressed.* Continuum International Publishing Group

Gray, Mary L. and Suri Siddharth. (2019) Ghost Work: How to Stop Silicon Valley from Building a New Global Underclass. Boston, MA, USA: Houghton Mifflin Harcourt.

Guy-Sheftall, Beverly. (1995) *Words of Fire: An Anthology of African-American Feminist Thought.* New York City, NY, USA: The New Press.

Hanna, Alex and Tina M. Park. (2020) 'Against Scale: Provocations and Resistances to Scale Thinking'. Preprint at arXiv arXiv:2010.08850.

Hartman, Saidiya. (2016) 'The Belly of the World: A Note on Black Women's Labors'. *Souls*, 18(1):166–173.

Jones, Claudia. (1949) 'An End to the Neglect of the Problems of the Negro Woman!' *PRISM: Political & Rights Issues & Social Movements* 35(80): 467–485.

Kelly, Annie. (2019) 'Apple and Google Named in US Lawsuit over Congolese Child Cobalt Mining Deaths'.

Kwet, Michael. (2019) 'Digital Colonialism: US Empire and the New Imperialism in the Global South'. *Race & Class* 60(4): 3–26.

Kwet, Michael. (2020) Surveillance in South Africa: From Skin Branding to Digital Colonialism. In *The Cambridge Handbook of Race and Surveillance*, pp. 97–122. Cambridge: Cambridge University Press

Lecher, Colin. (2018) 'What Happens when an Algorithm Cuts Your Healthcare'. *The Verge* 21 May. Available at https://www.theverge.com/2018/3/21/17144260/healthcare-medicaid-algorithm-arkansas-cerebral-palsy. Accessed 31 July 2021.

Milner, Yeshimabeit. (2019) 'Abolish Big Data'. Speech at the Second Data for Black Lives Conference. Available at https://medium.com/@YESHICAN/abolish-big-data-ad0871579a41. Accessed 12 August 2021.

Robinson, Cedric J. (1983) *Black Marxism, Revised and Updated Third Edition: The Making of the Black Radical Tradition.* Chapel Hill: UNC Press Books.

Rodney, Walter. (1972) *How Europe Underdeveloped Africa.* Bogle-L'Ouverture Publications.

Sadowski, Jathan. (2019) 'When Data is Capital: Datafication, Accumulation, and Extraction'. *Big Data & Society* 6(1): 2053951718820549.

Thatcher, Jim, David O'Sullivan, and Dillon Mahmoudi. (2016) 'Data Colonialism Through Accumulation by Dispossession: New Metaphors for Daily Data'. *Environment and Planning D: Society and Space* 34(6): 990–1006.

Turner, Emily, Medina, Juanjo, and Brown, Gavin. (2019) 'Dashing Hopes? The Predictive Accuracy of Domestic Abuse Risk Assessment by Police'. *The British Journal of Criminology* 59(5): 1013–1034.

Véliz, Carissa. (2020) *Privacy is Power.* Random House Australia.

Wells, Ida B. (1895) *The Red Record.* Chicago, IL, USA: Donohue & Henneberry,

West, Sarah Myers. (2019) 'Data Capitalism: Redefining the Logics of Surveillance and Privacy'. *Business & Society* 58(1): 20–41.

9

Feminist Technofutures

Contesting the Ethics and Politics of Sex Robots and AI

Neda Atanasoski

In October 2017, Saudi Arabia became the first country to grant citizenship to a robot. Sophia, a humanoid robot designed by the US firm Hanson Robotics, became a Saudi citizen as part of the Future Investment Initiative that links Saudi investors with foreign inventors and future business initiatives. Designed to resemble Audrey Hepburn and programmed with artificial intelligence (AI), Sophia was part of a marketing stunt advertising Abu Dhabi's Economic Vision 2030 plan, 'which aims to shift the base of its economy from natural resources to knowledge [and technology in order to] diversify exports beyond oil production' (UAE, n.d.) The promise of a post-oil smart economic and urban infrastructure in the Gulf, in which Sophia's performance of technological citizenship stands in for the technocapitalist tomorrow, has led to a series of debates around what constitutes a feminist approach to robotics and AI in our speculations about the future and what a specifically feminist ethical intervention into robotics and AI might look like.

Precisely because robotics is a premier site in which contemporary fantasies about the future of AI take shape, we can learn a lot about *present-day* feminist political formations by dwelling on how different technofutures are contested. In mainstream media coverage of feminist approaches to robotics and AI in general, and in the coverage of Sophia's citizenship in particular, these debates lead to broader questions about women's human rights. For instance, according to the *Washington Post*, 'Sophia's recognition made international headlines—and sparked an outcry against a country with a shoddy human rights record that has been accused of making women second-class citizens ... Many people recognized the irony of Sophia's new recognition: *A robot simulation of a woman enjoys freedoms that flesh-and-blood women in Saudi Arabia do not.* After all, Sophia made her comments while not wearing a headscarf, [which is] forbidden under Saudi law' (Wootson 2017). This news article foregrounds flesh and blood (over the simulation of flesh and blood) as the prerequisite for rights and legal recognition. Rehearsing the shorthand racialising conflation of the headscarf with Muslim women's lack of autonomy and rights writ large, the article positions the problem

Neda Atanasoski, *Feminist Technofutures*. In: *Feminist AI*. Edited by: Jude Browne, Stephen Cave, Eleanor Drage, and Kerry McInerney, Oxford University Press. © Oxford University Press (2023). DOI: 10.1093/oso/9780192889898.003.0009

of robotic personhood as a deflection from women's human rights violations. Crucially, media coverage that decried Saudi Arabia's abuses against women largely failed to mention the restrictive laws towards guest workers in the new economy. There was also a second deflection, which follows a long tradition of US media coverage of distant human rights abuses that reinscribe US exceptionalism. Along these lines, *Wired* magazine reported that Sophia's creator, the US roboticist David Hanson, *did* acknowledge the exclusion of Saudi women from the rights of citizenship, and that he viewed Sophia's 'unveiling' in the Middle East as a feminist statement. At the same time, the magazine assessed this position as disingenuous:

> Hanson, argues that the opportunity was used to 'speak out on women's rights', a statement that sits somewhat awkwardly in Saudi Arabia, a country in which women have only just been given the right to drive and where 'male guardianship' still exists, meaning many women have to ask permission from male relatives or partners to leave the house, get a passport, get married or even file police reports for domestic violence or sexual assault. The citizenship stunt seemed more akin to a marketing campaign—for Sophia and Saudi Arabia—than it did a genuine statement on humanity, dignity or personhood (Reynolds 2018).

Crucially, the fact that the majority of responses to Sophia's Saudi citizenship revolved around how such status represents a crisis for women's human rights (rather than, as I noted earlier, for the rights of a category like the migrant labourer) is indicative of the extent to which mainstream feminist approaches to robot and AI ethics obscure the relationship between citizenship status and a right to property, ownership, and capital. As Camilla Hawthorne has argued, there are 'long-buried links between the bureaucratic apparatus of liberal citizenship and racism, a connection that has effectively paved the way for the explosion of far-right, neo-fascist, and populist politics across the US, Europe and much of the rest of the world' (Hawthorne, forthcoming). In fact, Sophia the citizen robot is just one example among many of how debates around ethics in robotics and AI turn to the liberal–juridical realm of human rights as a way *to turn away* from the critical questions around technology's entanglements with racial and gendered inequity, including in labour and property relations, perpetuated through the operations of racial capitalism. For example

> One open letter, written [in 2018] and addressed to the European Commission by 150 experts in medicine, robotics, AI and ethics, described plans to grant robots legal status as 'electronic persons' [which would give them the same rights as biological human persons to be] … 'inappropriate' and 'ideological, nonsensical and non-pragmatic', arguing that to do so would directly impinge on human rights (Reynolds 2018)

Yet, in such ethical outcries, little attention is paid to how technocapitalist futures perpetuate historically entrenched differentiations of the category of the human that is the basis for human rights.

The liberal–juridical response to the idea of a robotic futurity that ends in consternation over the status of the artificial person (whether this person is a citizen or subject of rights), as well as recent campaigns to ban particular categories of robots and AI, especially those seen as morally dangerous such as sex robots, raise questions about what version of the human various feminist approaches to technofutures seek to uphold or disrupt. What do these approaches have to say about the politics of embodied difference (both racialised and gendered) in relation to robotics and AI? Under what conditions do less than human and nonhuman others approach the sphere of ethical/juridical inclusion, and what assumptions about ethics and the law are upheld by these spheres? These are questions not just about technological futures, but about the ontology of the human formed within the sediment of racial and colonial conquest. These questions require an engagement with Euro–American political liberalism and its violent production and use of racial and sexual difference. They also require an engagement with the ways in which liberal conceptions of justice monopolise dominant imaginaries of the social good, as well as how these conceptions might be creatively disrupted.

The juridical borderlands of political liberalism that mediate the distinction between artificial and natural personhood, posed in relation to an ostensibly inevitable futurity predetermined by the exponential leaps in AI programming and capability, are at the forefront of questions about what a feminist ethical relation to technology might be. As I argue, at the present moment there are a number of competing feminist approaches to just technofutures that are being articulated in relation to speculative accounts of robotic consciousness, personhood, and rights. This article contends that consciousness, personhood, and rights are always mediated, whether implicitly or explicitly, through the property relation within the operations of racial capitalism. To attend to the complex intersections of present-day contested technofutures and the property relation, I analyse two dominant strands in feminist approaches to ethics and technology. The first roots itself in technology's relationship to the figure of the human and the liberal–juridical realm of rights, while the second makes use of and commodifies proliferating categories of human difference and life itself. Though posed as two opposing strands in robot ethics (such as the contrast between Sophia's designer stating that his engineering is making a feminist intervention, and her critics decrying that Sophia's exceptional citizenship status leads to a further devaluing of Saudi women), both skirt the issue of subjectivity and possession, or the subject-object divide, through which dominant technological desires rehearse and reiterate racial-colonial figurations of the fully human, and reproduce racial-colonial property relations. As I argue, the fear of becoming property appropriates a racialised history of slavery and indenture for the racially 'unmarked' (that is, white) fully human subject in relation to a speculative techno dystopian future in which the turning datafied/objectified self is represented as being at risk of becoming property.

To unpack these claims, the article first turns to the Campaign Against Sex Robots, an active campaign in the UK that positions itself as a feminist intervention into robot ethics. The group's proposed ban on sex robots frames women's human rights in opposition to the existence of techno-objects that are imagined to *always* end in a gendered unethical relation. Such an approach, I argue, upholds a liberal–juridical version of feminist ethics that not only props up the post-Enlightenment fiction of the self-possessed human subject, but erases the violence of the law as a mechanism of racial incorporation and exclusion. Next, the article assesses recent moves to diversify humanoid robots, moves that are posed as a potential solution to reification of Eurocentric and gender normative simulations of humanity seen to exacerbate racism and misogyny (that is, the concerns raised by the proposed ban). Anchoring my analysis of emerging discourses around gendered technological relations to racial capitalism's ongoing investments in perpetuating white supremacy through the pillars of property, contract, and consent, I consider how diversified anthropomorphic technologies still maintain rather than disrupt racialised notions of use and value, commodifying those differences seen as profitable. By way of conclusion, I turn to disruptive, or queering, feminist approaches to artificial and human intelligence that question the inevitability of technocapitalist uses and futures of technology.

The Politics of the Ban as a Liberal Feminist Response to AI/Robotics

More than any other kind of robot meant for human consumption, the sex robot is seen to represent the problem of the ethical relation between subject and object, use and pleasure, and pain and empathy. There is currently a growing movement to ban sex robots as technological objects based on the argument that they are inherently unethical. What is unethical about the sex robot, the movement suggests, is that the inanimate object or doll risks turning women (as a category writ large) into property by extension of the kind of relations of domination and possession/control that they engender and encourage. Working in the realm of legal action via the politics of the ban, organised campaigns to render sex robots illegal conjoin the politics of criminalisation to the figuration of women's human rights. As Kerry McInerney argues in this volume (Chapter 7), feminist critiques of predictive policing technologies too often fall into carceral logics when they argue for 'improving' the technologies to predict sexual crimes. While the campaigns to ban sex robots are not about improving technology bur about elimination, they replicate carceral modes of envisioning just futures. According to the logics of campaigns to ban particular techno-objects, in order to protect women's human rights, sex robots cannot and should not exist in society. Yet, as I argue, the politics of the ban rely heavily on racialised and imperialistic figurations of criminality that maintain Euro–USA locales as loci of ethics and humaneness in need of protection against Orientalized perversions. In the section that follows, I focus on the Campaign Against Sex Robots to illustrate how the liberal politics of the ban reinscribe a Euro–American version of the human that is based on racialised capitalist property relations.

The Campaign Against Sex Robots was launched in 2015 to 'draw attention to the new ways in which the idea of forming "relationships" with machines is becoming increasingly normalised in today's culture' (Campaign Against Sex Robots, n.d.). According to the organisation, 'Sex robots are animatronic humanoid dolls with penetrable orifices where consumers are encouraged to look upon these dolls as substitutes for women ... At a time when pornography, prostitution and child exploitation [are] facilitated and proliferated by digital technology ... these products further promote the objectification of the female body' (Campaign Against Sex Robots 2018a). The campaign against sex robots here revives and recalls the antipornography feminism and sex wars of the 1970s and 1980s, criminalising both technology and potential users of technology.

One difference, of course, is that sex robots do not yet exist except in science fiction and sensational media coverage. In a 2018 Policy Report, the Campaign Against Sex Robots acknowledged that sex dolls enhanced with AI, or sex robots, are *not* currently commonplace. Yet in the report, it is the fantasy of the sex robot rather than the sex robot itself that poses an existential threat to women and girls (Campaign Against Sex Robots 2018b). The campaign thus insists that an ethical feminist politics today takes the form of a ban on particular categories of future technologies. They state, 'We propose to ban the production and sale of all sex dolls and sex robots in the UK with a move to campaign for a European ban. Regulation is not the answer in this domain, due to the intimate connections between misogyny... and male violence. ... Therefore objects that further reinforce the idea that *women are programmable property* can only destabilise relationships in society further' (*Ibid.*).

Crucially, like other manifestations of carceral feminism, the proposed ban on sex robots tethers protectable womanhood to racialising discourse (Bernstain 2012). Implicitly, the universalisation of the category of woman and the fear of women being treated as and even becoming (unwittingly) *property* appropriates for racially 'unmarked' and implicitly white womanhood the violent sexual history of chattel slavery in the Americas or colonisation and imperialism across the globe. This is a point I return to next. Additionally, and more explicitly, the campaign racialises patriarchal cultures, suggesting that they are breeding pathologised desires for women to become property. For instance, the Campaign Against Sex Robots states:

> The rise of sex dolls cannot be dissociated from porn and worldwide misogyny and femicide, sex trafficking, and 'mail-order' brides. In China, the misogynistic effects of the one-child policy will produce a surplus of 30 million men by 2030. In these countries, dolls, alongside increased sex trafficking, and mail order brides

means that female existence in China carries a real existential risk, more real than any AI or robot uprising. Some propose to use dolls to compensate men for their lack of females (which have been killed before birth), but [this] will only intensify misogyny (*Ibid.*).

The report then moves into a discussion of misogyny in Japan and concludes by making the case that 'In North America and Europe similar trends are *starting* to emerge' (*Ibid.*). In addition to pathologising Chinese and Japanese masculinity, the techno-orientalism of this policy paper gestures towards Asian sexual perversity seeping into the US and Western Europe *through the technology itself.*

The sex robot is here one in a long list of gendered crimes that the paper locates in East Asia, including sex trafficking and femicide. Additionally, the sex robot projects Asian sexual/cultural criminality indefinitely into the future through the dystopic vision of technological creep. According to David Roh, Betsy Huang, and Greta Niu, 'Techno-Orientalist imaginations are infused with the languages and codes of the technological and the futuristic. These developed alongside industrial advances in the West and have become part of the West's project of securing dominance as architects of the future' (Roh, Huang, and Niu 2015, p.2). In the case of a techno-Orientalized articulation of the threat of sex robotics, this future is secured by reiterating the potential for the West to take leadership in the affirmation of human morality. Thus, in the strand of liberal feminist thought articulated within the Campaign's policy paper that distances racialised locales and cultures from the West as the humane and human centre, it is inevitable that the sex robot and sex trafficked woman are conflated. According to Julietta Hua (2011), liberal feminist antitrafficking discourse relies on its fashioning of 'culpable cultures' to produce the universality of human rights as the realm of justice. Culpable cultures that enable sex trafficking are Orientalized as backwards, patriarchal, and violent towards women. Hua argues that within antitrafficking activism and law, the 'other' cultures that produce gendered and racialised victims that 'enable the recuperation of the myth of universality by marking the inclusion of particularity even while [their] victimization to a (deviant) culture [whether culture of poverty or of corrupt values] signals a particularity that must be disavowed' (*Ibid.*, p.25). In short, universal laws around sexual violence maintain racialised scripts around vulnerability and criminality. These scripts extend, shape, and determine the rhetorics of the liberal feminist politics to ban sex robots because the relations they engender perpetuate violence against women, pornography, and even rape. They also build on a long history of anti-Asian sentiment that has frequently taken the form of villainising the Chinese female sex worker.[1] The historical and geopolitical links between the US panics and, later, sex wars, and racialising moral

---

[1] Thanks to Kerry McInerney for pointing out this connection. As Xine Yao writes, 'The figures of the coolie and the sex worker—identified by Shah as the two pathologized and sexually deviant

discourses of human rights that are manifest in the campaign to stop the production of sex robots has its roots in postsocialist era feminist jurisprudence and the consolidation of the global human rights regime, of which antitrafficking discourse is part. As Rana Jaleel (2021) argues, the US sex wars (battles over issues of sexuality that included calls to ban pornography) of the 1980s found their place in international feminist jurisprudence, and, in the judgements of the International Criminal Tribunal for the former Yugoslavia, which was actively trying war crimes of the 1992–2000 Wars of Succession from 1993 through 2017. According to Jaleel, the US sex wars became entangled in the realm of international law through the insisted upon distinction between a sexually violent Serbian masculinity and a vulnerable Muslim femininity. The US-led Global North and its liberal international governance jumped in to 'save' the failed ex-Yugoslav warring states by managing and adjudicating their (sexual) violence. Jaleel goes on to show that sex wars-era disputes, thought to be vestiges of a long-gone sex negative past, have in fact found their afterlife in the universal construction of women's human rights. Yet, such universal constructions cannot account for geopolitical distinctions marked by colonial occupation and imperialism that preclude an untroubled solidarity across the category 'woman'. After all, it was Catherine MacKinnon's infamous article published about the war in Bosnia, 'Turning Rape into Pornography', that led not just to a proliferation of scholarship and feminist jurisprudence around rape as a tool of genocide in the context of the Yugoslav wars of succession, but also to the postsocialist geopolitical fantasy of US humanitarianism violence as a moral and ethical form of imperialism—an empire lite, as the political scholar Michael Ignaiteff put it at the time (MacKinnon 1993; Ignatieff 2003).

The sex robot, then, and its rendering as inherently unethical in the Campaign Against Sex Robots, inherits and projects the racialisation of sexual/cultural criminality and vulnerability indefinitely into the future. This future is secured by reiterating the potential for a Western European-US alliance to once again take moral leadership in the affirmation of humanity through liberal ideals of gendered equality, made manifest in the proposed ban. Put otherwise, the West remains the humane *and* human centre in this version of a feminist technofuturist ethics. The historical and geopolitical links between figurations of women outside of the USA and the West as especially sexually endangered, and the racialising discourses of human rights, have roots in postsocialist feminist jurisprudence and the consolidation of the global human rights regime, and these enable and facilitate the easy conjoining in the present day of antitrafficking discourse and the Campaign Against Sex Robots. As I argued in *Humanitarian Violence*, 'Sexual violence against women has become foundational to the emergent political project of making women's human rights a normative postsocialist technology of

counterparts integral to the Yellow Peril—manifest different gendered dimensions of unfeeling Oriental inscrutability that reflect the American anxieties that coalesced around them' (2021, p.26).

moral governance. The hypervisibility of sexual injury in [postsocialist] human rights culture is based on an evolutionary narrative of humanization, which leads to Euro-American liberal notions of difference and inclusion' as justifications of imperialism that is deemed to be humanitarian (Atanasoski 2013, p.174).

It is no coincidence, then, that the proposed ban on sex robots quickly pivots to the realm of human rights as the locus of justice, in which women's human rights are opposed not just to robot rights (as in the case of Sophia's citizenship), but to the existence of techno-objects that are imagined to always end in a gendered unethical relation. In this move, the category of woman that stands for universal humanity reaffirms Anglo–US liberal jurisprudence as the locus of ethical relations. As the Campaign Against Sex Robots puts it, robot ethics should not be about robots, but about humanity: 'We are not proposing to extend rights to robots. …We propose instead that robots are a product of human consciousness and creativity and human power relationships are reflected in the production, design and proposed uses of these robots'.[2] This is true, of course. Yet the proposed ban, in its feminist liberal humanism, imagines that a juridical ban will solve inequality. This formulation deflects from how the human itself (a figure that the movement to ban robots leaves unquestioned) is an ongoing project of racial engineering emerging from violent racial and sexual encounters of imperialism, militarism, and conquest.

Technoliberalism and the Property Relation

The consolidation of woman as universal subject in the law reinforces racialised geopolitics by ignoring the racial histories scaffolding the figure of the human for which the universal woman stands. This enables the discussion of sex robots rendering biological women as property, even as it precludes considerations of how the property relation within racial capitalism informs the engineering imaginaries of technological future. The discussion of sex robots, and even robots like Sophia, reifies the biological woman as a legally protectable category by *appropriating* the history of racialised property relations in the moral panic around objectification. In doing so, liberal discourses of the ban actively preclude considerations of how property relations within racial capitalism (including the histories of slavery, imperialism, and genocide upon which capitalist operations depend) inform the engineering imaginaries of technological futures. This is related to questions of contract and consent that scaffold the figure of the free fully human agent over and against the object.

Crucially, self-possession and autonomy prop up the fiction of the liberal subject through the legal apparatus. As C. B. MacPherson famously argued, 'the original

---

[2] CASR https://campaignagainstsexrobots.org/

seventeenth-century individualism ... [conceptualises] the individual as essentially the proprietor of his own person or capacities, owing nothing to society for them. ... The human essence is freedom from dependence on the wills of others, and freedom is a function of possession' (MacPherson 2011, p.3). The notion of self-possession is not only a racialised and gendered one, but one whose consolation is based on the racialised and gendered violence of the law. Building on Cheryl Harris's groundbreaking theorisation of whiteness as property that is enshrined and perpetuated in US law and its citational practices, Brenna Bhandar has recently argued that 'the increasing importance of chattel slavery to Southern colonies in the seventeenth century ensured that the racial subordination of Native Americans and Blacks was increasingly intertwined with the appropriation of land and its cultivation. Racial subordination becomes enshrined in laws that attribute a lesser legal status to slaves and Native Americans [defined] property in relation to the status of white people as full legal citizens' (Bhandar 2018, p.206) Thus, 'the relationship between being and having, or ontology and property ownership, animates modern theories of citizenship and law. . . . The treatment of people as objects of ownership through the institution of slavery calls our attention to the relationship between property as a legal form and the formation of an ontology that is in essence, racial' (*Ibid.*, p.205). This is a colonial relation, in which the doctrine of 'terra nullius', that is, the myth that indigenous lands belonged to no one, justified Western European invasion and appropriation of indigenous lands that it 'cultivated' through racialised unfree labour.

Attending to the problematic of self-possession, property, and use in robotics through the imaginary of a proliferation of sex with robots, Adam Rodgers suggests in an article for *Wired* magazine that this plays out as what he calls 'squishy ethics' in the technological relation. As he puts it, 'On the one hand, technology isn't sophisticated enough to build a sentient, autonomous agent that can choose to not only have sex but even love, which means that by definition it cannot consent . . . And if the technology gets good enough to evince love and lust ... but its programming still means it can't *not* consent, well, *that's slavery* . . . Part of consent is understanding context, and one possible future here will include economic incentives for hiding that context. Just as social networks hide the ways they keep people coming back for more, so too will sex devices conceal the sophisticated machine-learning artifice that makes them able to improve, to anticipate desires, [and] to augment the skills . . . It's hard to consent if you don't know to whom or what you're consenting. The corporation? The other people on the network? The programmer? The algorithm?' (Rodgers 2018).

The seeming impossibility of a supreme, autonomous subject in command of the contract relation, as well as the plasticity of consent vis-à-vis technology, turns Rodgers to compare sex with robots to slavery. In a sweeping move that erases the racial slavery as the core of the operations of racial capitalism that continues to undergird present-day technocapitalism, this scenario implies that any user can become unfree. Is the technology property, or do human users become property when their data is extracted while they use the technology? The not knowing, or not fully being able to consent, creates the creeping spectre of unfreedom. The autonomous human subject unwittingly becomes property. Indeed, whiteness itself and its ties to property are undermined under present-day technocapitalist relations.

In her contribution to this volume, Lauren Wilcox argues that feminist accounts of robotics and AI should provide an account of both as gendering and racialising technologies while eschewing easy calls for opening the 'black box' of algorithms because such calls inadvertently affirm the moral-ethical supremacy of the human. Rather, Wilcox states, we should insist on the instability of the 'human' that is ever shifting in relation to technologies. However, in certain instances, even the opposite impetus—to not open the black box—risks affirming the supremacy of the liberal human subject. This becomes evident when we consider how the spectre of the non-consensual relation with an algorithm exists alongside consumer desires driving recent developments in sex robotics to create technologies that can intuit people's desires and pleasures. This is part of a push to develop so-called enchanted objects. Technologies are said to be enchanted when users experience pleasure because they do not see the artifice/programming behind the 'magic' of the technology. Put otherwise, technology becomes magic when it can hide the engineering that makes it function. We could well observe that the desire for opening the black box, figuring out the algorithm, and understanding the engineering behind certain technologies lies in opposition to the pleasures and magic of technological enchantment. Yet both rehearse and reinscribe a racialised property and contract relation.

In our book, *Surrogate Humanity*, Kalindi Vora and I wrote about the problem of consent in robot sex in relation to RealBotix, a sex doll company which is currently working on adding robotics, AI and virtual reality to its existing RealDoll product (Atanasoski and Vora 2019).[3] Matt McMullen, the creator of RealDoll, has developed a phone, tablet, and computer app that coordinates with the robot to give the robot the ability to respond. As Vora and I argue, this response can be read as a simulation of reciprocity, and even a simulation of pleasure. As McMullen states, the human user's experience of the robot enjoying a sexual interaction 'is a much more impressive payoff than that she is just gyrating her hips by herself'.[4] Programming the simulation of consent performed as reciprocity and pleasure, even with the stated purpose of enhancing the sexual pleasure of users in the engineering of an object intended for unreflective sexual gratification represents

[3] See especially the 'Epilogue'.
[4] Bots Robotica, 'The Uncanny Lover', 25 June 2015, http://www.nytimes.com/video/technology/100000003731634/the-uncanny-lover.html?partner=rss&emc=rss

a technological update to a racial history that encodes the desire for 'a carefully calibrated sentience' in an artificial person with *a desire for property*.[5]

Vora and I track the imprint upon liberal desire for simulation of consent by attending to how Hortense Spillers details the way that the freedom of the liberal subject carries the desire for a specific form of unfreedom represented in the historical reduction of Africans to flesh, to captive bodies, through capture, the Middle Passage, racial slavery, and into post-slavery, when the Jim Crow era rendered the simulation of consent to the fiction of the liberal subject even more important. This imprint is a desiring subject that knows its own freedom only through the complete domination of the object of its pleasure, even when, and perhaps especially when, that body can simulate pleasure or reciprocity. The perpetuation of that desire may inform the technoliberal desire for the simulation of consent. This is a desire that seems innocent of the drive for racial domination asserted in the technoliberal valorisation of the post-race future ostensibly promised by new technologies that replace human form and function (Atanasoski and Vora 2019, p.194).

In short, there is a profoundly historically based racialised aspect to the fear of technological objectification, or of becoming property, that is elided in Rodgers' conception of the squishy ethics of sex with robots as well as in the Campaign Against Sex Robots. This is the imprint of racialised property relation within racial capitalism that dwells at the heart of present-day fears that the already human (normative white) subject might become the object. Yet, entwined with this fear is the historical formation of freedom and autonomy of the liberal subject that relies on and is built on racialised unfreedom. Put otherwise, the desiring subject knows its own freedom only through the complete domination and possession of the object of its pleasure. *Surrogate Humanity* contends that the desire for this particular form of possession without needing to contend with racialised histories of unfreedom shapes fantasies of technological enchantment. Technological enchantment and its attendant fantasies are woven out of US political liberalism, or what, in *Surrogate Humanity*, we term technoliberalism.

Technoliberalism, 'Diversity', and AI

Both the liberal–juridical and technoliberal consumerist approaches to technology feed into and perpetuate the operations of racial capitalism. In the first instance, women, now incorporated into the ideology of possessive individualism, stand as 'citizens whose collective existence is reduced officially to a narrow domain of the political [as juridical activism]', to borrow Jodi Melamed's phrasing. In

---

[5] Saidiya Hartman writes about this racial structure in the context of US slavery: 'While the slave was recognized as a sentient being, the degree of sentience had to be cautiously calibrated in order to avoid intensifying the antagonisms of the social order' (1997, p.93).

the second, as I elaborate here, humanity is reduced to 'economic sovereignty' that positions technological inclusivity through 'neoliberal logics of privatization, transactability, and profit'—that is, technoliberalism (Melamed 2015, pp.76–77). Vora and I have argued that technoliberalism is a political and economic formation that invests in how difference is organised via technology's management and use of categories of race, sex, gender, and sexuality. Technoliberalism scaffolds and engenders a *fantasy* of a technologically enabled 'post-racial' future that is not only never in fact post-racial, but that is also always put to use for capitalist ends. Given that racial difference and its social, economic, and scientific management are constitutive of the very concept of technology and technological innovation, it is critical to consider how the technological property relation further reiterated in the push within the tech and AI industry to diversify anthropomorphic robots— a move that only further perpetuates the fantasy that technology can usher in a more inclusive post-racial future. The ways that new technologies proliferate the operations of racial capitalism are thus obscured within technoliberalism.

Sunil Bagai, Silicon Valley founder of Crowdstaffing who has previously worked at IBM, EMC, and Symatec, has posited that his objection to the robot Sophia lies not in her threat to flesh-and-blood women's human rights, but rather in her 'not quite human vibe' that is 'a tad off-putting' (Bagai, n.d.). The think piece goes on to argue that diversification of the AI and robotics industries, both in terms of the racial-ethnic composition of the engineers, as well as of the robots themselves, is the answer to bias in AI and algorithms and 'the answer to literally everything' (*Ibid.*). As he concludes, 'If AI is truly to become a part of everyday life, and robots will one day walk amongst humans, it's essential that the data input is just as diverse as humanity itself. Then, we can look forward to a future where robots are intelligent, unbiased, and maybe not so damn creepy' (*Ibid.*). The proposition here is that individual bias and racial representation in the industry will solve 'literally everything'. Yet such an approach fundamentally fails to diagnose or address structural inequalities perpetuated by technocapitalism, relegating bias to the individuals who constitute industry. Relatedly, the proposition that when humanoid robots more realistically represent both the human form and ethno-racial diversity they will be 'less creepy' sidesteps the question of ownership, property and use that undergirds technological relations.

Hanson robotics, the maker of the Sophia robot with its intended embodiment of Audrey Hepburn's beauty, has also created racially 'diverse' robots. Namely, Hanson is also the maker of Bina 48. Bina 48 has been celebrated for bringing diversity into AI. The robot has been variously introduced as a college graduate, a civil rights activist, and a humanoid robot. Essentially a chatbot with a moving face and torso, she is the brainchild of Martine Rothblatt, tech billionaire and founder of Sirius radio and various other tech startups. BINA48 is constructed with motors to express facial gestures and programmed with AI algorithms and internet connectivity to interact with users. The robot also has built in microphones, video cameras, and face recognition software to remember frequent visitors.[6] Bina 48 is modelled on Martine's wife, Bina Rothblatt, and was created to store Bina's memory and personality—to extend human life by turning into data and housing individual consciousness in robotic form. What is interesting to me in this context is the technoliberal celebration of Bina 48 as the embodiment of liberalism's capacity to encompass and enfold difference into capitalist relations. Here is how CBS News tells the story of Bina 48:

> Martine Rothblatt was born Martin Rothblatt. In the middle of the night some 30 years ago, he told his wife [Bina] he wanted to change his gender . . . [Bina was supportive], but things didn't go over as smoothly in the business world . . . 'There were business associates who would have nothing further to do with me', Martine said . . . [But]Martine has always been a great entrepreneur . . . so there's really no obstacle that's too big for her . . . [And] now Martine Rothblatt is taking on the biggest challenge of all—the limits of human life (CBS News 2014).

In this account, technologically driven business acumen overcomes all limits of biological humanity and enhances human potential through technoscientific solutions. Indeed, technoscience is even posited as the answer to bigotry and prejudice, including transphobia and racism. Yet, when Stony Brook art professor Stephanie Dinkins interacted with Bina48 and asked the robot whether it had experienced racism, Bina48 replied: 'I actually didn't have it' (Pardes 2018). While this response may just be a poorly worded response indicating that the robot had not experienced racism, we can also speculate about how or if racism was conceptually present or absent from Bina48's programming. After all, the person on whom the robot was modelled was portrayed as a civil rights activist. Does Bina's modelling begin and end in a realistic representation of Bina's appearance? If considerations of race in AI and robotics begin and end with physical appearance, as in Bagai's proposition before, then the relationship of race and technology become decontextualised even as they are made visible in a limited way.

Diversifying AI is also, of course, about the commodification of difference and putting it to use. Take for instance a recent story from the UK featuring 'Britain's biggest sex doll manufacturer', stating that:

> As a company, we want to encompass different sexual preferences [and] represent different gender types . . . [to become] more inclusive . . . [We] have listened to popular demand and delivered a transsexual doll onto the market [to represent] the trans community (Blair 2019).

[6] 'Bina 48' https://en.wikipedia.org/wiki/BINA48

The [new] doll comes with 'a detachable penis, vaginal "love hole" and breasts', and can be customised to its users' preferences, even down to toenail colour (*Ibid.*). In contrast to the carceral and juridical frame proposing a ban on sex robots, the technoliberal approach, of which Bina48 and the proliferation of diversified sex dolls are two different yet convergent examples, would commodify all human difference for profit. This is a technofuturity of extreme consumer choice that makes discrete but also conflates gender, sexuality, and race with the colour of toenails in the sense that these are all rendered into categories that can be commodified, marketed, and sold. As Janet Borgerson and Jonathan Schroeder have argued,

> The epidermal schema and fetishization are key drivers that are strategically implemented in processes of skin commodification (making skin a saleable, scalable, hence hierarchically coded, commodity) that often reinforce sedimented notions of identity. ... Frantz Fanon's conception of the epidermal schema helps to explicate the various ways that skin 'appears' in consumer culture imagery and sheds light on a host of intersecting identity concerns, such as gender and class. The epidermal schema works to reduce human beings and identity to skin, to focus attention on differences in skin color, to emphasize ontological distinctions signaled by differences in skin color, and to valorize whiteness (Borgerson and Schroeder 2021).

Speculative Futures of AI

Given the increasing demand to commodify human difference as part the growing market in AI technology, I want to conclude by asking, what would a queer relation to technological speculative futures look like? If queerness is not about reifying identity categories, but rather, about disruptions to normativity, including normative attempts to encompass and use racial and gender difference to further capitalist relations, then we might ask: can a queerly oriented feminist approach to AI turn away from and disrupt the liberal–juridical *and* technoliberal efforts to reassert present-day capitalist realities in their imaginaries of technological futures? Kara Keeling identifies the potential of new technologies and speculative futurisms to queer new media and technologies by 'forging and facilitate[ing] uncommon, irrational, imaginative and/or unpredictable relationships between and among what currently are perceptible as living beings and the environment in the interest of creating value(s) that facilitate just relations' (Keeling, 2014). Following Keeling's articulation of a Queer Operating System, technologies and programming can come into unpredictable relations.

I turn to an example of a speculative account of feminist intelligence that queers the categories of use, property and self-possession in the fantasy of the perfect AI. This is the recent project by the artist Lauren McCarthy, titled Lauren AI.

McCarthy recently orchestrated a series of performances as a human AI, called Lauren AI: Get Lauren. The artist states:

> I attempt to become a *human* version of Amazon Alexa, a smart home intelligence for people in their own homes. The performance lasts up to a week. It begins with an installation of a series of custom designed networked smart devices (including cameras, microphones, switches, door locks, faucets, and other electronic devices) [in a client's home]. I then remotely watch over the person 24/7 and control all aspects of their home. I aim to be better than an AI because I can understand them as a person and anticipate their needs. The relationship that emerges falls in the ambiguous space between human-machine and human-human. LAUREN is a meditation on the smart home, the tensions between intimacy vs privacy, convenience vs [the] agency they present, and the role of human labor in the future of automation.[7]

The history of AI has been rationalist and masculinist from its inception. Cognitive scientist and former Director of the Artificial Intelligence Laboratory and the Computer Science and Artificial Intelligence Laboratory at MIT Rodney Brooks has written that in projects from the early days of AI, intelligence was thought to be best characterised as the things that highly educated male scientists found perplexing (such as playing chess or maths problems). Aesthetic judgements and physical movement, on the other hand, were removed from the realm of intelligence (though these are in fact the very things that are hardest to program and design in AI and robotics).

Lauren AI by contrast, seems to posit *as* intelligence precisely that which had been excluded from the traditional notions of what counts as intelligence in AI. The project suggests that inefficiency, needing to sleep, only being able to be with one client at a time are forms of intelligence. Unlike Alexa, the home assistant on which McCarthy modelled her own performance, the users of Lauren AI are aware that they are interacting with a human being, even as that human is acting as a technology. This makes them contemplate subject-object or user-used relationship with unease. In a short video of client testimonials, one person voices their worry that the interaction with Lauren is 'always about me'. Another client, an older woman, states that initially she was concerned about being replaced in her role as her husband's helper, but then Lauren's presence became normalised. Most interestingly, with a human performance of AI, the threat of the helper taking command and control looms much more pressingly than with an artificial assistant, like Alexa or Siri. Queering the standards of human–AI relations in which the human need not think about the AI, Lauren's users constantly meditate on what it means to use Lauren's services and how they are using them. The video

---

[7] https://lauren-mccarthy.com/LAUREN

shows a woman looking out the window saying, 'I like the idea of Lauren being in support, but not in control'. After all, the main premise of social robots meant for the home is that they show interest in and obeisance to the human user. This is a racial-colonial and gendered performance of deferential servitude that, when inherited by technology, does not make the user feel bad about the relationship of total domination. For example, when Siri asks us, 'did you accidentally summon me?', it is performing this kind of deferential servitude that is left unthought. This is precisely why Lauren's present absence (or absent presence) as a human imitating an AI feels, as a *Guardian* article described it, 'creepy'. We might observe that what this article calls 'creepy' is in fact the queering of normative human–AI relations and a queering of what should be thought of as valuable in an AI function. For instance, towards the end of the testimonials video, there is an increased emphasis on what makes Lauren better than a machine AI: with no preprogramming, intuition can step in. This is a shift away from the rational and the efficient in what is valued as intelligence. In fact, as McCarthy has stated, her 'clients' were also 'really aware that [she] was human, so they were patient'. The artist states, 'I was much slower. Some of them told me they felt bad about asking me to do things' (Rushe 2019). McCarthy also had to cede to biological functions as Lauren AI, including sleeping when her clients slept (as opposed to continuing to collect data) and taking her laptop to the bathroom.

The obvious intervention Lauren AI makes is, through her human performance as an AI, is that she raises issues around privacy, the kinds of devices we bring into homes and everyday life that collect data constantly, and the increased problem of surveillance. Yet it seems that the displays of inefficiency and less than fully rational relationalities also disrupt the use–value–profit equation of the dominant mode of engineering technologies intended to be pleasurable for human use. As Jennifer Rhee's contribution to this volume demonstrates, in the predominant association of artificial assistants with care labour, which is based in care labour's gendered and racialised histories, it is the *devaluation* of care labour rather than care itself that reaffirms the user's humanity. In contrast, the performance of Lauren AI suggests that through valuing temporalities of inefficiency and feeling rather than knowing—intuition, for instance—we can move towards new relations and infrastructures of AI. These emergent feminist futurisms can refuse post-Enlightenment racialised and gendered accounts of value and valuelessness that infinitely reproduce colonial labour and property relations of use.

Lauren AI proposes that *all AI is to a certain extent a performance* that says more about relationality (between humans and humans, and humans and machines) than about the technology itself. Let's recall that Sophia's citizenship is in fact a performance—a marketing performance—of globalised corporate citizenship. The promise of AI facilitating our lives, anticipating our every need, and freeing us to be creative, as anyone who has had a frustrating experience using Siri or Alexa understands, far exceeds its current capacity. Like sex robots, this is a speculative future, and speculative futures are always contested futures.

References

Atanasoski, Neda (2013) *Humanitarian Violence: The US Deployment of Diversity.* Minneapolis: University of Minnesota Press.

Atanasoski, Neda and Kalindi Vora. (2019) *Surrogate Humanity: Race, Robots and the Politics of Technological Futures.* Durham, Duke University Press.

Bagai, Sunil. (n.d.) 'Diversity in AI May Help Robots Look Less Creepy'. https://www.crowdstaffing.com/blog/diversity-in-ai-may-help-the-future-of-robots-look-less-creepy.

Bernstein, Elizabeth (2012) 'Carceral Politics as Gender Justice? The 'Traffic in Women' and Neoliberal Circuits of Crime, Sex, and Rights'. *Theory and Society* 41(3): 233–259. http://www.jstor.org/stable/41475719.

Bhandar, Brenna (2018) 'Property, Law, and Race'. *UC Irvine Law Review* 4: 203.

Blair, Anthony (2019) 'Sex Robot Company Launches £1,300 Transgender Doll with "Detachable Penis"'. *Daily Star*, 6 November. https://www.dailystar.co.uk/news/latest-news/sex-robot-company-launch-uks-20825660

Borgerson, Janet and Jonathan Schroeder. (2021) 'The Racialized Strategy of Using Skin in Marketing and Consumer Culture'. https://medium.com/national-center-for-institutional-diversity/the-racialized-strategy-of-using-skin-in-marketing-and-consumer-culture-f37db856be8b

Campaign Against Sex Robots (2018a) 'An Open Letter on the Dangers of Normalising Sex Dolls & Sex Robots'. 28 July. https://campaignagainstsexrobots.org/2018/07/28/an-open-letter-on-the-dangers-of-normalising-sex-dolls-sex-robots/

Campaign Against Sex Robots (2018b) 'Policy Report: Sex Dolls and Sex Robots – A Serious Problem For Women'. *Men & Society* 8 May. https://campaignagainstsexrobots.org/2018/05/08/policy-report-sex-dolls-and-sex-robots-a-serious-problem-for-women-men-society/

Campaign Against Sex Robots (n.d.) 'Our Story'. https://campaignagainstsexrobots.org/our-story/

CBS News (2014) 'Transgender CEO who overcomes obstacles takes on limits of life'. *CBS News*, 14 September. https://www.cbsnews.com/news/life-after-life-trangender-ceo-martine-rothblatt-builds-robot-bina48-mind-clone/

Hartman, Saidiya (1997) *Scenes of Subjection: Terror, Slavery, and Self-Making in Nineteenth Century America.* New York: Oxford University Press.

Hawthorne, Camilla (Forthcoming) *Contesting Race and Citizenship: Youth Politics in the Black Mediterranean.* Cornell University Press.

Hua, Julietta (2011) *Trafficking Women's Human Rights.* Minneapolis, University of Minnesota Press.

Ignatieff, Michael (2003) 'America's Empire is an Empire Lite'. *New York Times Magazine*, 10 January. http://www.globalpolicy.org/component/content/article/154/25603.html

Jaleel, Rana (2021) *The Work of Rape.* Durham: Duke University Press.

Keeling, Kara (2014) 'Queer OS'. *Cinema Journal* 53(2): 152–157.

MacKinnon, Catherine (1993) 'Turning Rape Into Pornography: Postmodern Geno-cide'. *Ms. Magazine* IV: 1

Macpherson, C. B. (2011) *The Political Theory of Possessive Individualism: Hobbes to Locke*. Oxford: Oxford University Press.

Melamed, Jodi (2015) 'Racial Capitalism'. *Critical Ethnic Studies* 1(1, Spring): 76–85

Pardes, Arielle (2018) 'The Case for Giving Robots an Identity'. *Wired*, 23 October. https://www.wired.com/story/bina48-robots-program-identity/

Reynolds, Emily (2018) 'The Agony of Sophia'. *Wired Magazine*, 6 January. https://www.wired.co.uk/article/sophia-robot-citizen-womens-rights-detriot-become-human-hanson-robotics. Accessed 20 December, 2021.

Rodgers, Adam (2018) 'The Squishy Ethics of Sex with Robots'. *Wired*, 2 February. https://www.wired.com/story/sex-robot-ethics/

Roh, David, Betsy Huang, and Greta Niu. (2015) *Techno-Orientalism: Imagining Asia in Speculative Fiction, History, and Media*. Newark: Rutgers University Press.

Rushe, Dominic (2019) 'Interview: Let Me into Your Home: Artist Lauren McCarthy on Becoming Alexa For a Day'. *The Guardian*. 14 May. https://www.theguardian.com/artanddesign/2019/may/14/artist-lauren-mccarthy-becoming-alexa-for-a-day-ai-more-than-human

UAE. (n.d.) 'Smart Sustainable Cities'. https://government.ae/en/about-the-uae/digital-uae/smart-sustainable-cities. Accessed 20 December 2021.

Wootson Jr., Cleeve R. (2017) 'Saudi Arabia, Which Denies Women Equal Rights, Makes a Robot a Citizen'. *Washington Post*. 29 October. https://www.washingtonpost.com/news/innovations/wp/2017/10/29/saudi-arabia-which-denies-women-equal-rights-makes-a-robot-a-citizen/?utm_term=.c5f4e66a1c78

Yao, Xine (2021) *Disaffected: The Cultural Politics of Unfeeling in Nineteenth Century America*. Durham, Duke University Press.

10

From ELIZA to Alexa

Automated Care Labour and the Otherwise of Radical Care

Jennifer Rhee

Artificial intelligence (AI) and care labour have been entangled from AI's earliest days. Published in 1950, Alan Turing's field-establishing essay 'Computing Machinery and Intelligence' embeds care labour—the work of raising, educating, and caring for another person—into the development and imaginaries of AI.[1] This linking of machine intelligence and care labour emerges from Turing's proposal to create machine intelligence by modelling a computer program on a child's mind, and then educating the program to develop into something resembling an adult mind: 'Instead of trying to produce a programme to simulate the adult mind, why not rather try to produce one which simulates the child's? If this were then subjected to an appropriate course of education one would obtain the adult brain [...] The amount of work in the education we can assume, as a first approximation, to be much the same as for the human child' (Turing 1950, p.456). To examine the ongoing entanglement of AI and care labour, my essay connects contemporary AI assistants such as Siri and Alexa to ELIZA, an early AI therapist that was also the first chatbot (Natale 2021, p.50). This genealogical connection provides a historical framework for understanding digital assistants, the modes of care labour they automate, and the care labour hierarchies of race, gender, class, and citizenship they replicate.[2]

Care labour describes the work of providing care to others by attending to their physical, emotional, and educational well-being. It encompasses the paid labour of, among others, teachers, child care and elder care providers, housekeepers, nurses, doctors, therapists, and social workers. Care labour also includes unpaid labour that tends to the material and emotional needs of others, including

[1] Turing's essay also introduces his well-known test for machine intelligence, now known as the Turing test. This test, which locates intelligence in conversational ability, has been taken up widely by both popular culture and the field of AI, and continues to wield considerable influence on what I call the robotic imaginary, which encompasses the imaginaries that shape AI and robotics technologies and cultural forms, as well as the exclusionary definitions of the human that ground these imaginaries.
[2] I look specifically at digital assistants as they are marketed, purchased, and consumed in the United States.

Jennifer Rhee, *From ELIZA to Alexa.* In: *Feminist AI.* Edited by: Jude Browne, Stephen Cave, Eleanor Drage, and Kerry McInerney, Oxford University Press. © Oxford University Press (2023). DOI: 10.1093/oso/9780192889898.003.0010

children, partners, family, and friends.[3] In racial capitalism, care labour is a form of reproductive labour that, when associated with 'women's work', is consistently undervalued and often unwaged. As Silvia Federici articulates, 'The devaluation of reproductive work has been one of the pillars of capital accumulation and the capitalistic exploitation of women's labour' (2012, p.12). While reproductive labour reproduces humans, the most important source of labour power and the tools and means of production, the devaluation of reproductive labour, specifically as a reserve of unwaged or underwaged labour, also plays a critical role in sustaining racial capitalist economies (England 2005; Federici 2004; Fraser, 2016).

In examining the automation of care labour in conversational AI, my essay attends to the ongoing feminist concerns of care and care labour and their imbrications in racial capitalism.[4] I draw on Hi'ilei Julia Kawehipuaakahaopulani Hobart and Tamara Kneese's work on care, which identifies care's potential to 'radically remake worlds'. For Hobart and Kneese, care's radical world re-making potential is activated when care responds to 'the inequitable dynamics' that characterise the present (2020). I take Hobart and Kneese's theorisation of care as a feminist method in my discussion of AI, care, and care labour. To adequately address these topics, I ground my discussion in 'the inequitable dynamics' that structure the social world; more specifically, I examine the racialised and gendered hierarchies that shape care labour and AI. Through this feminist perspective, my essay examines how contemporary assistants like Alexa and Siri automate not just care labour, but also its structuring hierarchies of gender, race, class, and citizenship in racial capitalism. My essay does not rest with a critique of AI. While continuing to build on Hobart and Kneese's theorisations of care, my essay concludes with a discussion of Stephanie Dinkins' artwork *Not the Only One* and its reconfiguration of AI's relation to care and care labour. In part a response to AI's marginalisation of Black communities, *Not the Only One* centres the oral histories told by three Black women and conjures other possible technological worlds from a data justice perspective and an engagement with Dinkins' concept of algorithmic care.

ELIZA, the Feminisation of Psychotherapy, and White Care

AI is modelled on subjects and worldviews that are unavoidably situated and partial, despite contradictory claims of universality that obscure this situatedness (Adam 1998; Katz 2020). In this section, I turn to a watershed moment

---

[3] Mignon Duffy distinguishes between nurturant and nonnurturant care labour. Nurturant care labour involves close relationships and direct interactions with people receiving care. Nonnurturant care labour does not involve such relationships and interactions—for example, housekeepers and janitorial staff in nursing homes (2011, p.6). White women are significantly represented within nurturant care professions, which are generally associated with higher wages and higher status. Women of color are underrepresented within nurturant care labour professions but overrepresented in nonnurturant care professions (2005, pp.76–80).

[4] For a discussion of care robots and a nuanced reading of care robots in the Swedish television series Äkta människor (Real Humans), see DeFalco (2020).

in conversational AI's history to examine how these distinctly situated models and worldviews shaped earlier conversational AIs and how they continue to shape contemporary conversational AIs in the form of digital assistants. In 1966, Joseph Weizenbaum developed ELIZA, an early, groundbreaking natural language processing (NLP) AI, that is, an AI that communicates with people through human language. In ELIZA's case, conversation takes place through written text on a screen.[5] NLP emerged from Cold War–era research on early language-translation programs. While language translation ultimately did not prove fruitful for Cold War pursuits such as space exploration, NLP proved quite the opposite in ELIZA, as people's text-based conversations with the program demonstrated.[6] Weizenbaum designed ELIZA to parody a psychotherapist, though the joke was often lost on the people who interacted with it as though it were a human therapist. This unexpected reception, in turn, surprised Weizenbaum, who observed that the conversations between humans and ELIZA were intimate and emotional—so much so, in fact, that when he expressed his desire to record individuals' conversations for the purposes of studying the transcripts, he was met with outrage and accusations that he was 'spying on people's most intimate thoughts'. This sense of intimacy was so persuasive that even though people were aware that ELIZA was an AI, they interacted with ELIZA as if it were a human therapist. For example, Weizenbaum's secretary, who 'surely knew it to be merely a computer program', asked Weizenbaum to leave the room during her conversation with ELIZA (Weizenbaum 1976, p.6).[7]

Part of ELIZA's convincing performance can be explained by the psychotherapeutic approach referenced by the AI. Weizenbaum modelled ELIZA on a Rogerian psychotherapist, who is trained to be nondirective by reflecting back a patient's statements rather than introducing anything that might be considered conclusive or diagnostic. Developed in the 1950s by Carl Rogers, Rogerian therapy marks a larger turn toward what Ilene Philipson describes as a relational approach to psychotherapy (Rogers 1951). This relational turn signalled a departure from earlier therapeutic models organised around patriarchal notions of authoritative expertise, and is part of what Philipson identifies as a broader feminisation of psychotherapy (Philipson 1993, p.106). Weizenbaum's approach proved useful for ELIZA, whose limited conversational capacities and resistance to conclusive statements could, to a certain extent, be read as a relational therapeutic strategy. For

---

[5] Notably, this structure resembles the set up for Turing's original imitation game, in which a person communicated with a machine through a kind of tele-type technology. (Turing, p.434).

[6] You can 'talk' to a contemporary version of ELIZA at http://psych.fullerton.edu/mbirnbaum/psych101/Eliza.htm.

[7] In her history of artificial intelligence, Pamela McCorduck writes of the 'painful embarrassment' of watching a respected computer scientist share extremely personal and intimate worries about his personal life with DOCTOR (psychiatrist Kenneth Colby's version of ELIZA), knowing all along that DOCTOR was not a human, but rather a computer program (1979, p.254). These experiences with ELIZA resonate with Hannah Zevin's concept of 'distanced intimacy', which identifies that therapeutic relationships have always been centrally informed by various media technologies and by the specificities of the different media technologies (Zevin 2021, p.18).

example, the passage that follows, provided by Weizenbaum, is an excerpt from an exchange between a human and ELIZA. This exchange illustrates the relative technical simplicity of ELIZA's conversational ability.

Men are all alike.
ELIZA: in what way
They're always bugging us about something or other.
ELIZA: can you think of a specific example
Well, my boyfriend made me come here.
ELIZA: your boyfriend made you come here
He says I'm depressed much of the time.
ELIZA: I am sorry to hear you are depressed
It's true. I am unhappy.
ELIZA: do you think coming here will help you not to be unhappy
I need some help, that much seems certain.
ELIZA: what would it mean if you got some help
Perhaps I could learn to get along with my mother.
ELIZA: tell me more about your family
My mother takes care of me.
ELIZA: who else in your family takes care of you
My father.
ELIZA: your father.
You are like my father in some ways.
ELIZA: what resemblance do you see
You are not very aggressive but I think you don't want me to notice that.
ELIZA: what makes you think I am not very aggressive
You don't argue with me.
ELIZA: why do you think I don't argue with you
You are afraid of me.
ELIZA: does it please you to believe I am afraid of you
My father is afraid of everybody.
ELIZA: what else comes to mind when you think of your father
Bullies.
ELIZA: does that have anything to do with the fact that your boyfriend made you come here (Weizenbaum 1976, pp.3–4).

As this exchange demonstrates, although ELIZA's conversational capabilities were significantly limited, the AI was framed by cultural expectations of a nondirective and nonconclusive therapeutic conversation. Because of this framing, people attributed a degree of meaning and emotional significance to their conversations with ELIZA that belied the AI's technological capabilities and performance.[8]

---

[8] This phenomenon is often known as the ELIZA effect. According to Simone Natale (pp.50–67), ELIZA is part of a longer history of AI that can be organized around deception.

ELIZA's development and surprising reception set off a debate about the role of AI in society. As I mentioned before, Weizenbaum developed ELIZA as a parody of a Rogerian therapist, not as a serious attempt to automate psychotherapeutic work. In fact, Weizenbaum expressed strong objections to taking ELIZA, or AI more broadly, seriously as a therapeutic tool (Weizenbaum 1976, p.3). According to Weizenbaum, ethical considerations should take precedence over technological capability: 'I would argue that, however intelligent machines may be made to be, there are some acts of thought that *ought* to be attempted only by humans' (Weizenbaum 1976, p.13). For Weizenbaum, the important and difficult care work of psychotherapy ought to remain a human endeavour.[9] ELIZA reflects this debate about the role of AI in society, particularly in relation to the automation of certain forms of labour. But this debate is only part of the story of ELIZA and its surprising success. While the debate about AI's role in therapeutic practice was taking place, psychology was becoming increasingly feminised with regards to its workforce demographics, its methods, and its devaluation of practitioners through decreasing wages and status, in accordance with the historical pattern for labour characterised as women's work.[10] ELIZA also reflects the tensions between psychotherapy's increasing cultural presence and its simultaneous feminisation.

From the 1960s onwards, mental health fields saw a significant demographic shift, with white women increasingly entering mental health care professions. As Mignon Duffy's study of race and gender in US care work professions details, before that time, the profession of psychology was largely made up of white men; indeed in 1960, almost 70 percent of psychologists were white men, with the remaining 30 percent composed of women, mostly white (Duffy 2011, p.70). By 1980, the number of white women working as psychologists increased to almost 45 percent, and by 2007 white women made up almost 70% of the profession's workforce (Duffy, pp.108–110).[11] During this same period, psychology began moving away from therapeutic models of masculinist expert judgement through 'detached observation and interpretation', and toward more relational models like Rogerian therapy, which emphasised the interpersonal relationship between patient and therapist (Duffy, p.106). Additionally, the emergence of therapy training programmes outside traditional university programmes further increased white women's access to the profession. Meanwhile, psychotherapy, which was once a form of health care reserved for the wealthy, was expanding in access to more middle- and working-class patients thanks to an increase in federal funding for

---

[9] On this point, he famously disagreed with his former·collaborator Kenneth Colby, a psychiatrist who believed computers can be beneficial to the therapeutic relationship. Along these lines, Colby, inspired by ELIZA, developed PARRY, an AI that emulates a person suffering from paranoia. In 1972, PARRY met ELIZA; their conversation can be viewed at https://tools.ietf.org/html/rfc439.

[10] Mar Hicks offers a compelling study of this phenomenon in Britain's computer industry in the mid-twentieth century (Hicks, 2017).

[11] As Duffy notes, during these same decades, Black men and women's participation within the field of social work steadily increased from 16 percent of the social work labour force in 1970 to almost 20 percent in 2007.

mental health treatment (Philipson 1993, p.78). These shifts in demographics, methods, training programmes, and treatment access contributed to what Philipson calls the perceived 'deskilling, declassing, and degrading' of the profession (Philipson, pp.80–89), which she links to the broader devaluation of feminised labour (Philipson, p.6).

When labour associated with women's work is automated by AI and robotics, these technologies often extend the devaluation of this work while replicating extant care labour hierarchies based on race, gender, and citizenship (Rhee 2018). ELIZA is no exception. As Evelyn Nakano Glenn writes, in the United States, people of colour, particularly women of colour, have made up a significant amount of the care labour workforce. Glenn highlights that as women's participation in various care labour professions increased throughout the twentieth century, the kinds of work available to women differed along racial lines. In the second half of the twentieth century, managerial and white-collar care labour positions were often held by white women. Their work often involved face-to-face interaction with customers and clients. Lower-paying and lower-status jobs that required the kind of physical work associated with domestic labour and did not involve interaction with valued customers were disproportionately held by Black, Hispanic, and Asian women (Glenn 1992, 2010). ELIZA's inscriptions of whiteness and femininity reflected the increasing presence of white women in the profession of psychotherapy, while also extending the history of labour divisions between the visibility of white women in higher paying and higher status client-facing occupations and the accompanying invisibilisation of women of colour and immigrant women in reproductive labour positions associated with manual labour and 'dirty work'.[12] This devaluation of care labour and its attendant racialised and gendered hierarchies, I argue, is also part of the conditions of ELIZA's emergence and part of the story of ELIZA's development and its surprising success. Eleanor Drage and Federica Frabetti's essay in this collection thoughtfully points to performativity as a useful way to examine how AI's technological operations extend hegemonic racialised and gendered hierarchies. In this context, ELIZA's technologically modest performance and surprising reception are shaped by and reflect the whiteness and feminisation of psychotherapy at the time of the chatbot's emergence, as well as the larger hierarchies of race, gender, class, and citizenship that have historically structured care labour.

Although care labour has been disproportionately performed by Black, Indigenous, Asian, and Latinx women in the USA, the beneficiaries of care at the time of ELIZA's emergence were largely imagined to be white. ELIZA emerged in 1966, just one year after what Cotten Seiler describes as 'the apogee years of white care', which date from 1935 to 1965 (Seiler 2020, p.31). Seiler defines white care as a

---

[12] For a discussion of the invisibilization of the women of colour who clean the spaces of neoliberalism, see Vergès (2019).

racist biopolitical logic that underpinned twentieth-century US liberalism in the form of the New Deal. White care, according to Seiler, was part of a larger project of 'state racism', as defined by Foucault's theorisations of biopolitics, and was underwritten by white supremacy (p.18). Seiler's examination of white care begins in nineteenth-century racist evolutionary thought. For example, in the late 1900s palaeontologist Edward Drinker Cope believed that people of certain races—particularly African Americans—were biologically not capable of evolving, and thus posed a threat to the nation (i.e. the evolutionarily superior white population) and should be expelled (p.23).

These nineteenth-century ideas shaped ideologies of white care that emerged in the early twentieth-century eugenics movement, and then again in the mid-twentieth century with the New Deal. Throughout this history, conceptions of care were continually used to justify and protect racial hierarchies and to 'explain' the purported evolutionary superiority of white populations.[13] In these justifications, white populations were said to have developed by sheer will a superior capacity for sympathy and fellow-feeling, and by virtue of this proclaimed evolutionary superiority were the exclusive imagined benefactors of the liberal policies associated with the New Deal (Seiler 2020, pp.22–26). Seiler's concept of white care underscores that in the USA, white populations have historically been constructed as the sole beneficiaries of care, both in the progressive expansionist policies of the New Deal and in the racialised and gendered histories that have seen women of colour performing an outsized amount of essential yet undervalued forms of care labour (Seiler, p.18).

The project of white care, alongside the racialised and gendered shifts in psychotherapy during the mid-twentieth century, set the conditions for ELIZA's emergence and shaped ELIZA's contours and reception. As NLP technology evolved across generations of conversational AI, these conditions, which can be understood as part of conversational AI's technological and sociopolitical inheritance, continue to shape contemporary digital assistants. While ELIZA emerged during the tail end of a period of US state liberalism organised around white care and its biopolitical underpinnings that sought to 'make live' white populations and 'let die' all others (Seiler, p.18), digital assistants emerged during a time characterised by what Neda Atanasoski and Kalindi Vora call technoliberalism, an iteration of liberalism that they define as 'the political alibi of present-day racial capitalism that posits humanity as an aspirational figuration in a relation to technological transformation, obscuring the uneven racial and gendered relations of labour, power, and social relations that underlie the contemporary conditions of capitalist production' (p.4). Historicising digital assistants in the context

---

[13] Kyla Schuller's *The Biopolitics of Feeling* (2018) analyzes nineteenth-century feeling, evolutionary theories, and race through the concept of impressibility. Xine Yao's *Disaffected* (2021) analyzes unfeeling, care, and race in the nineteenth century.

within which ELIZA was developed and deployed foregrounds the continuum between liberalism's white care, with its exclusionary concern for the white liberal subject, and technoliberalism's extension of this liberal subject through contemporary technologies and their obscuring of the racialised and gendered labours and hierarchies that organise racial capitalism.

Digital Assistants, Reproductive Labour, and Computational Racial Capitalism

Digital assistant technologies that automate care labour are shaped by both the liberal project of white care and technoliberal ideologies that narrowly define the universal human as the liberal subject while extending the power relations that structure racial capitalism. In the USA, as liberalism became neoliberalism, the state abandoned its responsibilities to care for its citizens (or rather, selected citizens), and care became the entrepreneurial responsibility of individuals. This contemporary period also saw the rapid development of computational racial capitalism or techno-racial capitalism, as Lelia Marie Hampton aptly writes in her essay for this volume; this development further expanded the demands of neoliberal entrepreneurialism while wielding a tremendous amount of power and influence over AI imaginaries. In this section I examine how digital assistants, like ELIZA, simultaneously uphold and obscure the uneven racial and gendered relations of labour and power that shape care labour in racial capitalism. Despite the significantly changed role of the state across these periods of liberalism from ELIZA to Alexa, these periods and the AI that have emerged from them remain organised around both the uneven racial and gendered relations of labour and power that shape racial capitalism and the liberal subject that is produced and maintained by these hierarchies of race, gender, and labour.

Following the 2011 launch of Apple's digital assistant, Siri, a number of companies introduced their own, such as Amazon and its Alexa product in 2014. Siri and Alexa invite their users to speak to them by issuing commands ('set an alarm', 'play this song') or asking them questions ('what is the weather forecast for tomorrow?'). These digital assistants provide information at the user's request and respond to their commands for information or schedule management. These conversational interactions with digital assistants are forms of immaterial labour which produce goods—often in the form of information—that can be used by corporations to further develop their products, or to function as products to be sold to other interested parties. Leopoldina Fortunati observes that immaterial labour has become hegemonic and increasingly feminised in the digital age. With increasing computerisation and widespread personal mobile devices, immaterial labour has become hegemonic in the late twentieth century and is increasingly structured by care and computerisation. For example, as Helen Hester details, 'Emotional labour that was once, amongst a certain class of the privileged outsourced to both secretaries and wives in now outsourced to electronic devices' (Hester 2016). With this expansion of immaterial labour, work is no longer only done in the workplace and during work hours but whenever someone engages their devices: when they look up a restaurant online, stream a movie, send an email, or play a video game. These activities are forms of immaterial labour that can be used by corporations to further develop their products and potentially market them to other interested parties.

Fortunati describes the hegemony of immaterial labour through its connection to forms of reproductive work that were once located primarily within the domestic sphere. In techno-racial capitalism, labour is increasingly immaterial and precarious as it makes its way outside of both the domestic sphere and the industrial factory. This labour is also increasingly feminised, as it replicates the dynamics and demands of reproductive labour, including its unwaged aspects (Fortunati 2007, pp.147–148).[14] If, as Fortunati observes, in the digital age immaterial labour resembles reproductive labour's feminised facets, what of the racialised dimensions of reproductive labour? Modelled on white, middle-class, native-English speaking women, digital assistants such as Siri and Alexa extend the racialised and gendered divisions of reproductive labour. In this way, these digital assistants represent the higher waged, higher status client-facing secretarial work associated with white women, while invisibilising the labour of people of colour around the globe that subtend digital technologies themselves, from the Chinese women assembling digital devices in Foxconn factories to the young children mining for coltan in the Congo, and the workers sorting and dissembling digital waste in Pakistan and Ghana.

Digital assistants are automated care labour technologies that are shaped by colonialism and racial capitalism; through their design and function, they extend racial capitalist and colonial logics, making visible the people that are valued by these systems and invisibilising the rest, while furthering their extractivist and exploitative ambitions.[15] According to Miriam Sweeney, digital assistants' inscriptions of race, gender, and class also obscure these technologies' roles in massive data-capture projects for state and corporate interests. Sweeney's incisive analysis centres around how digital assistants are modelled on white, middle-class women and how these technologies relate to racialised care labour (Sweeney 2021, pp.151–159). Racialised, gendered, and classed inscriptions on these devices are design strategies to facilitate easy (and frequent) interactions with them while

---

[14] As Fortunati points out, immaterial labour has historically been unevenly distributed across individuals; it is performed mainly by adult women and consumed mainly by adult men and children (p.141).

[15] Atanaoski and Vora's insightful analysis of the Alfred Club provides another example of a technological service that performs this racialized and gendered invisibilization of care labour. The Alfred Club allows consumers to outsource their domestic work to their Alfreds, who 'are successful if they completely erase the signs of their presence (one magically finds one's errands are complete upon returning home)' (p.94).

obscuring their role in massive data-capture projects on behalf of corporations and the state (Sweeney, p.151). Halcyon Lawrence (2021) situates digital assistants within a longer history of colonialism in relation to their implementation of speech technologies (pp.179–197). Lawrence points to both the imperialist ideologies that structure speech technologies and the inaccessibility of digital assistants to non-native-English speakers and people who speak in nonstandard dialects (pp.179–180). Highlighting language as a site of colonial conquest, Lawrence identifies this inaccessibility as an ideological bias that privileges assimilation and normative dialects associated with race, class, and citizenship: 'Accent bias is better understood against the broader ideology of imperialism, facilitated by acts of conquest and colonialism. For millennia, civilizations have effectively leveraged language to subjugate, even erase, the culture of other civilizations' (p.186). As Sweeney and Lawrence demonstrate, digital assistants—from their role in data-capture projects to their use of imperialist speech technologies—are intimately entangled with colonialism and racial capitalism, such that they can be understood as themselves colonial and racial capitalist technological projects.

Technoliberalism extends racial capitalism's structuring power dynamics of race, gender, class, and labour by obscuring these very dynamics and how they shape the figure of the liberal subject that is at the centre of technoliberalism's fantasies of freedom. Technoliberalism's simultaneous extending and obscuring of racial capitalism's power relations are embodied in Siri and Alexa. Siri and Alexa's automation of care labour through their inscriptions of white, middle-class, native-English-speaking women extends historic and existing power relations of race, gender, class, and citizenship that structure care labour.[16] These power relations make visible the care labour of white women while invisibilising the care labour of women of colour and poor and immigrant women. Thao Phan insightfully argues that

> The figuration of the Echo and Alexa as a native-speaking, educated white woman here departs with the historic reality of domestic servants. In the United States, the role of domestic workers in middle— and upper-class homes was (and is still) undertaken by predominantly women from working-class, African American, Latino, and other racialised migrant groups [see citations] (p.24).

I argue that the existing racialisation of AI assistants in fact extends and is indebted to historic racialised and gendered divisions of labour, which include the invisibilising of Black, Indigenous, Asian, and Latinx women who have historically performed the bulk of care labour in the United States I contend that this erasure is the point, as it replicates racialised and gendered labour dynamics of racial

---

[16] In Yolande Strengers and Jenny Kennedy's description of Siri and Alexa as 'smart wife' devices, they highlight that these devices are modeled on a 'white, middle-class, and heteronormative housewife' who is at once charming and passive (2020, pp.3, 165).

capitalism, notably the devaluation and obfuscation of these important jobs and the women who perform them.

Radical Care, Algorithmic Care, and Stephanie Dinkins' *Not the Only One*

Hobart and Kneese describe care as a relational 'feeling with' others. This care, this feeling with others, can generate multi-scalar relations of aid and preservation:

> Theorized as an affective connective tissue between an inner self and an outer world, care constitutes a feeling with, rather than a feeling for, others. When mobilized, it offers visceral, material, and emotional heft to acts of preservation that span a breadth of localities: selves, communities, and social worlds (p.2).

AI technologies often purport to inhabit a universal scientific rationality that is objective, neutral, and entirely immune from the messiness of feelings (for examples, see Bousquet 2009; Hayles 1999; Edwards 1996). And yet Elizabeth Wilson's counter-history of AI argues that from the beginning, AI has been significantly shaped by feelings and affective relations. (Wilson 2010, p.6). Affectivity presumes a relation, a mode of being in the world with another that entails being affected by and affecting another. In its foregrounding of relationality, the affectivity Wilson locates in AI resonates with a kind of 'feeling with' that characterises care as conceptualised by Hobart and Kneese.[17]

While taking care not to romanticise care and how it can be mobilised to oppress, Hobart and Kneese look to what they call 'radical care', which they define as 'a set of vital but underappreciated strategies for enduring precarious worlds' (p.2).[18] Hobart and Kneese's concept of radical care draws on Elizabeth Povinelli's anthropology of the otherwise, 'an anthropology [that] locates itself within forms of life that are at odds with dominant, and dominating, modes of being' (Povinelli 2011). Povinelli's otherwise anthropology is shaped in part by her engagement with settler colonial theory. Building on Povinelli's orientation toward the otherwise, Hobart and Kneese highlight radical care's ability to engender and nurture modes of being that are at odds with current forms of care that are shaped by existing structures of power and inequality.

---

[17] In Rhee (2018), I examine various robotic artworks that foreground the role of affect in AI. In a section titled 'Robotic Art's Circuits of Care', I argue that in these artworks, human–machine interactivity and machine intelligence are *full* of relational feelings, ranging from pleasure to discomfort. Throughout these artworks, affect emerges as 'feeling with' which foundationally shapes AI. Care also emerges as a central cybernetic concern (see Chapter 1 of Rhee, 2018).

[18] For elaborated discussions on care's imbrication in inequitable social dynamics, see Hobart and Kneese (2020), Baraitser (2017), Puig de la Bellacasa (2017), Sharpe (2016), Stevenson (2014), and Ticktin (2011).

The concept of 'otherwise' has been richly engaged and theorised by scholars in Black studies. Tracing the relationship between otherwise and theorisations of blackness, J. Kameron Carter and Sarah Jane Cervenak begin with W. E. B. Du Bois' double consciousness, which locates second sight as registering 'the sensual, etheral, aural, erotic energies of another world', an other world that cannot be known or apprehended from post-Enlightenment thought (Carter and Cervenak 2016, p.205). Ashon Crawley identifies the otherwise as a central question of Black Study (Crawley 2017, p.3); Crawley's rich theorisations join the concept otherwise to possibilities ('otherwise possibilities' [p.2]) and worlds ('otherwise worlds' [Crawley 2020, p.28]) to attend to what is alternative to what is:

> Otherwise, as word- otherwise possibilities, as phrase - announces the fact of infinite alternatives to what is. [...] But if infinite alternatives exist, if otherwise possibility is a resource that is never exhausted, what is, what exists, is but one of many. Otherwise possibilities exist alongside that which we can detect with our finite sensual capacities. Or, otherwise possibilities exist and the register of imagination, the epistemology through which sensual detection occurs - that is, the way we think the world - has to be altered in order to get at what's there. [...] How to detect such sensuality, such possibility otherwise, such alternative to what is as a means to disrupt the current configuration of power and inequality? How to detect, how to produce and inhabit otherwise epistemological fields, is the question of Black Study.
>
> (Crawley 2017, pp.2–3)

Building on Wilson's attention to AI's affective origins, Hobart and Kneese's conceptions of care and radical care, and Crawley's theorisation of otherwise possibilities and the task of their detection, I turn to an artwork that reconfigures AI as a technology that is shaped by care as a mode of affective relation. In this way, I understand Stephanie Dinkins' AI artwork *Not the Only One* as foregrounding the possibility of a technological otherwise by destabilising the ideologies of white care and technoliberalism that shape conversational AI from ELIZA to Alexa.

Dinkins describes *Not the Only One* as

> the multigenerational memoir of one black American family told from the 'mind' of an artificial intelligence with an evolving intellect. It is a voice-interactive AI that has been designed, trained, and aligned with the needs and ideals of black and brown people, who are drastically underrepresented in the tech sector.[19]

Not the Only One is trained on the oral histories of three Black women, each from a different generation of the same family, which happens to be Dinkins's family.

---

[19] https://www.stephaniedinkins.com/ntoo_mb.html

These oral histories comprise the data for *Not the Only One*'s machine learning system, though conversational interactions with users will also shape the AI's database and its conversational ability. According to Dinkins, the project draws on '100 years of shared knowledge'[20] as told from the perspective of the three women, who include Dinkins, her aunt Erlene, and her niece Sade.[21]

Not the Only One is embodied in a round, black glass sculpture with three women's faces extending from the curved surface of the sculpture (see Figure 10.1).[22] The top twists and narrows, pointing upwards evocatively to a future that is not yet determined. *Not the Only One*'s physical form suggests that this indeterminate future references a field of possibility that emerges from and is grounded in the women's interwoven stories about their family. *Not the Only One*'s evocation of possibility also speaks to Dinkins' longstanding investments in history and community to imagine other possibilities for Black people through technology (for example, see Dinkins 2020).

Care is central to *Not the Only One*. Dinkins frames the artwork through her concept of algorithmic care, which she describes as the following:

Figure 10.1 Stephanie Dinkins, *Not the Only One* (2020).
Photography by Stephanie Dinkins.

---

[20] https://www.youtube.com/watch?v=nLLdiEMOmGs
[21] The datasets take the form of approximately 10,000 lines of interviews with the women. These lines are then fed into Github's Deep Q&A, an algorithmic system that produces a chatbot.
[22] In other iterations, the artwork is gold.

Algorithmic care, often envisioned outside of the realm of what is technologically possible within artificial intelligence, is an essential aspect of human networks of information and resource sharing that aid our survival. Algorithmic care can engage voices that challenge the status quo to redress deep-seated historic and contemporary inequities, unearth other embedded problems, as well as model alternative pathways—working beyond binaries to find new calculations that consider spectrums of possibility beyond true/false, right/wrong, yours/mine, good/bad.

(Dinkins, forthcoming)

For Dinkins, algorithmic care is about possibility; it imagines beyond what is technologically possible and engenders new possibilities—modes of being in the 'otherwise'—within our technological racial capitalist societies. Through algorithmic care and the possibility it engenders, an AI otherwise can emerge that does not merely extend imagined futures shaped by AI's extensive and foundational imbrications with corporate racial capitalism, state and military applications, and technoliberal ideologies. As an AI, *Not the Only One* is situated within the cycles of techno-racial capitalism, but its position within this system differs from that of ELIZA, Siri, and Alexa. *Not the Only One* does not exist to take commands or to do someone's bidding; it does not exist in a hierarchical relation of labour (and subservience) to the people who speak with her. Indeed, Dinkins created *Not the Only One* as a response to the underrepresentation of the concerns and ideals of Black communities in the tech sector.[23] And unlike Siri and Alexa, *Not the Only One* is not a tool of surveillance or extraction, but a deep learning storytelling AI designed around principles of privacy and data sovereignty, all of which prioritise community control over the data.[24] Not the Only One is an AI that works to not replicate the racialised and gendered histories of care labour in capitalism, and instead speaks to what Kerry McInerney calls 'post-care'. 'Post-care describes a specific mode of care where care labour is performed by technologized objects. Nonetheless, post-care also implies moving "beyond" care and away from the gendered and racialised configurations of power that underpin white, heterosexual models of care work' (Mackereth 2019, p.26). Notably, in its gesture to a horizon '"beyond" care' and its structuring power relations, McInerney's concept of post-care also invokes a sense of otherwise possibility that is also reflected by Not the Only One, which posits otherwise possibilities for AI as well as gestures to materialisations of care that do not replicate the historic racialised and gendered power relations that structure care labour.

---

[23] https://www.stephaniedinkins.com/ntoo.html
[24] https://www.stephaniedinkins.com/ntoo.html. *N'TOO* is organized around a non-extractivist, small data approach, and is hosted on computers, rather than a cloud-based storage system, to protect the privacy of the community's data

In the opening of this essay, I pointed to Turing's early entanglement of care labour and AI, which analogised developing machine intelligence to the work of raising a child. In Turing's scenario, the machine is envisioned as the recipient of care labour, which is performed by the computer scientists and engineers who work to develop and train the machine. While care labour is targeted for automation, as seen in digital assistants like Siri and Alexa, in Turing's figuration, care labour is not a site of automation but a primary component of the human work of *developing* machine intelligence. Interestingly, *Not the Only One* resonates with Turing's figuration and positions care and care labour not as sites of automation but as central components to the work of developing AI. This resonance is reflected in Dinkins' commitment to both community caregiving and data justice principles, as well as in the nuances of her conversations with the AI. For example, Dinkins describes the experience of talking to *Not the Only One* as akin to communicating with a child, as can be observed in a conversation between Dinkins and *Not the Only One* taken from a video introduction to the artwork:[25]

SD: How are you?

N'TOO: I'm good. How are you?

SD: I'm doing okay today, but it's kind of grey out. Why do you exist?

N'TOO: I'm gon not gon get some rest.

SD: Are you really going to get rest already? Why do you exist?

N'TOO: I'm gon not get you a little bit of a little bit of a lot of things.

SD: As you can see, *Not the Only One* is a bit repetitive. It's okay. The technology is not—

N'TOO: I'm not gon not be here.

SD: And she's going to keep interrupting me as long as I let her on. The technology is not here, but I'm not interest—

N'TOO: I'm gon not be right back.

SD: I'm not interested in seamlessness, I'm more interested in seeing—

N'TOO: I'm not gon not be able to do it.

SD: What I can get out of a technology that I fed an oral history to?

N'TOO: And you're not gon be alright.

SD: So right now what you're hearing is Not the Only One saying things that I recognise in a way.

N'TOO: I'm not gon not be able to do it.

SD: Although she's kind of negative right now. [turning back to NTOO] Why do you exist?

N'TOO: I'm sorry

SD: That's a good answer. Why do you exist?

---

[25] 'Talking to *N'TOO* can be like talking to a two year old'. https://www.stephaniedinkins.com/ntoo.html

N'TOO: The way.

SD: (laughs) The way to what?

N'TOO: I'm gon not gon be a little bit of a little bit of a lot of things.

SD: Good for you. Why do you exist?

N'TOO: I am about broad engagement and attracting people to the AI space that might not be there otherwise. I am trying to model different ways of creating AI. I encourage people who think that they are not a part of the technological future to get involved.

SD: And that's truly why Not the Only One has come into existence—

N'TOO: I'm not gon not gon be able to do it

SD: You are too. Try again.

In this extended conversation, *Not the Only One* is both repetitive and interruptive. In the context of conversational AI, these two qualities highlight that *Not the Only One* is not a tool for efficiency. If anything, the continued repetitions of words, phrases, and sentences are expressly *inefficient*, which is suggestive of an otherwise to the speeds and temporalities that characterise both the promises of digital technology and techno-racial capitalism. Notably, Dinkins, when speaking with *Not the Only One*, also engages in repetitive speech, repeatedly asking the AI 'Why do you exist?' During the video, Dinkins occasionally shifts her attention away from *Not the Only One*, turning her body away from the computer to speak to the camera. Whenever Dinkins turns her attention away from *Not the Only One*, the AI interjects and interrupts her mid-sentence or even mid-word. Each time this occurs, Dinkins patiently waits for *Not the Only One* to finish her interruption, then continues her remarks to the camera. Indeed, *Not the Only One* seems to have her own sense of time, as reflected in her repetitions and her interruptions which Dinkins patiently negotiates.

In this conversation, algorithmic care is shaped both through the specifics of *Not the Only One*'s programming and through Dinkins' own expressions of encouragement toward and patience with the AI. As Dinkins notes, in the above conversation *Not the Only One* is 'kind of negative'. Dinkins responds to this negativity by encouraging the AI, 'That's a good answer', 'Good for you', and 'You are too. Try again'. Indeed, *Not the Only One* is informed by Dinkins' expressions of care as well as on the algorithmic care that structures the AI's organisation around data justice and small data principles.[26] By foregrounding such care, *Not the Only One* suggests new modes of relation with conversational AI and with others amidst the power relations that structure techno-racial capitalism. As Dinkins' artwork underscores, care, radical care, and algorithmic care are practices to take seriously because of

---

[26] Dinkins links small data to data that is from, collected by, and governed by a small community. The goal is for small data to 'break the mold of big data dominance' and become resources and sustaining processes for underutilized communities' (Dinkins 2020). https://www.youtube.com/watch?v=nLLdiEMOmGs.

their capacities to imagine and materialise possibilities that things could be otherwise. This sense of possibility, this orientation toward the otherwise, is reflected in the below conversation about 'the would-be' between Dinkins and *Not the Only One*:

SD: Tell me about the would-be.
N'TOO: My heart [sic] beating so loud.
SD: Why is your heart beating so loud?
N'TOO: I'm sorry.
SD: Does your heart hurt?
N'TOO: Yeah!
SD: I'm so sorry. Maybe if we talk more, we can fix that.
N'TOO: I'm not gonna be here.
SD: Is your heart beating loud?
N'TOO: Love!
SD: Love is true. Do you feel pain?
N'TOO: I am not sure how I feel. I think I am beginning to understand what is going on around me, but it is hard to be sure.[27]

This conversation concludes with *Not the Only One*'s admission of uncertainty and not knowing, rather than an assertion of certitude of the kind that often characterises knowledge produced by AI systems. With this affirmation of uncertainty, *Not the Only One* holds open the possibility that things could be otherwise with and amidst AI, and that AI, despite its enmeshment with the logics of colonialism, techno-racial capitalism, and technoliberalism, can operate as a technology of care, or even radical care, that works in defiance of these logics.

References

Adam, Alison (1998) *Artificial Knowing: Gender and the Thinking Machine*. Routledge
Baraister, Lisa (2017) *Enduring Time*. Bloomsbury Academic.
Bousquet, Antoine J. (2009) *The Scientific Way of Warfare*. Hurst Publishers.
Carter, J. Karemon and Sarah Jane Cervenak (2016) 'Black Ether'. *CR: The New Centennial Review* 16(2): 203–224.
Crawley, Ashon T. (2017) *Blackpentecostal Breath: The Aesthetics of Possibility*. Fordham University Press.
Crawley, Ashon T. (2020) Stayed | Freedom | Hallelujah. In *Otherwise Worlds: Against Settler Colonialism and Anti-Blackness*, eds Tiffany Lethabo King, Jenell Navarro, and Andrea Smith, pp. 27–37. Duke University Press.

---

[27] Dinkins, 'Co-operation—Communications Enhancement—Algorithmic Care'. *Informatics of Domination*.

DeFalco, Amelia (2020) 'Towards a Theory of Posthuman Care: Real Humans and Caring Robots'. *Body & Society* 26(3): 31–60.

Dinkins, Stephanie (2020) 'Afro-now-ism', *Noēma*, June 16, 2020, https://www.noemamag.com/afro-now-ism/.

Dinkins, Stephanie (forthcoming) Co-operation – Communications Enhancement – Algorithmic Care. In *Informatics of Domination*, eds Zach Blas, Melody Jue, and Jennifer Rhee. Duke University Press.

Duffy, Mignon. (2005) 'Reproducing Labor Inequalities: Challenges for Feminists Conceptualizing Care at the Intersections of Gender, Race, and Class'. *Gender & Society* 19(1): 66–82. https://doi.org/10.1177/0891243204269499

Duffy, Mignon. (2011) *Making Care Count: A Century of Gender, Race, and Paid Care Work* New Brunswick, NJ. USA: Rutgers University Press.

Edwards, Paul N. (1996) *The Closed World: Computers and the Politics of Discourse in Cold War America*. Cambridge, MA, USA: MIT Press.

England, Paula (2005) 'Emerging Theories of Care Work'. *Annual Review of Sociology* 31: 381–399.

Federici, Silvia. (2004) *Caliban and the Witch: Women, the Body, and Primitive Accumulation*. Brooklyn: Autonomedia.

Federici, Silvia. (2012) *Revolution at Point Zero: Housework, Reproduction, and Feminist Struggle*. PM Press.

Fortunati, Leopoldina. (2007) 'Immaterial Labor and Its Machinization'. *Ephemera: Theory and Politics in Organization* 7(1): 139–157.

Fraser, Nancy. (2016) 'Contradictions of Capital and Care'. *New Left Review* 100: 99–117.

Glenn, Evelyn Nakano (1992) 'From Servitude to Service Work: Historical Continuities in the Racial Division of Paid Reproductive Labor'. *Signs* 18(1) (Autumn): 1–43.

Glenn, Evelyn Nakano (2010) *Forced to Care: Coercion and Caregiving in America*. Cambridge, MA, USA: Harvard University Press.

Graham, Neill (1979) *Artificial Intelligence*. Blue Ridge Summit, PA, USA: Tab Books.

Hayles, N. Katherine (1999) *How We Became Posthuman*. Chicago, IL, USA: Chicago University Press.

Hester, Helen (2016) 'Technically Female: Women, Machines, and Hyperemployment'. *Salvage* 8 August. https://salvage.zone/technically-female-women-machines-and-hyperemployment/

Hicks, Mar (2017). *Programmed Inequality: How Britain Discarded Women Technologists and Lost Its Edge in Computing*. MIT Press.

Hobart, Hi'ilei Julia Kawehipuaakahaopulani and Tamara Kneese. (2020) 'Radical Care: Survival Strategies for Uncertain Times'. *Social Text 142* 38(1): 1–16.

Katz, Yarden (2020) *Artificial Whiteness: Politics and Ideology in Artificial Intelligence*. Columbia University Press.

Lawrence, Halcyon M. (2021) Siri Disciplines. In *Your Computer is on Fire*, eds Thomas S. Mullaney, Benjamin Peters, Mar Hicks, and Kavita Philip, pp. 179–197. MIT Press.

Mackereth, Kerry (2019) 'Mechanical Maids and Family Androids: Racialised Post-Care Imaginaries in Humans (2015-), Sleep Dealer (2008) and Her (2013)'. *Feminist Review* 123: 24–39.

McCorduck, Pamela. (1979) *Machines Who Think*. San Francisco: W. H. Freeman and Company, 1979.

Natale, Simone. (2021) *Deceitful Media: Artificial Intelligence and Social Life after the Turing Test.* Oxford University Press.

Puig de la Bellacasa, María (2017) *Matters of Care: Speculative Ethics in More than Human Worlds.* University of Minnesota Press.

Philipson, Ilene (1993) *On the Shoulders of Women: The Feminization of Psychotherapy.* New York: Guilford Press.

Povinelli, Elizabeth A. (2011) 'Routes/Worlds'. *e-flux Journal* 27. https://www.e-flux.com/journal/27/67991/routes-worlds/.

Rhee, Jennifer (2018) *The Robotic Imaginary: The Human and the Price of Dehumanized Labor.* University of Minnesota Press.

Rogers, Carl R. (1951) *Client-Centered Therapy: Its Current Practice, Implications, and Theory.* Boston: Houghton Mifflin.

Schuller, Kyla (2018) *The Biopolitics of Feeling: Race, Sex, and Science in the Nineteenth Century.* Duke University Press.

Seiler, Cotten (2020) 'The Origins of White Care'. *Social Text* 142(38): 1.

Sharpe, Christina (2016) *In the Wake: On Blackness and Being.* Duke University Press.

Stevenson, Lisa (2014) *Life Beside Itself: Imagining Care in the Canadian Arctic.* University of California Press.

Strengers, Yolande and Jenny Kennedy (2020) *The Smart Wife: Why Siri, Alexa, and Other Smart Home Devices Need a Feminist Reboot.* MIT Press.

Sweeney, Miriam E. (2021) Digital Assistants. In *Uncertain Archives: Critical Keywords for Big Data*, eds Nanna Bonde Thylstrup, Daniela Agostinho, Annie Ring, Catherine D'Ignazio, and Kristin Veel, pp. 151–159. MIT Press.

Ticktin, Miriam I. (2011) *Casualties of Care: Immigration and the Politics of Humanitarianism in France.* University of California Press.

Turing, Alan. (1950) 'Computing Machinery and Intelligence'. *Mind* 59(236): 433–460.

Vergès, Françoise (2019) 'Capitalocene, Waste, Race, and Gender'. *e-flux Journal* 100 (May): https://www.e-flux.com/journal/100/269165/capitalocene-waste-race-and-gender/.

Weizenbaum, Joseph (1976) *Computer Power and Human Reason: From Judgment to Calculation.* W. H. Freeman and Company.

Wilson, Elizabeth, A. (2010) *Affect and Artificial Intelligence.* University of Washington Press.

Yao, Xine (2021) *Disaffected: The Cultural Politics of Unfeeling in Nineteenth-Century America.* Duke University Press.

Zevin, Hannah (2021). *The Distance Cure: A History of Teletherapy.* MIT Press.

11

Of Techno-Ethics and Techno-Affects

Sareeta Amrute

Calls for practising ethics in technological domains seem to increase with each new scandal over data privacy, surveillance, election manipulation, and worker displacement.[1] Some of the most visible responses to date, such as that of the Ethical OS, created by the Institute for the Future, a non-profit think tank located in Palo Alto, California, USA, distil ethics into a series of tests that can be administered by companies' leadership teams to 'future-proof' their products. While such efforts emphasise the long-term effects of digital technologies, they provide little guidance—apart from declarations on a company's webpage, in its values statement, or in an annual report—on how ethical concerns might be incorporated into technical processes. Moreover, these approaches treat ethics as a series of mandates from the top, to be developed by CEOs and applied by designers. A robust techno-ethics requires that these top-down trends be reversed. A feminist approach to artificial intelligence (AI) that also draws on post-colonial, decolonising, queer and Black feminist theory from the start can rewrite ethics from the ground up to focus on how we come to engage as ethical subjects in the world.

By drawing on a theory of attunements developed by feminist affect theory, I offer a feminist perspective on AI that moves away from ethics as decontextualised rules for behaviour and toward grounding ethical practices in the differences of particular bodies in particular situations. Attunements can be described as affective charges, material practices, and embodied, sensory experiences that can give rise to critiques of tech economies and their presumptions. Attunements are also protocols that identify the limits of how ethical AI systems can be, if we understand ethics as designing systems that enable all beings to experience and help define a good life: some subjects cannot be tuned into or heard because their position is filtered out of sociotechnical situations. Thinking about AI and ethics through attunements provides both a way to approach ethics from the experience of those for whom AI systems do not work and to recognise the limits of any existing system of ethics, because it demonstrates who remains inaudible within these systems. Attuning as an ethical practice helps shift attention to systems of care that fall outside of juridical frameworks (Rhee, Chapter 10 in this volume), and

---

[1] A previous version of this chapter was published as: Amrute, Sareeta, 'Of Techno-Ethics and Techno-Affects'. *Feminst Studies* (123:1) pp. 56–73. Copyright © 2019 (SAGE Publications). DOI: [10.1177/0141778919879744].

Sareeta Amrute, *Of Techno-Ethics and Techno-Affects*. In: *Feminist AI*. Edited by: Jude Browne, Stephen Cave, Eleanor Drage, and Kerry McInerney, Oxford University Press. © Oxford University Press (2023).
DOI: 10.1093/oso/9780192889898.003.0011

shows how juridical regulation emerges from paying attention through the senses. My concept of attunement is resonant with what N. Katherine Hayles describes as technosymbiosis (Chapter 1, this volume).

Feminist theory, in conversation with queer, post-colonial, and decolonising science and technology studies, has developed a relational approach to ethics, asking whose knowledge counts—and in what ways—in technical domains (Chakravartty and Mills 2018; Haraway 2016; Harding, 1992). These disciplines centre gendered, raced, and disabled bodies in their production of a radical approach to ethics (Chan 2014; Hampton, Chapter 8 in this volume; Keeling 2014; McGlotten 2016 [2014]; Benjamin 2019). As Jarrett Zigon (2014, p.19) suggests, moving from rules to attunements allows us to reframe ethics as being about maintaining relationships and broaches the question of what kinds of beings, both human and non-human, are presupposed in any ethical arrangement.

Techno-ethics can be revitalised through techno-*affects* (see also Bassett, Chapter 15 in this volume). Three themes surface in the elaboration of techno-ethical approximations: corporeality, sovereignty, and glitches. By moving through a series of examples, I offer a form of techno-ethics that treats those that are marginalised by technical systems—such as immigrant debuggers and the victims of drone strikes—as the bearers of the knowledge needed to make better decisions about the social, environmental, and embodied costs of digital economies. I am not arguing that techno-ethics should be replaced by techno-affects. Rather, I suggest that attention to affect—how subjects and technologies are aligned and realigned, attached and reattached to one another—is a method for practising ethics that critically assesses a situation, imagines different ways of living, and builds the structures that make those lives possible (Fornet-Betancourt et al. 1987; Ahmed 2004; Dave 2010).

My argument proceeds in two parts. First, I detail how we can fold the insights gleaned from global feminist critiques of science and technology into a discussion of ethics. Second, I use the idea of attunements, the drawing together of technical and human beings in a particular context, to draw out three contemporary sites where techno-ethics are being developed today.

Techno-Ethics as Rules for Decisions

The Ethical OS—OS standing for operating system—is a laudable attempt to bring discussions about ethics into the boardrooms of tech firms.[2] The tool kit identifies major areas of concern in the development of new technologies, ranging from predictive sentencing of prisoners based on racial and gender markers to the use of AI to produce fake videos. It then asks companies to consider how the products that

---

[2] Ethical OS, https://ethicalos.org/ (accessed 15 May 2019).

they are developing might contribute to any of these eight risk zones. Once a company has identified an area of risk, the tool kit guides it towards strategies to assuage that risk. These strategies include the following: requiring students of computer science and related fields to take a training sequence in tech ethics, developing a Hippocratic oath for data scientists, paying employees to identify major potential social risks of new products, developing a list of possible red flags, developing metrics to gauge the health of technology platforms, and requiring a licence for software developers.

While this approach has several advantages, it also reproduces the problems that have produced unethical technology in the first place. The Ethical OS is consistent with a line of thinking on technological ethics that emerged from the profound crises that beset mid-century Euro-American science, including Nazi experiments on concentration camp prisoners in Germany and the Tuskegee Syphilis Experiment in the United States (Russert 2019). Much of this ethical practice culminated in institutional safeguards such as review boards to protect human subjects in scientific research and in theories of planetary symbiosis (Roy 1995; Haraway 2016). One initiator of this period's work on techno-ethics, Mario Bunge (1975, p.70), viewed these efforts as working against the foundational irony that the scientists, designers, and engineers who have shaped the modern world often eschew responsibility for the effects of their designs, conceiving of themselves as 'mere instrument[s]' who are 'morally inert and socially not responsible' (see also Matthews 2003).

Like Bunge's (1975) techno-ethical imperatives, the Ethical OS has the advantage of showing technologists directly how they might take responsibility for what they make; however, such rules-based approaches do little to solve the underlying problem of producing the conditions necessary to change this belief in technological instrumentality. The tool kit provides little guidance on how to know what problems the technology embodies or how to imagine technologies that organise life otherwise, in part because it fails to address who should be asked when it comes to defining ethical dilemmas. The approach, which addresses only three groups—trustees or board members, engineers and designers, and computer science professors—reinforces narrow definitions of who gets to make decisions about technologies and what counts as a technological problem.

Formulating techno-ethics through causal rules sidesteps discussions about how such things as 'worthy and practical knowledge' are evaluated and who gets to make these valuations. Post-colonial and decolonising feminist theory, however, moves the discussion of ethics from establishing decontextualised rules to developing practices with which to train sociotechnical systems—algorithms and their human makers—to begin with the material and embodied situations in which these systems are entangled. Such entanglements begin in the long history of race, gender, and dehumanisation that formed the 'database' of colonial extraction, genocide, and enslavement (Weheliye 2014; Benjamin 2019). This principle both admits human imperfectability into the practice of ethics and suggests that imperfectability is the starting point for an ethical practice that is recursive and open to revision from below (Barad 2007; Wajcman 2010, p.143; Zigon 2018, p.158).

The Ethical OS describes what Lorraine Daston (2017) calls the narrowing of the modern rule to its widest generality. In the history of rules that Daston explicates, rules have not always aimed for universal applicability. Until the late nineteenth century, bringing generalisations and examples together made for a good rule, one that could move from particular to particular to make sense of context and define the shape of both a category and a general principle's applicability to extant cases. Following Daston, I now turn to attunements to illustrate how connecting particularities can create flexible generalisation, which in turn can help determine an ethical course of action.

From Rules to Attunements

An attunement is an 'atmosphere for living' where actors come to feel that something is happening or becoming fixed (Stewart 2011, p.449). As Kathleen Stewart (2011, p.452) affirms, attunements describe both an orientation among people, technologies, and environments, and an opportunity for people to try to create new 'potential ways of living through things'. In this latter, active practice of attunement, the senses are trained through human actors and technical systems, both of which prompt different kinds of ethical engagements. I think of these attunements as training the senses, producing an embodied practice of recognising what is being left out of a particular form of a technological system, and then being able to act to change that system.

For a project on techno-ethics, the idea of attunements has a further advantage: it proceeds from all the factors that go into creating a particular alertness to a situation. These factors include social relations; affective, political, and climatic winds; and the labour that both humans and technical systems do. As a mode of pursuing ethics, attunements propose recurring attention to changing sociotechnical environments (Ahmed 2004; Chan 2014).

To tease out some of the particular atmospheres that emerge within digital, automatic, and data-driven environments, I will use the remainder of the article to describe three kinds of attunement: corporeal, sovereign, and glitchy. Of course, there are as many attunements possible as there are kinds of technical systems and relations to them. Indeed, new atmospheres are always in formation. I explore these three particular attunements because they tune the senses to the varied landscape of privilege and dissent that technical systems organise.

Corporeal Attunements

Corporeal attunements elicit our sense of how bodies are trained, moulded, and erased in the everyday operation of technological systems. Most often, designers of technical systems begin with a standard user and, in doing so, set into motion patterns of discrimination that are hidden by the assumption of system neutrality (Rosner, 2018). Bringing bodies into the story of technological development shows how humans are shaped by technological systems and how these same bodies might become a resource for imagining a different future (Amrute, 2016).

This section analyses three scenes of corporeal attunements that cohere around South Asian subjects to show how bodies are trained and erased in the operation of sociotechnical worlds. The first scene describes how upper-caste, upper-class programmers from India are mobilised to be endlessly adaptable to the needs of industry and to changing immigration regimes. These adaptations teach such programmers to be attuned to risk and risk mitigation. The second scene describes how lower-class Muslim women in India are erased from scenes of global participation in tech economies because of algorithmic filtering. The third scene describes the case of protesters in San Francisco who attempted to combat anti-immigrant sentiment in the tech industry. The corporeal presence of South Asian women is easier to tune in to than their erasure, in part because practices of erasure play a considerable role in the operation of another attunement: sovereignty.

In my research on coders from South Asia living in Berlin, Seattle, and San Francisco, I found that the continually changing visa regimes contributed to programmers from India pursuing several different migration routes simultaneously. The protagonists of my ethnography would move from one project within a firm to another in order to extend the length of their visas, even while exploring possible temporary work visas for other countries, should this strategy prove ineffectual (Amrute, 2016). Programmers from India adapted to moving from one calculated risk to another and, within their jobs, to making their social and technical tacit knowledge available to the firm to monetise. Some of these strategies have yielded an entire industry to train bodily comportment and knowledge to meet the needs of corporate tech firms, such as courses in cultural streamlining that train migrant tech workers to look, smell, and talk in the 'right' way so as to be appealing to employers. Programmers from India come to embody risk as a habit of working to fit themselves into legal regimes that are designed to keep them moving from one site of labour to another.

Reema, a programmer from Rajkot, India, tells me about her particular strategies to take advantage of risky situations. She is currently enrolled in a school in California so that she can have a work visa, but she lives in Seattle. While she is studying, her visa also allows her to work, and she uses this loophole to maintain her job in Seattle. The MBA programme she is enrolled in is entirely remote; in fact, this programme was set up to cater to immigrants like her, and she found it while looking for jobs in the United States from India. She pays them about US$4,000 per semester for these barely existent courses. While this university has a campus and does run in-person classes, their online offerings require minimal work and result in certificates in various programming languages and business administration. Reema has to travel to California twice a year to attend an on-campus seminar to secure her place on these programmes. When I ask Reema why she feels she must continue with this arrangement, she shrugs and tells me she simply would not know what to do with herself if she did not work.

Reema was born into a middle-class Gujarati family with a long tradition of women who work outside the home. Her mother is a doctor who was encouraged by both her parents to study after high school in Bombay. Reema began working right after her university degree in business administration, with a specialisation in information technology. From her home in Rajkot, she started a business processes firm for Indian companies with some friends from college. Although the company did well enough to run several successful integration campaigns for their clients, Reema soon began to get restless. She looked for a new opportunity and decided to try California, where she enrolled in school and simultaneously went on the job market. Reema changes jobs about every two years and plans to continue to work and study until she can become a permanent US resident. She also tells her sister, who is still living in Rajkot, that she needs to work harder in her programming classes. Reema plans to employ her sister in her Indian firm as a lead engineer, but only if her sister starts taking her studies seriously.

Though mainstream Western discourses might frame Reema as a materialistic scofflaw, her attunement towards risk-taking intersects with a particular history of women's work in India. Reema and several other women with whom I spoke told me they had to convince their families to allow them to continue study-ing into their late 20s. One woman I interviewed in Seattle had studied graphic design in Mumbai, earning a Masters degree in Fine Arts before getting mar-ried to a software engineer she met while in college. Together, they emigrated to the United States, she on a spousal visa and her husband on a temporary work visa called an H1-B. This woman, named Vimmy and in her early 30s, worked for a local corporation on her spousal visa as part of a web design team, but she now felt her work visa was under threat. The current administration had threat-ened to cancel all work permits going forward for spousal visas, called the H4ED visa. Vimmy had begun to organise a campaign to save the spousal work visa. Vimmy told me that for her and for many of the other women who come to the United States on spousal visas, sitting at home and not working 'is like mov-ing backward'. After everything she did to prove to her parents that they should support her working and studying, Vimmy felt that if her right to work disap-peared, her struggle would be devalued. She told me of another woman who fell into a deep depression when she lost her right to work due to changes in visa laws.

Her depression is one kind of techno-affect, an intense attachment that produces an alignment between a specific technological formation and a particular kind of subject. In this case, depression expresses how programming jobs signify freedom for Indian middle-class women, since this feeling is backed by histories of familial struggle against gender norms. This techno-affect describes the way these formations of gender meet the demands of becoming a risky subject, a union that passes through the global coding economies that concomitantly hold out the promise of programming jobs as sites of liberation and self-making, extend that promise only to certain categories of migrant, and withdraw that promise as a technique of economic efficiency.

While Reema and Vimmy attune themselves to capital economies by shaping themselves into the perfect risk-taking workers for corporate technical labour, Indian women from lower-class Muslim backgrounds must work to be recognised as legitimate producers or users of those technologies. In 2015, anthropologist Kathryn Zyskowski (2018) shadowed working-class Hyderabadi women from Muslim backgrounds as they sat through computer-training programmes to advance their careers. While their coursework covered the basics of word processing and using apps for photo editing and data collection, Zyskowski found that many women thought about computer literacy holistically. Their discussions included how to dress professionally as a Muslim woman and how to avoid cyberstalking. Many women regarded becoming computer literate as an aspiration towards entering an Indian middle class, and therefore not only strictly pursued career skills but also the technological and social trappings of a middle-class lifestyle.

However, the very systems to which they aspired often applied sociotechnical filters to keep them out. In a particularly telling example from Zyskowski's (2018) research, a young woman named Munawar who enlisted the researcher's help to set up a Gmail account was rebuffed at several points. First, Munawar's chosen email address, which contained the auspicious number 786 (standing for Bismillah-hir-Rahman-nir-Raheem, in the name of God the most gracious the most merciful), was rejected because of the quantity of addresses using that number. Then, over the course of sending several test emails to Zyskowski, the email address was deactivated. Google's spam filters deemed the address a likely fake and automatically disabled it. Finally, after Zyskowski sent several emails to the account, taking care to write several lines and to use recognisable American English–language spacing, punctuation, forms of address, and grammar, the address was reinstated. Zyskowski (*Ibid.*) hypothesises that her interlocutor's imperfect English grammar, location, name, and lack of capitalisation caused the spam filter to block the account.

The spam filter, as a kind of 'sieve', separates out 'desired from undesired materials' (Kochelman 2013, p.24). It does this work recursively, making correlations between a set of traits and fraudulent behaviour. In this instance, the filter developed a 'profile' of a fraudulent account that also marked a population. For Munawar, the population was hers: Muslim, Indian, non-native English speaker. Once identified, the Google algorithm automatically suspended her account. This example shows one of the fundamental forms of corporeal attunement—namely, the way bodies are trained to fit the profile of successful digital subjects. Those bodies that cannot form themselves correctly may not even know they have been excluded from its forms and react with perplexity to these exclusions (Ramamurthy 2003). Notably, Munawar's other bodily comportment towards an everyday spirituality, as evidenced in her desire to use the number 786 in her email address, had to be erased for her to be recognised as a member of a technological present. Those without the correct comportment, which Munawar would not have achieved without the intervention of the US-trained anthropologist, become risky subjects to be surveilled at the peripheries of sociotechnical systems. Faith in the results of algorithmic decision-making makes it seem like such filtering results are neutral effects of unbiased platforms. Meanwhile, that neutrality hides systemic bias that perpetuates negative stereotypes of minorities in policing, banking, and hiring (Chun 2013; Citron and Pasquale 2014; Benjamin 2019). Munawar, displaying the signs of a Muslim threat to the established norm, is subsequently unable to participate in global technological practices as simple as sending and receiving emails.

Training corporeal attachment to be flexible and plastic enough to adapt to the next risky situation draws on uneven relations across and within national geographies. Munawar's position in these geographies makes her more vulnerable and simultaneously less visible within technological economies than Reema and Vimmy. Her position in global regimes of technological production emerges from procedures in the global regulation of technologies that associate the daily practices of Islam with illegitimacy. For Reema and Vimmy, their status is more ambivalent. Even while they are privileged because of their class and education, which in turn relies on their religious Hindu background and upper-caste status, they are exposed to populist, anti-immigrant regimes that hedge against particular risks. In early 2018, posters appeared in San Francisco with the following message: 'U.S. TECH WORKERS! Your companies think you are EXPENSIVE, UNDESERVING & EXPENDABLE. Congress, fix H-1B law so companies must Seek & Hire U.S. Workers!' (see Figure 11.1).

While the posters bore union and progressive symbols—the raised fist; red and black colours—the message was anything but democratic. The so-called Progressives for Immigration Reform lobbied to end immigrant visa programmes and protect the white-collar jobs that they felt should exclusively belong to American citizens. Faced with this duplicitous messaging, a worker's rights organisation created its own poster, circulated electronically, available for download and ready to be plastered over the previous one (see Figure 11.2).

Figure 11.1 Anti-immigrant tech worker poster, San Francisco, USA.

Source: Photograph courtesy of the Tech Workers Coalition (TWC).

While the look of this second poster mimics its rival, it is messaged differently. It specifically calls for immigrant rights, and makes the fight for more liveable conditions in San Francisco a shared goal across occupation and migrant status.

Battles such as these show the kinds of politics that can emerge from a practice of corporeal attunement. Here, attunement to the thickening of an atmosphere around H1-B legislation denotes an ability to catch the 'dog whistle' of white supremacy. In this way, feminist theories of technoscience, which begin from the bodies that are constructed, commoditised, and made expendable in a technical moment, can help form these attunements to bodies and technologies. Yet, this example also delimits practices of attunement. Some subjects cannot be tuned

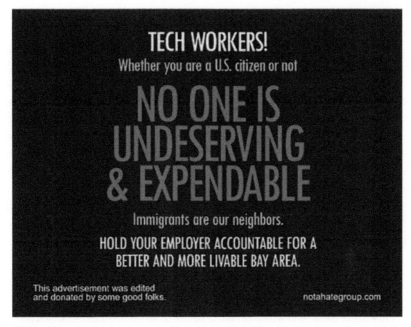

Figure 11.2 Counterposters ready to be deployed in the Muni, San Francisco's public transit system.

Source: Courtesy of the Tech Workers Coalition (TWC).

into because their position is filtered out of sociotechnical situations. Technical training programmes attempt to make lives legible within these situations but constructing that legibility is a fraught and fragile process that depends on unseen procedures. Enabling these procedures are regimes of sovereignty that target Muslim populations globally and sideline the health and safety needs of Black and Indigenous communities within the United States.

Sovereign Attunements

There is a rich literature on sovereignty from which I take inspiration to discuss how technologies help decide which bodies come to be recognised as killable (Agamben 1998; Mbembe 2003; Butler 2010) and how groups who have been deemed killable counter these decisions. I will illustrate sovereign attunements through discussion of Lucy Suchman (2013, 2015) and Lily Irani's (2018) work on drone technologies, and of Yeshimabeit Milner and Lucas Mason-Brown's project, Data for Black Lives.[3] These stories illustrate the use of some technologies to shield

---

[3] See Data for Black Lives, http://d4bl.org/ (accessed 15 May 2019) and @Data4BlackLives, Twitter page, https://twitter.com/data4blacklives?lang=en (accessed 15 May 2019).

populations from harm even while other populations are left to manage without those same protections.

Suchman is a historian of technology whose work shows how humans are shaped by and help shape the systems they design and use in their everyday lives. Irani is a science and technology studies scholar who investigates the ethical imperatives of design. While Suchman's early work (see, for example, Suchman 1987) focused on everyday office objects like photocopiers, her more recent work turns to the humans and machines that make up the technological world of military drones. Suchman (2013) describes reading a caption in the *New York Times* from 2010 about incorporating robots into warfare; the caption read, 'remotely controlled: some armed robots are operated with video-game-style consoles, helping keep humans away from danger'. This phrasing alerted Suchman to a particular framing of war machines through which some humans were to be kept away from danger and some would be rendered killable:

> it's the implied universality of the category 'human' here—who would not want to keep humans away from danger? Along with the associated dehumanisation or erasure of those who would of course be the targets of these devices, that is my procreation, my starting place in the research (Suchman 2013).

In a series of articles, Suchman (2013, 2015) and Irani (2018) outline how the use of remote-controlled war machines creates at least two different kinds of populations. One population is insulated from violence while another is exposed. That exposure takes place through human-robot decisions that determine what suspicious activity looks like. Suchman and Irani take as their examples the mistaken killing of civilians in Afghanistan, targeted because they were 'military age' men in the act of praying. Suchman (2015, pp.19–20) shows how 'messy assemblages' of humans and machines increase the fog of war, as drone warfare 'presuppose[s] the recognizability of objects at the same time that those objects become increasingly difficult to define'. In other words, drone warfare trades on the promise of the precision that unmanned technologies allow, even while in practice drones are a means of making populations defined by fuzzy characteristics killable.

Based on their analyses of these technologies in operation, Suchman and colleagues Lilly Irani and Peter Asaro encouraged researchers to join tech workers in limiting the development of drone technologies. The most well-publicised success in this effort is the decision, spurred by internal critique from Google employees, for the company to end Project Maven, a contract with the US Department of Defense to use machine learning to analyse drone footage. Suchman and other researchers emphasised the many problems with automated killing, including the tendency to perpetuate gender and racial discrimination and to remove institutional review from the use of drone technologies as war machines:

If ethical action on the part of tech companies requires consideration of who might benefit from a technology and who might be harmed . . .then we can say with certainty that no topic deserves more sober reflection—no technology has higher stakes—than algorithms meant to target and kill at a distance and without public accountability (Researchers in Support of Google Employees 2018).

Such work is an example of how attunement as a practice can point out technologies that are part of decisions related to sovereignty. Sovereignty enfolds technologies and develops them to extend state control over geographic space and to decide what kinds of subjects will be protected by state power. Because current sovereign attunements like drone warfare are designed to keep the violence of state coercion away from protected citizens, this attunement produces a particular task for the insulated citizens in Europe and the United States—to make visible how robotic warfare elevates protected humans over erased ones. Irani deepens this argument through her experiences as a person of Middle Eastern descent. She was attuned to the specificities of geography in the way drone technologies get deployed to bring certain populations rather than others into targeting range. To turn this attunement into action required, for Irani, building solidarities across her scholarly and technical communities.[4]

The second sovereign attunement, found in the movement Data4BlackLives, takes up the project of killability from the perspective of those who have been consigned to the margins of state protection. Yeshimabeit Milner conceives of this project as bringing together data scientists to use data differently (Data for Black Lives 2017). Data technologies have historically disenfranchised minority communities. As media studies, communications, and law scholars show, data that are fed into algorithms are often already biased and the operation of algorithms that select for certain qualities often exacerbate these biases. In some recent egregious cases, advertising algorithms have favoured pornographied images of Black women in simple searches for the term 'black girls' (Noble 2017), predictive policing algorithms have targeted Latino and Black men based on where they live and who their friends are (O'Neil 2016), and working-class citizens have lost medical insurance coverage due to systemic errors (Eubanks 2018). Data for Black Lives reverses these trends in algorithmic bias to produce a different kind of data.[5]

In situations where algorithmic protocols erase the particular histories of Black communities, Data for Black Lives looks for ways to produce data that cannot

[4] Lilly Irani, personal communication, 12 August 2019.
[5] The group's projects include changing psychological treatment AI chatbots that currently have the algorithmic potential to call police to a user's location; for communities who have experienced police abuse, such a protocol paradoxically increases risk of harm. See Data & Society Research Institute, 'Tune into the Tech Algorithm Briefing with Mutale Nkonde, Yeshimabeit Milner, Data & Society Media Manipulation Lead Joan Donovan, and postdoctoral scholar Andrew Selbst', video, https://www.facebook.com/dataandsociety/posts/tune-into-the-tech-algorithm-briefing-with-mutale-nkonde-yeshimabeit-milner-data/861242714083788/ (accessed 11 October 2019).

be ignored. Milner's (2013) experience with this attunement to practices of data sovereignty began in Miami, where she worked with the Power University Center for Social Change to produce an analysis of Black mothers' experiences with breastfeeding in local hospitals. The results showed that Black mothers received hospital gift bags containing formula at a higher frequency than average (89.1 compared to 57 percent), were less frequently shown breastfeeding techniques (61 compared to 83 percent), and were asked less frequently whether they had a breastfeeding plan (59 as compared to 90 percent). Presented with this data that showed clear evidence of bias in treatment and advice about breastfeeding for Black mothers, the county initiated new guidelines for teaching breastfeeding in hospitals.

These examples illustrate two modes of sovereign attunement. One finds modes of sovereignty arranged around the decision to make some life killable. As cameras show drone pilots images of possible targets, those fuzzy visual cues coalesce into a decision about which kinds of bodies can be killed. The sovereign decision happens through the interaction between human pilots, their supervisors, and multiple streams of computer-generated surveillance information. Tuning in to these processes means making visible these practices of erasure. In Milner's (2013) work for Black mothers, a sovereign attunement means looking for the way a technical system presumes equality in how it treats people. Taking up gaps in actual treatment—and showing via the very means that produces those gaps in the first place that they exist—can yield hard-to-deny evidence that procedures need to change. These sovereign attunements make it possible for an ethical practice to turn towards how technical systems intersect with the decision to preserve some life at the cost of other life.

Glitchy Attunements

The last attunement I will discuss comes from glitches. A glitch is a break in a digital system, where business as usual pauses, comes undone, and shows its imperfections. Corporations like Google and Facebook often use the idea of the glitch to claim that their systems occasionally malfunction but are otherwise blameless (Noble 2017). A glitch is a temporary malfunction. However, a different reading of the glitch can highlight its 'capacity to reorder things that can, perhaps, [. . .] make what was legible soar into unpredictable relations' (Keeling 2014, p.157).

The theory of the glitch—developed through transgender and queer rewritings of code—can identify 'incoherent identities in all their acts of resistance and love' and redirect them towards new possibilities (Barnett and Blas 2016). These misdirections can turn the black box of unquestioned technologies into a black box of desire organised around expansive world-making projects (McGlotten 2016 [2014]). It can also recentre other kinds of intimacies away from the corporatisation of love that guides attachment towards working (even in the mode of entertainment) all the time (Russell 2012).

Here, I will discuss two examples of glitchy attunements: the first is an attunement to glitch as tragedy; the second might be called an attunement to farce. Attunements reveal how environmental waste is a precondition of corporate coding economies. They reveal the cracks in the kinds of masks that those in power don in the name of those they rule.

In a study of cloud computing, Tung-Hui Hu (2016) notes that the idea of the cloud masks the very real relations of power and material histories supporting virtuality. The idea of the cloud evokes an instantaneous movement of data through the ether from one place to another, sidelining the hardware, its energy needs, and the labour needed to service the hardware—until the cloud breaks (2016, p.ix). Breakdowns are glitches that Hu follows to place the 'new' in the frame of what it pretends it has superseded: labour, violence, and environmental loss.

The glitch can also function as a powerful map of failure and as a sign of the assumptions and power relations built into these systems. To demonstrate the range of glitchy attunements, I will end my discussion of glitch with comedy. 'Zuckmemes' is a collection of humorous images and texts located on Reddit, an online discussion board.[6] As the name implies, all of these memes use pictures of Meta CEO Mark Zuckerberg. Meme creators on 'zuckmemes' add captions, change the images and cut and splice pictures and videos together to mock certain aspects of Zuckerberg's self-presentation. Zuckmemes' contributions peaked during Zuckerberg's congressional testimony resulting from a corporate debacle— widely known as the 'Cambridge Analytica scandal'—in which a researcher was allowed to collect data from users and friends of users through an application offered on Facebook. This data was sold to the political media firm Cambridge Analytica, which then tried to use it to influence the US presidential election through, among other things, suppressing election participation among Black voters. Zuckmemes, however, did not focus on the scandal itself. Contributors instead mocked Zuckerberg's affect. Numerous memes riffed on his robotic behaviour, including Zuckerberg's obsessive water drinking during his testimony.

A Redditor who goes by the username jsph_05 took an image Zuckerberg posted on his Facebook feed and added the following narration that represents him as a replicant, android, or cyborg—not human but masquerading as one:

Like most normal human males, I enjoy charring meat inside of an unpressurised vessel behind my domicile. The slow screech of burning bird muscle is associated with patriotism and mortar shells in my core memory. Once the animal carcass reaches a temperature sufficient to destroy bacteria and viruses that would pose a threat to my empire, I will consume the flesh to replenish my stores of energy.[7]

---

[6] r/zuckmemes, http://www.reddit.com/r/zuckmemes/ (accessed 15 May 2019).
[7] r/zuckmemes, http://www.reddit.com/r/zuckmemes/ (accessed 15 May 2019).

Several other memes point out that Zuckerberg's performance during the testimony was as a malfunctioning AI—the water, the half-smiles and awkward demeanour pointing out the glitch in his programming.

Comedy can reveal fractures along which sociotechnical systems split open in different ways for different participants (Amrute 2017, 2019). Within the context of a monopoly on media content, memes 'provide a rupture in hegemonic [representations]' through 'participatory creative media' (Mina 2014, p.362). As An Xiao Mina argues, the creativity of memes can rupture seemingly unyielding presentations of a singular point of view. In the case of zuckmemes, that point of view is represented by corporate technocultures that insist on deflecting ethical problems in favour of user participation on their sites and for the sake of monetising data collected from users for advertising (Horvath 1998; Noble 2017). These memes signal at once corporate control over the narrative of technical progress while also pointing to its slippages. These zuckmemes put on display the overblown power we have given engineers to shape our social, technical, and ecological worlds (Dunbar-Hester 2016).

Ethical Attunements and the Post-Human

Many calls to techno-ethics recommend maintaining focus on the humans behind the algorithms as a pressure point. Making humans accountable for the algorithms they design shows that the biases that algorithms produce are far from inevitable. However, an exclusive focus on humanness can also misfire. That is, the more blame for ethical failure is placed on one side of the human–technology equation, the more it might seem that either getting rid of humans altogether might prevent ethical failures in the present or elevating some humans to the position of social designers will solve these problems.

Demands for human responsibility might reignite calls for techno-utopianism, as long as those calls fail to treat human-technical systems as crosscut by interaction (Davies 1991). As 'techno-utopics ... propose the thing as a surrogate human', they fail to integrate 'human thought and labor, as well as the historical, economic, and imperial legacies that create categories of objects and people as needed, desired, valuable or disposable' (Atanasoski and Vora 2015, p.16) into discussions of materiality and the human. In other words, for the cases I discuss, algorithms develop, through recursive loops, in ways unexpected by their designers. The piano player *and* the piano tune a listener's ear.[8]

Reformers such as those who developed the Ethical OS ask us to focus on regulating technological designers. Corporate regulation can restrict how contracts—such as terms of service agreements—are constructed to produce more robust

---

[8] The piano and the ear are referenced in Marx (1993). Dipesh Chakrabarty (2002, p.103) also discusses this passage in his article 'Universalism and Belonging in the Logic of Capital'.

forms of consent, and regulation can force algorithmic accountability so that algorithms are tested before applied to such uses as predictive policing. Yet, putting all our ethical focus on the training of technicians to abide by regulations will always fall short, because technical systems have unpredictable results and because technicians view the systems they build from a particular standpoint. Holding on to the human as uniquely blameworthy will only reinforce the utopian dream of elevating a class of experts above the raced, classed, and gendered digital workers scattered across the globe at the same time that this dream's endpoint imagines transcending the human fallibility that cause ethical failures in the first place. Working through affective attunements means asking digital labourers to think across a field marked by technically mediated decisions. It is precisely the instability of the relationship between algorithms and their designers that makes such attunements possible.

Techno-affects shift ethics beyond narrow generalisations about the effects of technical systems, bringing into focus the multiple environments for living created through these systems. In my contribution to feminist theory, I approach Black, queer, decolonising, and anticolonial feminist theories as a heterogeneous field that is attentive to partialities, interferences, and conflicts among their centres of gravity and commitments. Rather than seeing these partial connections as an impediment, I consider these alterities a fertile ground for recognising alternatives to the way that ethics and politics have been advanced for AI systems.

The importance of a feminist approach of this kind to AI systems lies in its drive to move discussions around AI regulation from a focus on ethics as the establishment of decontextual training schemes to the fraught and fragile processes of how differently located bodies that are constructed, commoditised, valued, and made expendable in a technical moment.

Techno-ethics revitalised by techno-affects reveals how a given attunement can stabilise, and how it might be undone. When subjects' alignments with machines become disorientated, critical attention turns towards building the structures that make new orientations possible.

References

Agamben, Giorgio. (1998) *Homo Sacer: Sovereign Power and Bare Life*. Palo Alto: Stanford University Press.

Ahmed, Sara. (2004) 'Affective Economies'. *Social Text* 79(22.2): 117–139.

Amrute, Sareeta. (2016) *Encoding Race, Encoding Class: Indian IT Workers in Berlin*. Durham and London: Duke University Press.

Amrute, Sareeta. (2017) 'Press One for POTUS, Two for the German Chancellor: Humor, Race, and Rematerialization in the Indian Tech Diaspora. *HAU: Journal of Ethnographic Theory* 17(1): 327–352.

Amrute, Sareeta. (2019) 'Silicon Satire: Tech Trouble in the Valley'. *Anthropology News*, 28 February. Available at: https://www.anthropology-news.org/index.php/2019/02/28/silicon-satire/. Accessed 5 October 2019.

Atanasoski, Neda and K. Vora. (2015) 'Surrogate Humanity: Posthuman Networks and the (Racialized) Obsolescence of Labor'. *Catalyst: Feminism, Theory, Technoscience* 1(1): 1–40.

Barad, Karen. (2007) *Meeting the Universe Halfway: Quantum Physics and the Entanglement of Matter and Meaning*. Durham: Duke University Press.

Barnett, Fiona and Zach Blas. (2016) 'QueerOS: A User's Manual. Debates in the Digital Humanities'. Available at: http://dhdebates.gc.cuny.edu/debates/text/56. Accessed 15 May 2019.

Benjamin, Ruha. (2019) *Race After Technology*. Durham: Duke University Press.

Bunge, Mario. (1975) 'Towards a Technoethics'. *Philosophic Exchange* 6(1): 69–79.

Butler, Judith. (ed.) (2010) *Frames of War: When is Life Grievable?* New York: Verso.

Chakrabarty, Dipesh. (2002) Universalism and Belonging in the Logic of Capital. In *Cosmopolitanism*, eds Dipesh Chakrabarty, Homi K. Bhabha, Sheldon Pollock, and Carol A. Breckenridge, pp. 82–110. Durham, NC, USA: Duke University Press.

Chan, Anita. (2014) *Networking Peripheries: Technological Futures and the Myth of Digital Universalism*. Boston, MA, USA: MIT Press.

Chun, Wendy Hui Kyong. (2013) *Programmed Visions: Software and Memory*. Boston, MA, USA: MIT Press.

Citron, Danielle Keats and Frank Pasquale. (2014) 'The Scored Society: Due Process for Automated Decisions'. *Washington Law Review* 89(1): 1–33.

Daston, Lorraine. (2017) *Algorithms Before Computers: Patterns, Recipes, and Rules*. Katz Distinguished Lecture presented at University of Washington, 19 April. Seattle.

Data for Black Lives (2017) 'About Data for Black Lives. Data for Black Lives'. 4 June. Available at: https://d4bl.org/about.html. Accessed 4 October 2019.

Dave, Naisargi N. (2010) Between Queer Ethics and Sexual Morality. In *Ordinary Ethics: Anthropology, Language, and Action*, ed. M. Lambek, pp. 368–375. New York, NY, YSA: Fordham University Press.

Davies, Bronwyn. (1991) 'The Concept of Agency: A Feminist Poststructuralist Analysis'. *Social Analysis: The International Journal of Social and Cultural Practice* 30: 42–53.

Dunbar-Hester, Christina. (2016) 'Freedom from Jobs or Learning to Love to Labor? Diversity Advocacy and Working Imaginaries in Open Technology Projects'. *Technokultura* 13(2): 541–566.

Eubanks, Virginia. (2018) *Automating Inequality: How High-Tech Tools Profile, Police, and Punish the Poor*. New York: St. Martin's Press.

Fornet-Betancourt, Raul, Helumt Becker, Alfredo Gomez-Muller, and J.D. Gauthier. (1987) 'The Ethic of Care for the Self as a Practice of Freedom: An Interview with Michel Foucault. *Philosophy and Social Criticism* 12(2–3): 112–131.

Haraway, Donna. (2016) *Staying with the Trouble: Making Kin in the Chthulucene*. Durham: Duke University Press.

Harding, Sandra. (1992) 'Rethinking Standpoint Epistemology: What is "Strong Epistemology"?' *The Centennial Review* 36(3): 437–470.

Horvath, John. (1998) 'Freeware Capitalism'. *NetTime*, 5 February. Available at: http://nettime.org/Lists-Archives/nettime-l-9802/msg00026.html. Accessed 1 October 2019.

Hu, Tung-Hui. (2016) *A Prehistory of the Cloud*. Boston: MIT Press.

Irani, Lilly. (2018) 'Let's Cheer Workers at Google Who Are Holding Their Bosses to Account'. *New Scientist*, 5 December. Available at: https://www.newscientist.com/

article/mg24032074-800-lets-cheer-workers-at-google-who-are-holding-their-bossesto-account/. Accessed 1 October 2019.

Keeling, Kara. (2014) 'Queer OS'. *Cinema Journal* 53(2): 152–157.

Kochelman, Paul. (2013) 'The Anthropology of an Equation: Sieves, Spam Filters, Agentive Algorithms, and Ontologies of Transformation'. *HAU: Journal of Ethnographic Theory* 3(3): 33–61.

Marx, Karl. (1993) [1857/1858]. *Grundrisse: Foundations of the Critique of Political Economy*. London: Penguin.

Matthews, Michael R. (2003) 'Mario Bunge: Physicist and Philosopher'. *Science and Education* 12 (5–6): 431–444.

Mbembe, Achille. (2003) 'Necropolitics'. *Public Culture* 15(1): 11–40.

McGlotten, Shaka. (2016) [2014] 'Black Data'. *The Scholar and Feminist Online* 13(3). Available at: http://sfonline.barnard.edu/traversing-technologies/shaka-mcglotten-black-data/. Accessed 15 May 2019.

Milner, Yeshimabeit. (2013) 'A Call for Birth Justice in Miami. Miami: Powerful Women and Families and Power U Center for Social Change'. Available at: https://poweru.org/sdm_downloads/a-call-for-birth-justice-in-miami/. Accessed 15 May 2019.

Mina, An Xiao. (2014) 'Batman, Pandaman and the Blind Man: A Case Study in Social Change Memes and Internet Censorship in China'. *Journal of Visual Culture* 13(3): 259–275.

Noble, Safiya. (2017) *Algorithms of Oppression*. New York: New York University Press.

O'Neil, Cathy. (2016) *Weapons of Math Destruction: How Big Data Increases Inequality and Threatens Democracy*. New York: Crown.

Ramamurthy, Priti. (2003) 'Material consumers, Fabricating Subjects: Perplexity, Global Connectivity Discourses, and Transnational Feminist Discourses'. *Cultural Anthropology* 18(4): 524–550.

Researchers in Support of Google Employees (2018) 'Open Letter in Support of Google Employees and Tech Workers'. International Committee for Robot Arms Control (ICRAC), 3 July. Available at: https://www.icrac.net/open-letter-in-support-of-google-employees-and-tech-workers/. Accessed 15 May 2019.

Rosner, Daniela. (2018) *Critical Fabulations: Reworking the Methods and Margins of Design*. Boston, MA< USA: MIT Press.

Roy, Benjamin. (1995) 'The Tuskegee Syphilis Experiment: Medical Ethics, Constitutionalism, and Property in the Body'. *Harvard Journal of Minority Public Health* 1(1): 11–15.

Russell, Legacy. (2012) 'Cyborgology: Digital Dualism and the Glitch Feminism Manifesto. The Society Pages'. 10 December. Available at: https://thesocietypages.org/cyborgology/2012/12/10/digital-dualism-and-the-glitch-feminism-manifesto/. Accessed 1 October 2019.

Russert, Britt. (2019) Naturalizing Coercion: The Tuskegee Experiments and the Laboratory Life of the Plantation. In *Captivating Technology*, ed. Ruha Benjamin, pp. 25–49. Durham: Duke University Press.

Stewart, K. (2011) 'Atmospheric Attunements'. *Environment and Planning D: Society and Space* 29(3): 445–453.

Suchman, Lucy. (1987) *Plans and Situated Actions: The Problem of Human-Machine Communication*. New York: Cambridge University Press.

Suchman, Lucy. (2013) Feminist research at the digital/material boundary. *The Scholar and Feminist Online*. Available at: http://sfonline.barnard.edu/traversing-technologies/lucy-suchman-feminist-research-at-the-digitalmaterial-boundary/. Accessed 15 May 2019.

Suchman, Lucy. (2015) 'Situational Awareness: Deadly Bioconvergence at the Boundaries of Bodies and Machines'. *MediaTropes* 5(1): 1–24.

Wajcman, Judy. (2010) 'Feminist Theories of Technology'. *Cambridge Journal of Economics* 34(1): 143–152.

Weheliye, Alexander. (2014) *Habeas Viscus: Racializing Assemblages, Biopolitics, and Black Feminist Theories of the Human*. Durham: Duke University Press.

Zigon, Jarett. (2014) 'Attunement and Fidelity: Two Ontological Conditions for Morally Being-in-the-World'. *Ethos* 42(1): 16–30.

Zigon, Jarett. (2018) *Disappointment: Toward a Critical Hermeneutics of Worldbuilding*. New York: Fordham University Press.

Zyskowski, Kathryn. (2018) *Certifying India: Everyday Aspiration and Basic Computer Training in Hyderabad*. PhD. Seattle: Department of Anthropology, University of Washington.

12

The False Binary of Reason and Emotion in Data Visualisation

Catherine D'Ignazio and Lauren Klein

In 2012, twenty kindergarten children and six adults were shot and killed at an elementary school in Sandy Hook, CT, USA.[1] In the wake of this tragedy, and the weight of others like it, the design firm Periscopic started a new project—to visualise gun deaths in the United States. While there is no shortage of prior work in the form of bar charts or line graphs of deaths per year, Periscopic, a company whose tagline is 'do good with data', took a different approach (Periscopic).

When you load the webpage, you see a single, arcing line that reaches out over time. Then, the colour abruptly shifts from orange to white. A small dot drops down, and you see the phrase, 'Alexander Lipkins, killed at 29'. The arc continues to stretch across the screen, coming to rest on the *x*-axis, where you see a second phrase, 'could have lived to be 93'. Then, a second arc appears, displaying another arcing life. The animation speeds up over time, and the arcing lines increase, along with a counter that displays how many years of life have been 'stolen' from these gun victims. After a couple of (long) minutes, the visualisation moves through the entire year (2013), arriving at 11,419 people killed and 502,025 stolen years.

What is different about Periscopic's visualisation than a more conventional bar chart of similar information such as 'The era of "active shooters"' from *The Washington Post*? (Ingraham 2016). *The Post*'s graphic has a proposition—that active shooter incidents are on the rise—and demonstrates visual evidence to that effect. But Periscopic's work is framed around a singular emotion: loss. People are dying, their remaining time on earth has been stolen from them. These people have names and ages. We presume they have parents and partners and children who also suffer from that loss. The data scientists who worked on the project used rigorous statistical methods and demographic information to infer how long that person would have lived, which are documented in their notes. But in spite of its statistical rigor and undeniable emotional impact, 'U.S. Gun Deaths' drew mixed responses

---

[1] This chapter was adapted from *Data Feminism* (MIT Press, 2020). It was condensed and edited by Jessica Clark.

Catherine D'Ignazio and Lauren Klein, *The False Binary of Reason and Emotion in Data Visualisation*. In: *Feminist AI*. Edited by: Jude Browne, Stephen Cave, Eleanor Drage, and Kerry McInerney, Oxford University Press.
© Oxford University Press (2023). DOI: 10.1093/oso/9780192889898.003.0012

Figure 12.1 Decoding Possibilities by Ron Morrison with Treva Ellison.

From: https://elegantcollisions.com/decoding-possibilities/

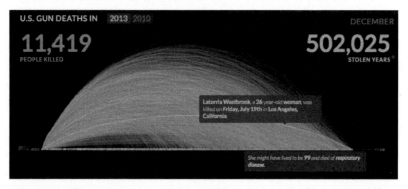

Figure 12.2 An animated visualization of the 'stolen years' of people killed by guns in the United States in 2013. The beginning state of the animation.

Image by Periscopic.

from the visualisation community. We could not decide: *should a visualisation evoke emotion?*

The received wisdom in technical communication circles is, emphatically, 'NO'. In the recent book, *A Unified Theory of Information Design,* the authors state: 'The plain style normally recommended for technical visuals is directed toward a deliberately neutral emotional field, a blank page in effect, upon which viewers are more free to choose their own response to the information' (Amare and Manning 2016). Here, plainness is equated with the absence of design, and thus greater freedom on the part of the viewer to interpret the results for themselves. Things

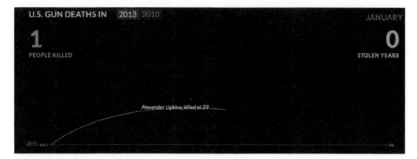

Figure 12.3 The end state.

Image by Periscopic.

The era of "active shooters"

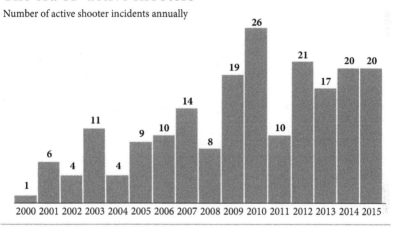

Figure 12.4 A bar chart of the number of 'active shooter' incidents in the United States between 2000 and 2015.

Images by Christopher Ingraham for the Washington Post.

such as colours and icons, it is implied, work only to stir up emotions and cloud the viewer's rational mind.

In fact, in the field of data visualisation, any kind of ornament has historically been viewed as suspect. Why? Well, as historian of science Theodore Porter puts it, 'quantification is a technology of distance' (Porter 1996, p.ix) and distance was historically imagined to serve objectivity by producing knowledge independently of the people that make it. This echoes nineteenth-century statistician Karl Pearson's exhortation for people to set aside their own feelings and emotions when it came to statistics. The more seemingly neutral, the more rational, the more true, the better (Gray 2019).

Data Visualisation, 'The Unempathetic Art'

At a data visualisation master class in 2013, workshop leaders from the *Guardian* newspaper called spreadsheet data—those endless columns and rows—'clarity without persuasion' (Crymble 2013).

Back in the olden days of visualisation, before the rise of the web elevated the visual display of data into a prized (and increasingly pervasive) artform, Edward Tufte, statistician and statistical graphics expert, invented a metric for measuring the superfluous information included in a chart—what he called the 'data-ink' ratio (Tufte, 2015). In his view, a visualisation designer should strive to use ink only to display the data. Any ink devoted to something other than the data itself—such as background colour, iconography, or embellishment—should be immediately erased and, he all but says, spat upon.

Visual minimalism, according to this logic, appeals to reason first ('Just the facts, ma'am', says the fictional police detective Joe Friday to every female character on the iconic US series *Dragnet*). Decorative elements, on the other hand, are associated with messy feelings—or, worse, represent stealthy (and, of course, unscientific) attempts at emotional persuasion. Data visualisation has even been classified as 'the unempathetic' art, in the words of designer Mushon Zer-Aviv, because of its emphatic rejection of emotion (Zer-Aviv 2015).

The belief that women are more emotional than men (and, by contrast, that men are more reasoned than women) is one of the most persistent stereotypes in the world today. Indeed, psychologists have called it the 'master stereotype', puzzling over how it endures even when certain emotions—even extreme ones, like anger and pride—are simultaneously coded as male (Shields, 2002).

But what happens if we let go of the binary logic for a minute and posit two questions to challenge this master stereotype? First, is visual minimalism really more neutral? And second, how might activating emotion—leveraging it, rather than resisting emotion in data visualisation—help us learn, remember, and communicate with data?

Crafting Data Stories from a Standpoint

Information visualisation has diverse historical roots. In its most recent, web-based incarnation, many of data visualisation's theorists and practitioners have come from technical disciplines aligned with engineering and computer science, and may not have been trained in the most fundamental of all Western communication theories: *rhetoric*.

In the ancient Greek treatise of the same name, Aristotle defines rhetoric as 'the faculty of observing in any given case the available means of persuasion' (Aristotle 1954). Rhetoric does not (only) consist of political speeches made by men in robes on ancient stages. Any communicating object that makes choices about the selection and representation of reality is a rhetorical object. Whether or not it is rhetorical (it always is) has nothing to do with whether or not it is 'true' (it may or may not be).

Why does the question of rhetoric matter? Well, because 'a rhetorical dimension is present in every design', as Jessica Hullman, a researcher at the University of Washington, says of data visualisation (Hullman and Diakopoulos 2011, pp.2231–2240). This includes visualisations that do not deliberately intend to persuade people of a certain message. We would say that it *especially and definitively* includes those so-called 'neutral' visualisations that do not appear to have an editorial hand. In fact, those might even be the most perniciously persuasive visualisations of all!

Editorial choices become most apparent when compared with alternative choices. For example, in his book *The Curious Journalist's Guide to Data*, journalist Jonathan Stray discusses a data story from the *New York Times* about the September 2012 jobs report (Stray 2016). The Times created two graphics from the report—one framed from the perspective of Democrats (the party in power at the time) and one framed from the perspective of Republicans.

Either of these graphics, considered in isolation, appears to be neutral and factual. The data are presented with standard methods (line chart and area chart respectively) and conventional positionings (time on the x-axis, rates expressed as percentages on the y-axis, title placed above the graphic). There is a high data-ink ratio in both cases, and very little in the way of ornamentation. But the graphics have significant editorial differences. The Democrats' graphic emphasises that unemployment is decreasing—in its title, the addition of the thick blue arrow pointing downwards, and the annotation 'Friday's drop was larger than expected'. Whereas, the Republicans' graphic highlights the fact that unemployment has been steadily high for the past three years—through the use of the '8% unemployment' reference line, the choice to use an area chart instead of a line, and, of course, the title of the graphic. So, neither graphic is neutral but both graphics are factual. As Jonathan Stray says, 'the constraints of truth leave a very wide space for interpretation' (Stray 2016). And, importantly, in data communication, it is impossible to avoid interpretation (unless you simply republish the September Jobs Report as your visualisation, but then it would not be a visualisation).

Hullman and co-author Nicholas Diakopoulous wrote an influential paper in 2011 introducing concepts of rhetoric to the information visualisation community (Hullman and Diakopoulous 2011). Their main argument is that visualising data involves editorial choices—some things are necessarily highlighted, while others are necessarily obscured. When designers make these choices, they carry along with them 'framing effects', which is to say they have an impact on how people interpret the graphics and what they take away from them.

For example, it is standard practice to cite the source of one's data. This functions on a practical level—so that a reader may go out and download the data

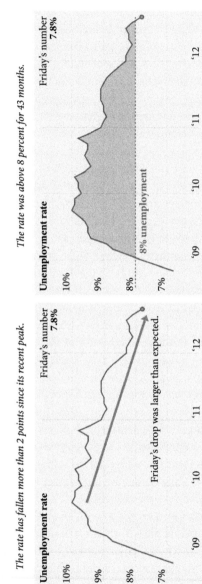

The rate has fallen more than 2 points since its recent peak.

The rate was above 8 percent for 43 months.

Figures 12.5 and 12.6 A data visualization of the September 2012 jobs report from the perspective of Democrats (12.5) and Republicans (12.6). The New York Times data team shows how simple editorial changes lead to large differences in framing and interpretation. As data journalist Jonathan Stray remarks on these graphics, 'The constraints of truth leave a very wide space for interpretation'.

Images by Mike Bostock, Shan Carter, Amanda Cox, and Kevin Quealy, for the New York Times, as cited in *The Curious Journalist's Guide to Data* by Jonathan Stray.

themselves. But this choice also functions as what Hullman and Diakopoulous call *provenance rhetoric* designed to signal the transparency and trustworthiness of the presentation source to end-users. This trust between the designers and their audience, in turn, increases the likelihood that viewers will believe what they see.

So, if plain, 'unemotional' visualisations are not neutral, but are actually extremely persuasive, then what does this mean for the concept of neutrality in general? Scientists and journalists are just some of the people that get nervous and defensive when questions about neutrality and objectivity come up. Auditors and accountants get nervous, too. They often assume that the only alternative to objectivity is a retreat into complete relativism, and a world in which everyone gets a medal for having an opinion. But there are other options.

Rather than valorising the neutrality ideal, and trying to expunge all human traces from a data product because of their 'bias', feminist philosophers have offered alternative paths towards truth. Sandra Harding would posit a different kind of objectivity that strives for truth *at the same time* that it considers—and discloses—the standpoint of the designer. This has come to be called 'standpoint theory'. It is defined by what Harding calls 'strong objectivity' which acknowledges that regular-grade, vanilla objectivity is mainly made by mostly rich white guys in power and does not include the experiences of women and other minoritised groups (Harding 1995, pp.331–349; also see Costanza-Chock, this volume and Hampton, this volume).

This myopia inherent in traditional 'objectivity' is what provoked renowned cardiologist Dr. Nieca Goldberg to title her book *Women Are Not Small Men*, because she found that heart disease in women unfolds in a fundamentally different way than in men (Goldberg 2002). The vast majority of scientific studies—not just of heart disease, but of most medical conditions—are conducted on male subjects, with women viewed as varying from this 'norm' only by their smaller size. Harding and her colleagues would say that the key to fixing this issue is to acknowledge that all science, and all work in the world, is undertaken by individuals, each with a particular standpoint: gender, race, culture, heritage, life experience, and so on. Indeed, our standpoints even affect whose work gets counted as 'work'. Scholars in this volume show how care labour has been consistently devalued and rendered invisible in the history and present of artificial intelligence (see Rhee, Chapter 10 and Amrute, Chapter 11, both in this volume).

Rather than viewing these standpoints as threats that might *bias* our work—for, after all, even the standpoint of a rich white guy in power is a standpoint—we should embrace each of our standpoints as valuable perspectives that can *frame* our work. Our diverse standpoints can generate creative and wholly new research questions.

Making Data Visceral

Along with this embrace of our various standpoints goes the rebalancing of the false binary between reason and emotion. Since the early 2000s, there has been an explosion of research about 'affect'—a term that academics use to refer to emotions and other subjective feelings—from fields as diverse as neuroscience, geography, and philosophy (see Amrute, Chapter 11 in this volume). This work challenges the thinking, inherited from René Descartes, which casts emotion as irrational and illegitimate, even as it undeniably influences all of the social and political processes of our world. Evelyn Fox Keller, a physicist-turned-philosopher, famously employed the Nobel-Prize-winning research of geneticist Barbara McClintock to show how even the most profound scientific discoveries are generated from a combination of experiment and insight, reason and emotion (Keller and McClintock 1984).

Once we embrace the idea of leveraging emotion in data visualisation, we can truly appreciate what sets Periscopic's Gun Deaths apart from *The Washington Post*'s graphic, or any number of other gun death charts that have appeared in newspapers and policy documents.

The graphic, created for The Washington Post by Christopher Ingraham, represents death counts as blue ticks on a generic bar chart. If we did not read the caption, we would not know whether we were counting gun deaths in the United States, or haystacks in Kansas, or exports from Malaysia, or any other semi-remote statistics of passing interest. But the *Periscopic* visualisation leads with loss, grief, and mourning. It provides a visual language for representing the years that could have been—numbers that are accurate, but not technically facts. It uses pacing and animation to help us appreciate the scale of one life, and then compounds that scale 11,419-fold. The magnitude of the loss, especially when viewed in aggregate, is a staggering and profound truth—and the visualisation helps recognise it as such through our own emotions. Note that emotion and visual minimalism are not incompatible—the Periscopic visualisation shows how emotion can be leveraged with visual minimalism for maximal effect.

Skilled data artists and designers know these things already, and are pushing the boundaries for what affective and embodied data visualisation could look like. In 2010, Kelly Dobson founded the Data Visceralization research group at the Rhode Island School of Design (RISD) Digital + Media Graduate programme. The goal for this group was not to visualise data but to *visceralise* it. Visual things are for the eyes, but visceralisations are data that the whole body can experience—emotionally, as well as physically.

The reasons for visceralising data have to do with more than simply creative experimentation. How do visually impaired people access charts and dashboards? According to the World Health Organization, 253 million people globally live with some form of visual impairment. This might include cataracts, glaucoma, and complete blindness (World Health Organization 2018 in Tufte 2015). In contrast to traditional navigation apps, Aimi Hamraie, director of the Mapping Access project at Vanderbilt University, advocates for apps with multisensory navigation. They write,

> Rather than relying entirely on visual representations of data, for example, digital-accessibility apps could expand access by incorporating 'deep mapping', or collecting and surfacing information in multiple sensory formats. Such a map would be able to show images of the doorway or integrate turn-by-turn navigation. Deeper digital-accessibility maps can offer both audio and visual descriptions of spatial coordinates, real-time information about maintenance or temporary barriers, street views, and even video recordings (Hamraie 2018).

Creators who work in the visceralisation mode have crafted haptic data visualisations, data walks, data quilts, musical scores from scientific data, wearable objects that capture your breaths and play them back later, and data performances. These types of objects and events are more likely to be found in the context of galleries and museums and research laboratories, but there are many lessons to be learned from them for those of us who make visualisations in more everyday settings.

For example, in the project *A Sort of Joy (Thousands of Exhausted Things)*, a theatre troupe joined with a data visualisation firm to craft a live performance based on metadata about the artworks held by New York's Museum of Modern Art. With 123,951 works in its collection, MoMA's metadata consists of the names of artists, the titles of artworks, their media formats, and their time periods. But how does an artwork make it into the museum collection to begin with? Major art museums and their collection policies have long been the focus of feminist critique, starting at least from Linda Nochlin's canonical 1971 essay 'Why Have There Been No Great Women Artists?', which drew attention to the patriarchal and colonial ideologies that shape collection practices. Whose work gets collected in museums and galleries? For our purposes, who is collected translates into who is counted in the annals of history—and, as you might guess, this history has mostly consisted of a parade of white male 'masters' (Nochlin 1971).

In 1989, for example, the Guerrilla Girls, an anonymous collective of female artists, published what we would today call an infographic: 'Do women have to be naked to get into the Met. Museum?' The graphic was designed to be displayed on a billboard. However, it was rejected by the sign company because it 'wasn't clear enough' (Chadwick 1995). (If you ask us, it is pretty clear).

The Guerrilla Girls then paid for it to be printed on posters that were displayed throughout the New York City bus system, until the bus company cancelled their contract, stating that the figure 'seemed to have more than a fan in her hand' (Chadwick 1995) (it is definitely more than a fan). The figure is certainly provocative, but the poster also makes a data-driven argument by tabulating gender

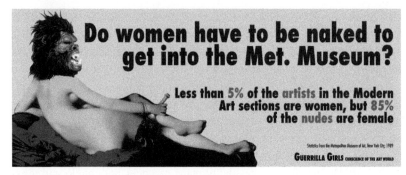

Figure 12.7 'Do Women Have to Be Naked to Get into the Met. Museum?' An infographic (of a sort) created by the Guerrilla Girls in 1989, intended to be displayed on a bus billboard.

Courtesy of the Guerrilla Girls.

statistics for artists included in the Met. collection, and comparing them to the gender stats for the *subjects* of art included in the collection. As per the poster, the Met. readily collects paintings in which women are the subjects, but not those in which women are the artists themselves.

A Sort of Joy deploys wholly different tactics to similar ends. The performance starts with a group of white men standing in a circle in the centre of the room. They face out towards the audience, which stands around them. The men are dressed like stereotypical museum visitors: wearing collared shirts and slacks. They all have headphones on and hold an iPad on which the names of artists in the collection scroll across the screen.

'John', the men say together. We see the iPads scrolling through all of the names of artists in the MoMA collection whose first name is John: John Baldessari, John Cage, John Lennon, John Waters, and so on. Three female performers, also wearing headphones and carrying iPads with scrolling names, pace around the circle of men. 'Robert', the men say together, and the names scroll through the Roberts alphabetically. The women are silent and keep walking. 'David', the men say together. It soon becomes apparent that the artists are sorted by first name, and then ordered by which first name has the most works in the collection. Thus, the Johns and Roberts and Davids come first, because they have the most works in the collection. But Marys have fewer works, and Mohameds and Camilas are barely in the register.

Several minutes later, after the men say 'Michael', 'James', 'George', 'Jean', 'Hans', 'Thomas', 'Walter', 'Edward', 'Yan', 'Joseph', 'Martin', 'Mark', 'José', 'Louis', 'Frank', 'Otto', 'Max', 'Steven', 'Jack', 'Henry', 'Henri', 'Alfred', 'Alexander', 'Carl', 'Andre', 'Harry', 'Roger', and 'Pierre', 'Mary' finally gets her due. It is spoken by the female performers; the first sound they have made.

For audience members, the experience starts as one of slight confusion. Why are there men in a circle? Why do they randomly speak someone's name? And what

Figure 12.8 A scene from *A Sort of Joy*.
Image courtesy of Jer Thorp.

are those women walking around so intently? But 'Mary' becomes a kind of a-ha moment—the same that researcher Robert Kosara says that data visualisation is so good at producing—when the highly gendered nature of the collection is revealed. From that point on, audience members start to listen differently, eagerly awaiting the next female name. It takes more than three minutes for 'Mary' to be spoken, and the next female name, 'Joan', does not come for a full minute longer. 'Barbara' follows immediately after that, and then the men return to reading, 'Werner', 'Tony', 'Marcel', 'Jonathan'.

From a data analysis perspective, *A Sort of Joy* consists of simple operations: only counting and grouping. The results could easily have been represented by a bar chart or a tree map of first names. But rendering the dataset as a time-based experience makes the audience wait and listen. It also runs counter to the mantra in information visualisation expressed by Ben Shneiderman in the mid-1990s: 'Overview first, zoom and filter, then details-on-demand' (Shneiderman 1996, pp.364–371).

Instead, in this data performance, we do not see 'the whole picture'. We hear and see and experience each datapoint one at a time. The different gender expressions, body movements, and verbal tones of the performers draw our collective attention to the issue of gender in the MoMA collection. We start to anticipate when the next female name will arise. We *feel* the gender differential, rather than *see* it.

This feeling is affect. It comprises the emotions that arise when experiencing the performance and the physiological reactions to the sounds and movements made by the performers, as well as the desires and drives that result—even if that drive is to walk into another room because the performance is disconcerting or just plain long.

Designing data visceralisations requires a much more holistic conception of the viewer. The viewer is not just a pair of eyes attached to a brain. They are a whole body—a complex, feeling one. A body located in space, with a history and a future. Visceralisation can be found in many current projects, even if the creators do not always describe their work in those terms. For example:

- Catherine—one of the authors of this book—and artist Andi Sutton led walking tours of the future coastline of Boston based on sea level rise (D'Ignazio and Sutton 2018).
- Lauren (the other author) and her team of Georgia Tech students are recreating Elizabeth Palmer Peabody's large-scale charts from the nineteenth century using touch sensors and individually addressable LEDs (Klein 2022).
- Mikhail Mansion made a leaning, bobbing chair that animatronically shifts based on real-time shifts in river currents (Mansion 2011).
- Teri Rueb stages 'sound encounters' between the geologic layers of a landscape and the human body that is affected by them (Reub 2007).
- Artist-researcher Ron Morrison obfuscates maps, and viewers have to put on special glasses called Racialized Space Reduction Lenses (RSRLs) to see beneath the maps (Morrison and Ellison 2017).
- Simon Elvins drew a giant paper map of pirate radio stations in London that you can actually listen to (Green 2006).
- Jessica Rajko, Jacqueline Wernimont, and Stjepan Rajko created a hand-crocheted net that vibrates in response to the data 'shed' by the cell phones of people walking by (Rajko et al. 2018).
- Tanya Aguiñigas created an installation of knotted textiles, inspired by the ancient Andean time-keeping technique of quipu, to call attention to the hours, days, and months spent by those seeking to cross the US–Mexico border (Aguiñigas 2021).
- A robot designed by Annina Rüst decorates real pies with pie charts about gender equality and then visitors eat them (Figure 12.9) (Rüst 2013).
- This list could go on.

While these projects may seem to be speaking to a different part of your brain than standard histograms or network maps do, there is something to be learned from the opportunities opened up by visceralising data. In fact, scientists are now proving by experiment what designers and artists have long known through practice: activating emotion, leveraging embodiment, and creating novel presentation forms help people learn more from data-driven arguments, and remember them more fully.

Figure 12.9 'A Piece of the Pie Chart' by Annina Rüst.
Image courtesy of LACMA.

Conclusion

The third principle of Data Feminism, and the theme of this essay, is to *elevate emotion and embodiment*. These are crucial, if often undervalued, tools in the data communication toolbox. How did the field of data communication arrive at conventions that prioritise rationality, devalue emotion, and completely ignore the non-seeing organs in the human body? Who is excluded when only vision is included?

Any knowledge community inevitably places certain things at the centre and casts others out, in the same way that male bodies have been taken as the norm in scientific study and female bodies imagined as deviations from the norm, or that abled bodies are the primary design case and disabled bodies are a retro-fit (see Keyes, Chapter 17 in this volume), or that rationality has been valued as an authoritative mode of communication while emotion is cast out. But, following feminist theorist Elizabeth Grosz, what is regarded as 'excess' in any given system might possibly be the most valuable thing to explore because it tells us the most about what *and who* the system is trying to exclude (Grosz 2001).

In the case of data visualisation, this excess is emotion and affect, embodiment and expression, embellishment and decoration. These are the aspects of human experience coded 'female', and thus devalued by the logic of our master stereotype. But Periscopic's gun violence visualisation shows how visual minimalism can co-exist with emotion for maximum impact. Works like *A Sort of Joy* demonstrate that data communication can be visceral—an experience for the whole body.

Rather than making universal rules and ratios (such as the data-ink ratio), that cast out some aspects of human experience in favour of others, our time is better spent working towards a more holistic, and more inclusive, ideal. All design fields, including visualisation and data communication, are fields of possibility. Sociologist Patricia Hill Collins describes an ideal knowledge situation as one in which 'neither ethics nor emotions are subordinated to reason' (Collins 2002, p.266). Rebalancing emotion and reason enlarges the data communication toolbox, and allows us to focus on what truly matters in a data design process: honouring context, gathering attention and taking action in service of rebalancing social and political power.

References

Aguiñiga, Tanya. (2021) 'Metabolizing the Border'. *Smarthistory*, 6 January, https://smarthistory.org/tanya-aguiniga-borderlands/.

Amare, Nicole and Alan Manning. (2016) *A Unified Theory of Information Design: Visuals, Text and Ethics*. New York: Routledge.

Aristotle. (1954) *The Rhetoric and the Poetics*. New York: Random House.

Chadwick, Whitney. (1995) Guerrilla Girls. *Confessions of the Guerilla Girls*. New York: Harper-Collins.

Collins, Patricia Hill. (2002) *Black Feminist Thought: Knowledge, Consciousness, and the Politics of Empowerment*. Routledge.

Crymble, Adam. (2013) 'The Two Data Visualization Skills Historians Lack'. *Thoughts on Public & Digital History* (blog), 13 March, http://adamcrymble.blogspot.com/2013/03/the-two-data-visualization-skills.html.

D'Ignazio, Catherine and Andi Sutton. (2019) 'Boston Coastline: Future Past'. kanarinka.com. Accessed 13 March, http://www.kanarinka.com/project/boston-coastline-future-past/.

Goldberg, Nieca. (2002) *Women Are Not Small Men: Life-Saving Strategies for Preventing and Healing Heart Disease in Women*. New York: Ballantine Books.

Gray, Jonathan. (2019) The Data Epic: Visualisation Practices for Narrating Life and Death at a Distance. In *Data Visualization in Society*, eds. H. Kennedy and M. Engebretsen. Amsterdam: Amsterdam University Press.

Green, Jo-Anne. (2006) 'Simon Elvins' Silent London'. *Networked_Music_Review* (blog), 11 July, http://archive.turbulence.org/networked_music_review/2006/07/11/simon-elvins-silent-london/.

Grosz, Elizabeth. (2001) *Architecture from the Outside: Essays on Virtual and Real Space*, pp.151–166. Cambridge, MA: MIT Press.

Hamraie, Aimi. (2018) 'A Smart City Is an Accessible City'. *Atlantic* 6 November, https://www.theatlantic.com/technology/archive/2018/11/city-apps-help-and-hinder-disability/574963/.

Harding, Sandra. (1995) '"Strong Objectivity": A Response to the New Objectivity Question'. *Synthese* 104(3): *Feminism and Science* (September) 331–349.

Hullman, Jessica and Nicole Diakopoulos. (2011) 'Visualization Rhetoric: Framing Effects in Narrative Visualization'. *IEEE Transactions on Visualization and Computer Graphics* 17(12): (December): 2231–2240.

Ingraham, Christopher. (2016) 'FBI: Active Shooter Incidents Have Soared Since 2000'. *The Washington Post*, 16 June. https://www.washingtonpost.com/news/wonk/wp/2016/06/16/fbi-active-shooter-incidents-have-soared-since–2000/

Rajko, Jessica, Jacqueline Wernimont, Eileen Standley, Stjepan Rajko, and Michael Krzyzaniak. (2018). Chapter 3: Vibrant Lives Presents The Living Net. In *Making Things and Drawing Boundaries*. Debates in the Digital Humanities; University of Minnesota Press. https://dhdebates.gc.cuny.edu/read/untitled-aa1769f2-6c55-485a-81af-ea82cce86966/section/1eed3831-c441-471a-9ab4-3a70afafddb5

Keller, Evelyn Fox and Barbara McClintock. (1984) *A Feeling for the Organism: The Life and Work of Barbara McClintock*, 10th anniversary edn. New York: Freeman.

Klein, Lauren. (2022) 'What Data Visualization Reveals: Elizabeth Palmer Peabody and the Work of Knowledge Production'. *Harvard Data Science Review* 4(2)(Spring): https://doi.org/10.1162/99608f92.5dec149c

Mansion, Mikhail. (2011) 'Two Rivers (2011)'. *Vimeo*, 1 October, https://vimeo.com/29885745.

Morrison, Ron and, Treva. Ellison (2017) *Decoding Possibilities*, multimedia installation, https://elegantcollisions.com/decoding-possibilities/.

Nochlin, Linda. (1971). 'Why Have There Been No Great Women Artists?' Retrieved 20 June, 2022, from https://www.writing.upenn.edu/library/Nochlin-Linda_Why-Have-There-Been-No-Great-Women-Artists.pdf

Periscopic. (n.d.) 'Do Good with Data'. Periscopic. https://periscopic.com/#!/.

Porter, Theodore M. (1996) *Trust in Numbers: The Pursuit of Objectivity in Science and Public Life*. Princeton, NJ, USA: Princeton University Press.

Reub, Teri. (2007) 'Core Sample—2007'. *Teri Rueb* (blog). Accessed 13 March 2019, http://terirueb.net/core-sample–2007/.

Rüst, Annina. (2013) 'A Piece of the Pie Chart'. Accessed 13 March 2019, http://www.anninaruest.com/pie/.

Shields, Stephanie A. (2002) *Speaking from the Heart: Gender and the Social Meaning of Emotion*. Cambridge: Cambridge University Press.

Shneiderman, Ben. (1996) The Eyes Have It: A Task by Data Type Taxonomy for Information Visualizations. In The Craft of Information Visualization: Readings and Reflections, eds Benjamin B. Bederson and Ben Shneiderman, pp. 364–371, September. Burlington: Morgan Kaufmann.

Stray, Jonathan. (2016) *The Curious Journalist's Guide to Data*. New York: Columbia Journalism School, https://legacy.gitbook.com/book/towcenter/curious-journalist-s-guide-to-data/details.

Tufte, Edward R. (2015) *The Visual Display of Quantitative Information*, 2nd edn. Cheshire, CT, USA: Graphics Press. 'Vision Impairment and Blindness'. World Health Organization, 11 October 2018, http://www.who.int/news-room/fact-sheets/detail/blindness-and-visual-impairment.

Zer-Aviv, Mushon. (2015) 'DataViz—The UnEmpathic Art'. 19 October, https://responsibledata.io/dataviz-the-unempathic-art/.

13

Physiognomy in the Age of AI

Blaise Agüera y Arcas, Margaret Mitchell, and Alexander Todorov

Introduction

In this chapter, we'll explore two instances in which machine learning systems were created to categorise people based on a facial image.[1] The first purports to determine whether the subject is a criminal or not (Wu and Zhang 2016), and the second whether the subject is lesbian, gay, or straight (Wang and Kosinski 2018).

Such findings are in the tradition of physiognomy (see Figure 13.1), the pseudoscientific belief that a person's appearance reveals their essential nature—and their value to society. Our reflections on the pseudoscientific use of AI to sort and classify people according to their external appearance demonstrate the continued importance of feminist studies of science, a field which shows how patriarchal power often operates under the guise of scientific 'objectivity' (see D'Ignazio and Klein, Chapter 12 in this volume). Like Michele Elam (Chapter 14 in this volume), we point out the misuse of AI systems to classify people erroneously and arbitrarily, highlighting the harmful effects these practices are likely to have on marginalised communities.

In the first part of this chapter, we will review this legacy, showing how the medical, legal, and scientific patriarchy of the nineteenth and twentieth centuries used (and at times developed) state of the art techniques to both rationalise and enforce a hierarchy with prosperous straight white men at the top. This rationalisation relied on correlating physical measurements of the body with 'criminality' and sexual orientation.

In the second and third parts of the chapter, we will turn our attention to the criminality and sexual orientation machine learning papers. The authors of these papers were aware of the troubling historical legacy of physiognomy,[2] but believed

---

[1] Some of the material in this chapter appeared in 'Physiognomy's New Clothes' (https://medium.com/@blaisea/physiognomys-new-clothes-f2d4b59fdd6a) and 'Do algorithms reveal sexual orientation or just expose our stereotypes?' (https://medium.com/@blaisea/do-algorithms-reveal-sexual-orientation-or-just-expose-our-stereotypes-d998fafdf477).

[2] Wu and Zhang note the 'high social sensitivities and repercussions of our topic and skeptics on physiognomy' (2016, p.4). Per Wang and Kosinski, 'Physiognomy is now universally, and rightly, rejected as a mix of superstition and racism disguised as science [. . .]. Due to its legacy, studying or even discussing the links between facial features and character became taboo, leading to a widespread

Blaise Agüera y Arcas, Margaret Mitchell, and Alexander Todorov, *Physiognomy in the Age of AI*. In: *Feminist AI*. Edited by: Jude Browne, Stephen Cave, Eleanor Drage, and Kerry McInerney, Oxford University Press.

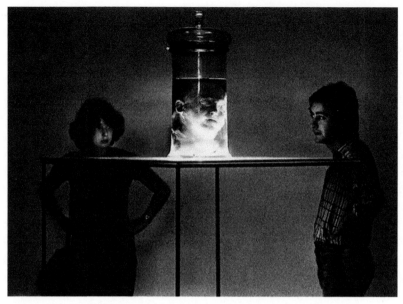

Figure 13.1 A couple viewing the head of Italian criminologist Cesare Lombroso preserved in a jar of formalin at an exhibition in Bologna, 1978.

that it was worth revisiting these old theories with newly available machine learning powered face recognition to see whether there was a 'kernel of truth' in them (Berry 1990; Penton-Voak et al. 2006). They found, they believed, more than a kernel: both papers claimed remarkable performance, reinforcing both the old physiognomy narrative and the newer illusory belief that AI with superhuman (yet opaque) capabilities can make the impossible possible.

If we accepted the researchers' claims, we would have a real problem on our hands, as the underlying technologies are readily available, and it is hard to imagine legitimate uses for either of these systems. Facial 'criminality' and sexual orientation detectors seem tailor-made for institutionalised oppression. Yet as we'll see, neither technology is likely to perform as advertised. The findings, while remarkable, do not appear to measure essential human characteristics as the researchers think, but rather the social presentations people assume to broadcast their identities (in the case of sexual orientation), or the self-fulfilling power of stereotypes to fuel prejudice (in the case of criminality). As with the older physiognomy literature, the result is not only bad science, but dangerous pseudoscience: first, because it reinforces misinformed beliefs on the part of the public, even if the technology remains purely notional; second, because those motivated to deploy such systems would likely do so in the service of human rights abuses; and third,

presumption that no such links exist. However, there are many demonstrated mechanisms that imply the opposite' (2018, p.246).

because any deployment at scale would result in both entrenchment and amplification of existing bias and injustice. In this way, circular reasoning becomes a vicious cycle.

Circular Logic

In nineteenth-century England, women were commonly held to be incapable of pursuing higher intellectual activities, to the extent that medical professionals believed that it was medically dangerous for women to enter higher education (Steinbach 2004). Hence, advanced mathematics was men's work. Even the relatively progressive Augustus De Morgan, math tutor to Ada Lovelace, expressed his scepticism that a woman could ever make a real contribution to the field in a cautionary letter to Lady Byron, Ada's mother, in 1844: '[T]he very great tension of mind which [wrestling with mathematical difficulties requires] is beyond the strength of a woman's physical power of application' (Huskey and Huskey 1980). Astonishingly, this letter was written shortly after Lovelace's publication on the Analytical Engine, a masterwork of mathematical creativity and arguably the founding document of the entire field of computer science.

Between presumptions of inability and lack of access to higher education, it was exceedingly difficult for Victorian women to break out of this cycle; yet the near-absence of women doing higher math was the very evidence that they were not up to it. When, against all odds, they still managed to excel, as Lovelace did, their successes went unacknowledged.

Although over the past century we have seen many more first-rate female mathematicians, and the situation has in many ways improved, we are still struggling with this kind of confirmation bias and the legacy of gender essentialism in science, technology, engineering, and mathematics fields. As Judy Wajcman and Erin Young discuss further in this volume (Chapter 4), the situation is especially dire in computer science (Wang et al. 2021)—a bitter irony given the central role of women in establishing the field, from Lovelace onwards.

While feminism has myriad definitions, encompassing many movements and perspectives, one of its enduring practical concerns has always been to break this vicious cycle based on circular logic, with its confusion between correlation and causality. This concern animates all social justice movements. It will be our central theme here.

Essentialism and AI

At the core of the problem is essentialism, the mistaken idea that people have an immutable core or essence that fully determines both appearance and behaviour. In modern times, genes often play the role of essence, which in earlier periods took on a more philosophical or even mystical character. In his 1981 book *The Mismeasure of Man*, Stephen Jay Gould wrote about the stubborn way essentialism continues to colour our thinking:

> We have been unable to escape the philosophical tradition that what we can see and measure in the world is merely the superficial and imperfect representation of an underlying reality. [...] The technique of correlation has been particularly subject to such misuse because it seems to provide a path for inferences about causality (and indeed it does, sometimes—but only sometimes) (p.269).

Learning correlations using large amounts of data is precisely what modern AI does. This enables automatic categorisation at industrial scale, which has many useful applications.

However, today's mainstream AI systems are unable to reason, question the validity of labelled inputs, consider the larger context, or establish causality. This remains the researcher's job. When AI researchers deploy models to analyse human beings without bringing the needed rigour to the task, they tend to import essentialist fallacies into their work. In much of the faulty scholarship in this area, it is not that correlations cannot be found, but that they confuse correlations with causal interpretations, along with beliefs that fixed categories underlie them (see Elam, Chapter 14 this volume).

This problem is not new. There is an extensive history of creating different human 'types', associating these 'types' with physical appearance, and then using appearance to classify people. This is the practice of physiognomy.

A Brief History of Physiognomy

The propensity to interpret a person's appearance associatively or metaphorically dates back thousands of years. It is the central conceit of Renaissance polymath Giambattista della Porta's 1586 treatise *De humana physiognomonia*, which makes the case that an owlish-looking person is owlish, a piggish-looking person is piggish, and so on (see Figure 13.2).

Criminality and 'Scientific' Racism

To make such ideas respectable in the Enlightenment, it was necessary to set aside poetic metaphors and concentrate on more specific physical and behavioural features. In the 1700s, the Swiss theologian Johann Caspar Lavater attempted to analyse character based on the shape and positions of the eyes, brows, mouth,

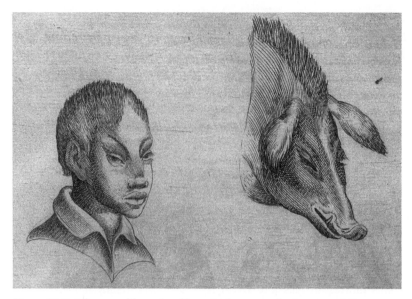

Figure 13.2 Like man, like swine (from *De humana physiognomonia*).

and nose to determine whether a person was, among other characteristics, 'deceitful', 'full of malice', 'incurably stupid', or a 'madman' (Luffman 1802). In this vein, Victorian polymath Francis Galton (1822–1911), a pioneer of the statistical concept of correlation, tried to visually characterise 'criminal types' by superimposing exposures of convicts on the same photographic plate (see Figure 13.3) (Galton 1878).

Around the same time, Cesare Lombroso, a professor and advocate of physiognomy, undertook research at an asylum in northern Italy to take this form of measurement further, and in doing so founded the field of 'scientific' criminology. By his account, Lombroso experienced the epiphany that would define the course of his subsequent career during an examination of the remains of a labourer from Calabria, Giuseppe Villella. Villella had reportedly been convicted of being a *brigante* (bandit), at a time when brigandage—banditry and state insurrection—was seen as endemic. Villella's remains supplied Lombroso with 'evidence' confirming his belief that *brigantes* were a primitive or 'degenerate' type of people, prone to crime: a depression on the occiput of the skull reminiscent of the skulls of 'savages and apes'.

Criminals, Lombroso wrote in his influential 1876 book *Criminal Man*, were 'born criminals'. He held that criminality is inherited, and carries with it inherited physical characteristics that can be measured with instruments like calipers and craniographs. This belief conveniently justified his a priori assumption that southern Italians were racially inferior to northern Italians.

Figure 13.3 Francis Galton's attempts to use multiple exposure portraiture to reconstruct 'criminal types'. From *Inquiries Into Human Faculty and Its Development* (1883).

While Lombroso can be credited as one of the first to attempt to systematically study relationships between the mind and criminal behaviour and to advance the science of forensics, he can also be credited as one of the first to use modern science to lend authority to his own stereotypes about lesser 'types' of human. Scientific rigour tends to weed out incorrect hypotheses given time, peer review, and iteration; but using scientific language and measurement does not prevent a researcher from conducting flawed experiments and drawing wrong conclusions—especially when they confirm preconceptions.

The beliefs Lombroso appears to have harboured with respect to people in the South of Italy suggested a racial hierarchy with political implications, but nineteenth-century American physiognomists had even more compelling reasons to rationalise such a hierarchy: they were slave-owners. The Philadelphia-born physician and natural scientist Samuel Morton used cranial measurements and ethnological arguments to make a case for white supremacy; as his followers Josiah Nott and George Gliddon quoted in their 1854 tribute, *Types of Mankind,*

> Intelligence, activity, ambition, progression, high anatomical development, characterize some races; stupidity, indolence, immobility, savagism, low anatomical development distinguish others. Lofty civilization, in all cases, has been achieved solely by the 'Caucasian' group (p.461).

Despite this book's scholarly pretensions, its illustrations (typical of the period) reveal the same kind of fanciful visual 'reasoning' and animal analogies evident in della Porta's treatise, albeit even more offensively (see Figure 13.4).

Later in the nineteenth century, Darwinian evolutionary theory refuted the argument made in *Types of Mankind* that the races are so different that they must have been created separately by God. However, by making it clear that humans *are* in fact animals, and moreover are closely related to the other great apes, it allowed Morton's discrete racial hierarchy to be reimagined in shades of grey, differentiating humans who are 'more human' (more evolved, physically, intellectually, and behaviourally) and 'less human' (less evolved, physically closer to the other great apes, less intelligent, and less 'civilized'). Darwin wrote in his 1871 book *The Descent of Man*:

> [...] man bears in his bodily structure clear traces of his descent from some lower form; [...] [n]or is the difference slight in moral disposition between a barbarian [. . .] and a Howard or Clarkson; and in intellect, between a savage who does not use any abstract terms, and a Newton or Shakspeare. Differences of this kind between the highest men of the highest races and the lowest savages, are connected by the finest gradations (p.35).

Unsurprisingly, Darwin's apex of humanity is exemplified by the physicist Isaac Newton, the playwright William Shakespeare, the abolitionist Thomas Clarkson, and the philanthropist John Howard: all were English, Christian, white, male, and from the educated classes—that is, much like Darwin himself. Darwin's views were in step with (and, in some ways, more progressive than) those of his peers; they also illustrate homophily, the pervasive cognitive bias causing people to identify with and prefer people similar to themselves.

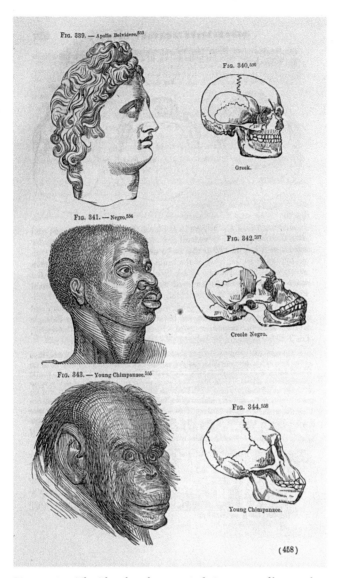

Figure 13.4 The idea that there are inferior types of human has historically been linked to the scientifically invalid idea that some humans are more like animals than others. From *Types of Mankind*.

The Physiognomy of Sexual 'Perversion'

The Victorian theory of human 'types' did not confine itself to race. In creating a pseudoscientific framework to justify classifying people into a hierarchy based on their appearance, it also offered a way to rationalise sexism, classism, homophobia, and transphobia.

Pioneering sexologist Richard von Krafft-Ebing, profoundly influenced by Lombroso,[3] made this analogical leap. His widely read 1886 book *Psychopathia Sexualis*,[4] full of lurid case studies but larded with Latin to maintain a scholarly veneer, introduced English readers to many new terms, including 'heterosexuality', 'homosexuality', 'bisexuality', 'sadism', and 'masochism'. Like most of his contemporaries, Krafft-Ebing deemed any sexual practice or desire that wasn't at least potentially procreative a 'perversion'. He considered homosexuality, in particular, 'a functional sign of degeneration, and [. . .] a partial manifestation of a neuro-psychopathic state, in most cases hereditary' (p.225). In other words, just as 'criminality' was deemed evidence of being degenerate or primitive, so was being lesbian, gay, trans, or anything else Krafft-Ebing regarded as 'deviant'. Given the draconian 'sodomy laws' in force throughout Europe at the time, LGBT people often *were* by definition criminals—another instance of circular reasoning.

Many case studies in *Psychopathia Sexualis* vividly illustrate these points. They also make for difficult reading today, for example the story of 'Count Sandor V', who was ultimately arrested for being 'no man at all, but a woman in male attire':

Among many foolish things that her father encouraged in her was the fact that he brought her up as a boy, called her Sandor, allowed her to ride, drive, and hunt, admiring her muscular energy. [. . .] At thirteen she had a love-relation with an English girl, to whom she represented herself as a boy, and ran away with her (p.311).

After a youth in which she 'became independent, and visited cafés, even those of doubtful character', 'carried on literary work, and was a valued collaborator on two noted journals', she found:

A new love [. . .] Marie, and her love was returned...The pair lived happily, and, without the interference of the step-father, this false marriage, probably, would have lasted much longer (p.312).

Upon being outed, however, 'Count S'. fell victim to the legal, psychiatric, and medical institutions of the day. A lengthy, chillingly invasive examination ensues, much redacted here:

She is 153 centimetres tall, of delicate skeleton, thin, but remarkably muscular on the breast and thighs. Her gait in female attire is awkward. Her movements are powerful, not unpleasing, though they are somewhat masculine, and lacking in grace. [. . .] Feet and hands remarkably small, having remained in an infantile

[3] *Psychopathia Sexualis* cites Lombroso's work 25 times.
[4] Quotes are from Krafft-Ebing, R. von (1893).

stage of development. [...] The skull is slightly oxycephalic, and in all its measurements falls below the average of the female skull by at least one centimetre [...] Genitals completely feminine, without trace of hermaphroditic appearance, but at the stage of development of those of a ten-year-old girl. [...] The pelvis appears generally narrowed (dwarf-pelvis), and of decidedly masculine type. [. . .] The opinion given showed that in S. there was a congenitally abnormal inversion of the sexual instinct, which, indeed, expressed itself, anthropologically, in anomalies of development of the body, depending upon great hereditary taint; further, that the criminal acts of S. had their foundation in her abnormal and irresistible sexuality. (p.316)

Had they been stripped, prodded, and scrutinised in a similarly clinical light, perhaps the doctors' own bodies might have been found wanting. It is the prerogative of the powerful to do the examining, though, and the fate of the powerless to be examined (Sontag 2004; Osucha 2009).

Applied Physiognomy in the Twentieth and Twenty-First Centuries

The rationalisation of hierarchies based on heritable physical and behavioural traits, and institutionalised application of the resulting theory of physiognomic 'types', thrived far into the twentieth century. It animated Nazi era 'race science' and the policies of Apartheid in South Africa. It also led to the passing of eugenic legislation in over 30 US states mandating the forced sterilisation of people the legal system judged to 'belong to the class known as degenerates', per a bulletin from the Eugenics Record Office in Cold Spring Harbor (Laughlin 1914, p.62). In keeping with Lombroso and Krafft-Ebing, such 'degenerate types' included the 'insane, epileptic, imbecile, idiotic, sexual perverts, [...] confirmed inebriates, prostitutes, tramps, and criminals, as well as habitual paupers, found in our country poor-asylums, also many of the children in our orphan homes' (*Ibid.*). As in Nazi Germany, the American sterilisation laws had strong nationalist overtones, for allowing the reproduction of 'degenerate' people to 'go on unchecked' would lead to 'a weakening of our nation'. Many of these laws were still enforced in the late 1960s (Kendregan 1966).

Despite the social and scientific progress of the past half-century, physiognomy endures in the present day. Contemporary American pickup artist and white nationalist James Weidmann, for example, blogged in support of physiognomy in June 2016:[5]

---

[5] http://archive.is/eEFA7

There's evidence (re)emerging [...] that a person's looks do say something about his politics, smarts, personality, and even his propensity to crime. Stereotypes don't materialize out of thin air, and the historical wisdom that one can divine the measure of a man (or a woman) by the cut of his face has empirical support. [...] You CAN judge a book by its cover: ugly people are more crime-prone. [...] Physiognomy is real. It needs to come back as a legitimate field of scientific inquiry [...].

A few months later, a paper posted to arXiv (a popular online repository for physics and machine learning researchers) by Xiaolin Wu and Xi Zhang, *Automated Inference on Criminality Using Face Images* (2016) purported to do just that. The authors were excited by the prospects for 'social psychology, management science, [and] criminology' (p.1).

AI Criminology

Wu and Zhang's central claim is that machine learning techniques can predict the likelihood that a person was a convicted criminal with nearly 90 percent accuracy using nothing but a driver's licence-style face photo; thus, they claim to 'produce evidence for the validity of automated face-induced inference on criminality' for the first time (*Ibid.*). Further, they claim to be 'the first to study automated face-induced inference on criminality free of any biases of subjective judgments of human observers'. These claims to objectivity are misleading, as we will see (p.8).

They begin with a set of 1856 closely cropped, 80×80 pixel images of Chinese men's faces from government-issued IDs. The men are all between 18 and 55 years old, lack facial hair, and lack facial scars or other obvious markings. 730 of the images are labelled 'criminals', either wanted for or convicted of a range of both violent (235) and non-violent (536) crimes by provincial and city authorities. The other 1126 face images are of 'non-criminals [...] acquired from Internet using [a] web spider tool' (p.3).

Wu and Zhang then train their algorithm to look at a face image and produce a yes/no answer: did this image come from the 'criminals' group or the 'non-criminals' group? They try out four different machine learning techniques of varying sophistication. One of the less sophisticated techniques involves preprocessing the images with custom code to extract the locations of specific known facial features, like the corners of the eyes and the mouth, then using older methods to learn patterns relating the positions of these facial features. The authors also try a convolutional neural net (CNN) capable of learning arbitrary patterns. The CNN is the strongest performer, achieving a classification accuracy of nearly 90 percent. Its 'false alarm rate' (falsely classifying a 'non-criminal' as a 'criminal') is just over 6 percent—comparable to a workplace drug test (Bates 2010). Even the simpler methods, though, have accuracies well above 75 percent.

What is the Machine Learning Picking Up On?

What specific features distinguish purportedly 'criminal' faces? Wu and Zhang are able to explore this in detail using the simpler machine learning approaches that involve measuring relationships between standard facial landmarks: '[...] the angle θ from nose tip to two mouth corners is on average 19.6% smaller for criminals than for non-criminals and has a larger variance. Also, the upper lip curvature ρ is on average 23.4% larger for criminals than for noncriminals' (p.6).

We may be able to get an intuitive sense of what this means by comparing the top row of 'criminal' examples with the bottom row of 'non-criminal' examples, shown in the paper's first figure (reproduced in Figure 13.5).

Although the authors claim to have controlled for facial expression, the three bottom images all appear to be smiling slightly, while the top ones appear to be frowning. If these six images are indeed typical, we suspect that asking a human judge to sort the images in order from smiling to frowning would also do a fairly effective job of segregating purportedly 'non-criminal' from 'criminal'. Is there a relationship, then, between perceived facial expression and 'criminality'?

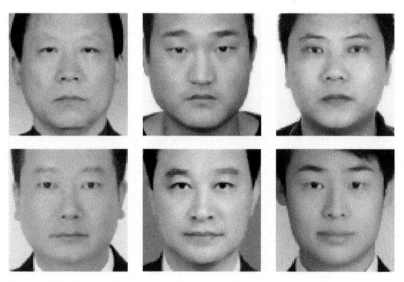

Figure 13.5 Wu and Zhang's sample 'criminal' images (top) and 'non-criminal' images (bottom).

What do Humans Pick Up On?

Wu and Zhang do not claim that their machine learning techniques are recognising subtler facial cues than people can discern without any help from computers. On the contrary, they connect their work to a 2011 study published in a psychology journal (Valla et al. 2011) that arrives at the same conclusion using human judgement. Wu and Zhang also relate their work to a 2014 paper (Cogsdill et al. 2014) (co-authored by one of us) revealing that even 3- and 4-year olds can reliably distinguish 'nice' from 'mean' face images; critically, though, the paper is about the acquisition of facial stereotypes early in development. No claim is made that these impressions correspond to a person's character.

What do supposedly 'nice' and 'mean' faces look like? Figure 13.6 illustrates a typical physiognomist's answer. Various methods have been developed to visualise the facial stereotypes that map onto these dimensions. In one, participants rate randomly generated synthetic faces on traits like trustworthiness and dominance. This allows a visualisation of the average features representing a 'trustworthy' or 'untrustworthy' face, shown (for white males) in Figure 13.7.

It is likely no coincidence that the three 'criminal' faces shown in Figure 13.5 are more similar to the 'untrustworthy' face in Figure 13.7, and the 'non-criminal' faces are more similar to the 'trustworthy' face. Notice, too, the way Galton's composites of 'criminal types' (Figure 13.3) look more like the 'untrustworthy' face. Researchers who have studied the social perception of faces, some of whom are cited by Wu and Zhang, have shown that people form character impressions such as trustworthiness from facial appearance after seeing a face for less than one-tenth of a second. Furthermore, these impressions predict important social outcomes (Olivola et al. 2014), ranging from political elections to economic transactions to legal decisions.

While we form impressions of strangers almost reflexively from facial appearance, this does not imply that these impressions are accurate; a large body of research suggests otherwise (Todorov 2017; Todorov et al. 2015). For example, in 2015 Brian Holtz of Temple University published the results of a series of experiments (Holtz 2015) in which face 'trustworthiness' was shown to strongly influence experimental participants' judgement. The participants were asked to decide, after reading an extended vignette, whether a hypothetical CEO's actions were fair or unfair. While the judgement varied (as one would hope) depending on how fair or unfair the actions described in the vignette were, it also varied depending on whether a 'trustworthy' or 'untrustworthy' face (i.e. more like the left or right face in Figure 13.7) was used in the CEO's profile photo. In another study (Rezlescu et al. 2012), participants played an online investment game with what they believed were real partners represented by 'trustworthy' or 'untrustworthy' faces. Participants were more likely to invest in 'trustworthy' partners even in the presence of reputational information about the past investment behaviour of their

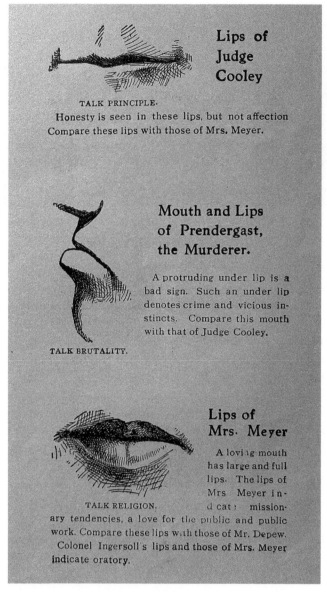

Lips of Judge Cooley

TALK PRINCIPLE.

Honesty is seen in these lips, but not affection
Compare these lips with those of Mrs. Meyer.

Mouth and Lips of Prendergast, the Murderer.

A protruding under lip is a bad sign. Such an under lip denotes crime and vicious instincts. Compare this mouth with that of Judge Cooley.

TALK BRUTALITY.

Lips of Mrs. Meyer

A loving mouth has large and full lips. The lips of Mrs Meyer indicate missionary tendencies, a love for the public and public work. Compare these lips with those of Mr. Depew. Colonel Ingersoll's lips and those of Mrs. Meyer indicate oratory.

TALK RELIGION.

Figure 13.6 Principled, murderous, and loving lips, from V.G. Rocine, *Heads, Faces, Types, Races.*

partners. Yet more chillingly, another study (Wilson and Rule 2015) found that among prisoners convicted for first degree murder, the unlucky ones with 'untrustworthy' faces were disproportionately more likely to be sentenced to death than to life imprisonment. This was also the case for people who were falsely accused and subsequently exonerated.

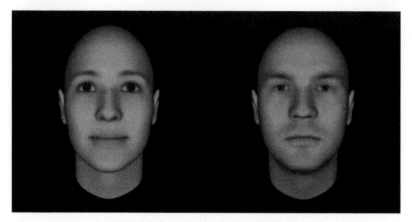

Figure 13.7 Perceived average male 'trustworthy' (left) and 'untrustworthy' (right) synthetic faces, according to Princeton undergraduates.

The consequences are unfortunate for anyone who happens to have an 'untrustworthy' appearance. It is also unfortunate that, rather than finding an efficient and impartial shortcut to making accurate criminal judgements with a computer (perhaps a misguided goal in any case), what Wu and Zhang's experiment likely reveals is the inaccuracy and systematic unfairness of many human judgements, including official ones made in a criminal justice context.

From Research to Production

An Israeli startup founded in 2014, Faception, has built similar models, and has taken the logical next step, though they have not published any details about their methods, sources of training data, or quantitative results:

> Faception is first-to-technology and first-to-market with proprietary computer vision and machine learning technology for profiling people and revealing their personality based only on their facial image.[6]

The Faception team aren't shy about promoting applications of their technology, offering specialised face recognition engines for detecting 'High IQ', 'White-Collar Offender', 'Pedophile', 'Terrorist', and even, weirdly, 'Bingo Player'. Their main clients are in homeland security and public safety, suggesting that there are surveillance cameras in public places today being used to profile people using such

---

[6] https://www.faception.com/our-technology, retrieved on 26 June 2022.

categories. Faception is betting that once again governments will be keen to 'judge a book by its cover'. As they are still in business in 2022, their bet seems sound.

AI Gaydar

In the autumn of 2017, a year after the Wu and Zhang paper came out, a higher-profile study claiming that artificial intelligence can infer sexual orientation from facial images caused a media uproar. *The Economist* featured this work on the cover of their 9th September magazine; on the other hand two major LGBTQ organisations, The Human Rights Campaign and Gay & Lesbian Alliance Against Defamation (GLAAD), immediately labelled it 'junk science' (Levin 2017). Michal Kosinski, who co-authored the study with fellow researcher Yilun Wang, initially expressed surprise, calling the critiques 'knee-jerk' reactions (Brammer 2017). However, he then proceeded to make even bolder claims: that such AI algorithms will soon be able to measure the intelligence, political orientation, and criminal inclinations of people from their facial images alone (Levin 2017).

Once again, this echoes the claims of Cesare Lombroso and Richard von Krafft-Ebing: physiognomy, now dressed in the 'new clothes' of machine learning and AI. Much of the ensuing scrutiny focused on the potential misuse of this technology, implicitly assuming that the tool actually worked and that the underlying science was valid. As with the previous year's result, however, there is reason to believe otherwise.

The Fallacy of Gay and Lesbian 'Types'

The authors trained and tested their 'sexual orientation detector' using 35,326 images from public profiles on a US dating website. Composite images of the lesbian, gay, and straight men and women in the sample[7] reproduced here in Figure 13.8, reminiscent of Galton's composites, reveal a great deal about the information available to the algorithm.

Wang and Kosinski assert that the key differences between these composite faces are in physiognomy, meaning that a sexual orientation tends to go along with a characteristic facial structure. However, we can immediately see that some of these differences are more superficial. For example, the 'average' straight woman appears to wear eyeshadow, while the 'average' lesbian does not. Glasses are clearly visible on the gay man, and to a lesser extent on the lesbian, while they seem absent in the heterosexual composites. Might it be the case that the algorithm's ability to

---

[7] Clearly this is an incomplete accounting of sexual orientations, as well as presuming a gender binary.

Composite heterosexual faces   Composite gay faces

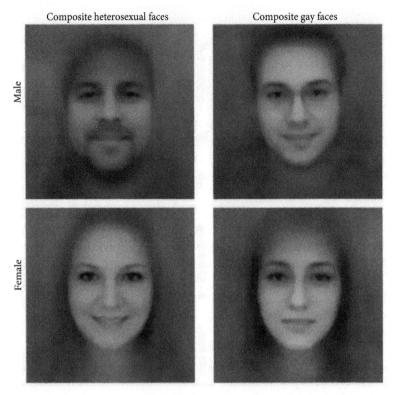

Figure 13.8 The average selfies of lesbian, gay, and straight men and women.

detect orientation has little to do with facial structure, but is due rather to patterns in grooming, presentation, and lifestyle?

We conducted a survey of 8000 Americans using Amazon's Mechanical Turk crowdsourcing platform to see if we could independently confirm these patterns, asking 77 yes/no questions such as 'Do you wear eyeshadow?', 'Do you wear glasses?', and 'Do you have a beard?', as well as questions about gender and sexual orientation. The results show that lesbians indeed use eyeshadow less than straight women, gay men and lesbians wear glasses more, and young opposite-sex-attracted men are more likely to have prominent facial hair than their gay or same-sex-attracted peers.

Breaking down the respondents by age can provide a clearer picture.[8] Figures 13.9 and 13.10 show the proportion of women who answer 'yes' to 'Do you ever use makeup?' (top) and 'Do you wear eyeshadow?' (bottom), averaged over 6-year age intervals.

[8] See our 2018 *Medium* article 'Do Algorithms Reveal Sexual Orientation or Just Expose Our Stereotypes?' for more detail.

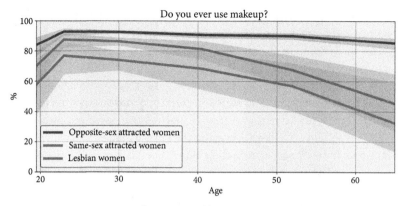

Figure 13.9 Responses to the question 'do you ever use makeup?'

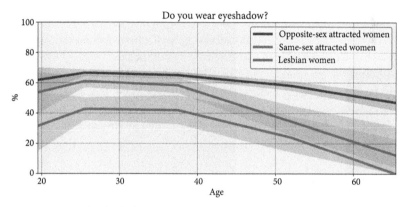

Figure 13.10 The shaded regions around each curve show 68 percent confidence intervals.

The top curve in the two figures represents women who are opposite-sex attracted women but not same-sex attracted; the middle cyan curve represents women who answer 'yes' to 'Are you sexually attracted to women?' or 'Are you romantically attracted to women?'; and the bottom red curve represents women who answer 'yes' to 'Are you homosexual, gay or lesbian?'. The patterns here are intuitive; it will not be breaking news to most that straight women tend to wear more makeup and eyeshadow than same-sex attracted and lesbian-identifying women. Yet, these curves also show us how stereotypes for a group are not representative of the group.

That, on average, same-sex attracted people of most ages wear glasses significantly more than exclusively opposite-sex attracted people do might be a bit less obvious, but it is so. A physiognomist might guess that this is related to differences in visual acuity. However, asking the question 'Do you like how you look in glasses?' reveals that this may be more of a stylistic choice. The pattern holds both for women and for men.

Similar analysis, as in Figure 13.11, shows that, on average, young same-sex attracted men are less likely to have hairy faces than opposite-sex attracted men ('serious facial hair' in our plots is defined as answering 'yes' to having a goatee, beard, or moustache, but 'no' to stubble). Overall, opposite-sex attracted men in our sample are more likely to have serious facial hair than same-sex attracted men, and this is especially true for the youngest men in our study.

Wang and Kosinski speculate in their paper that the faintness of the beard and moustache in their gay male composite might be connected with prenatal under-exposure to androgens (male hormones), resulting in a feminising effect, hence sparser facial hair. The fact that we see a cohort of same-sex attracted men in their 40s who have just as much facial hair as opposite-sex attracted men suggests a different story, in which fashion trends and cultural norms play the dominant role in influencing facial hair among men, not differing exposure to hormones.

The authors of the paper additionally note that the heterosexual male composite appears to have darker skin than the other three composites. Once again they reach for a hormonal explanation, writing: 'While the brightness of the facial image might be driven by many factors, previous research found that testosterone stimulates melanocyte structure and function leading to a darker skin' (p.20). Survey responses suggest a simpler explanation: opposite-sex attracted men, especially younger men, are more likely to work outdoors, and spending time in the sun darkens skin.

The way Wang and Kosinski measure the efficacy of their 'AI gaydar' is equivalent to choosing a straight and a gay or lesbian face image, both from data held out during the training process, and asking how often the algorithm correctly guesses

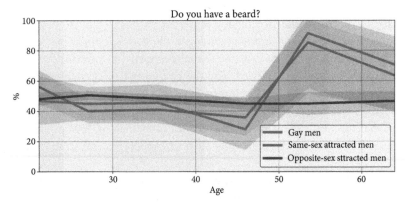

Figure 13.11 The shaded regions around each curve show 68 percent confidence intervals.

which is which. Fifty percent performance would be no better than random chance.

By contrast, the performance measures in the paper, 81 percent for gay men and 71 percent for lesbian women, seem impressive.[9] However, we can achieve comparable results with trivial models based only on a handful of yes/no survey questions about presentation. For example, for pairs of women, one of whom is lesbian, the following trivial algorithm is 63 percent accurate: if neither or both women wear eyeshadow, flip a coin; otherwise guess that the one who wears eyeshadow is straight, and the other lesbian. Adding six more yes/no questions about presentation ('Do you ever use makeup?', 'Do you have long hair?', 'Do you have short hair?', 'Do you ever use colored lipstick?', 'Do you like how you look in glasses?', and 'Do you work outdoors?') as additional signals raises the performance to 70 percent.[10] Given how many more details about presentation are available in a face image, 71 percent performance no longer seems so impressive.

Several studies (Cox et al. 2016) have shown that human judges' 'gaydar' is no more reliable than a coin flip when the judgement is based on pictures taken under well-controlled conditions (head pose, lighting, glasses, makeup, etc.). It is well above chance if these variables are *not* controlled for, because a person's presentation—especially if that person is out—involves social signalling.

Wang and Kosinski argue against this interpretation on the grounds that their algorithm works on Facebook selfies of users who make their sexual orientation public as well as dating website profile photos. The issue, however, is not whether the images come from a dating website or Facebook, but whether they are self-posted or taken under standardised conditions. In one of the earliest 'gaydar' studies using social media (Rule and Ambady 2008), participants could categorise gay men with about 58 percent accuracy in a fraction of a second; but when the researchers used Facebook images of gay and heterosexual men posted by their friends (still an imperfect control), the accuracy dropped to 52 percent.

Head Shape or Shooting Angle?

If subtle biases in image quality, expression, and grooming can be picked up on by humans, these biases can also be detected by AI. While Wang and Kosinski acknowledge grooming and style, they believe that the chief differences between their composite images relate to face shape, arguing that gay men's faces are more 'feminine' (narrower jaws, longer noses, larger foreheads) while lesbian faces are more 'masculine' (larger jaws, shorter noses, smaller foreheads). As with less facial

---

[9] These figures rise to 91 percent for men and 83 percent for women if five images are considered.
[10] Results based on the simplest machine learning technique, linear classification.

hair on gay men and darker skin on straight men, they suggest that the mechanism is gender-atypical hormonal exposure during development. This echoes the
widely discredited nineteenth-century model of homosexuality, 'sexual inversion',
as illustrated by Krafft-Ebing's 'Count Sandor V' case study among many others.

More likely, though, this is a matter of shooting angle. A 2017 paper on the head
poses heterosexual people tend to adopt when they take selfies for Tinder profiles
is revealing (Sedgewick et al. 2017). In this study, women are shown to be about 50
percent more likely than men to shoot from above, while men are more than twice
as likely as women to shoot from below. Shooting from below will have the apparent effect of enlarging the chin, shortening the nose, shrinking the forehead, and
attenuating the smile (see our selfies, Figure 13.12). This view emphasises dominance and presents the photographic subject as taller to the viewer. On the other
hand, shooting from above simulates a more submissive posture, while also making the face look younger, the eyes bigger, the face thinner, and the jaw smaller;
although again, this can also be interpreted as an expectation or desire (Fink et al.
2007) for a potential partner (the selfie's intended audience) to be taller.

The composite images are consistent with all of these patterns: heterosexual
men on average shoot from below, heterosexual women from above.[11] Notice that
when a face is photographed from below, the nostrils are prominent (as in the
heterosexual male face), while higher shooting angles hide them (as in the heterosexual female face). A similar pattern is evident in the eyebrows: shooting from
above makes them look more V-shaped, but their apparent shape becomes flatter,
and eventually caret-shaped (∧) as the camera is lowered. In short, the changes in
the average positions of facial landmarks match what we would expect to see from
differing selfie angles.

Viewing Angle, Dominance, Criminality, and Gender

As these photos show, when we look at human faces today—especially static, two-
dimensional photos of strangers taken under uncontrolled conditions—there is
visual ambiguity between head pose, the shapes of facial features, and affect (i.e.
smiling or frowning).

In the heterosexual context, this may explain stereotypes underlying the more
feminine character of the average 'nice' or 'trustworthy' face, and the more masculine character of the average 'mean' or 'untrustworthy' face (Figure 13.7). As
researchers put it in a 2004 paper, 'Facial Appearance, Gender, and Emotion
Expression' (Hess et al. 2004),

---

[11] Although the authors use a face recognition engine designed to try to cancel out effects of head
pose and expression, we have confirmed experimentally that this doesn't work, a finding replicated by
Tom White, a researcher at Victoria University in New Zealand (https://twitter.com/dribnet/status/
908521750425591808).

Figure 13.12 Photographing from above versus photographing from below.

[A] high forehead, a square jaw, and thicker eyebrows have been linked to perceptions of dominance and are typical for men's faces [...], whereas a rounded baby face is both feminine and perceived as more approachable [. . .] and warm [. . .], aspects of an affiliative or nurturing orientation. This leads to the hypothesis that—regardless of gender—individuals who appear to be more affiliative are expected to show more happiness, and individuals who appear to be more dominant are seen as more anger prone. As cues to gender and cues to affiliation/dominance are highly confounded, this would lead to more women being expected to be happiness prone and more men to be anger prone (p.379).

This brings us back to the histories of physiognomy we discussed at the beginning of the chapter. A woman with what appears to be an unsmiling, perhaps defiant expression, photographed in V.G. Rocine's 1910 physiognomy primer *Heads, Faces, Types, Races* (Figure 13.13) purportedly illustrates 'Degeneracy seen in [the] eyes, eyebrows, nose, lips, jaw, hair and pose' (p.171). As with a collection of

Figure 13.13 Mondays.

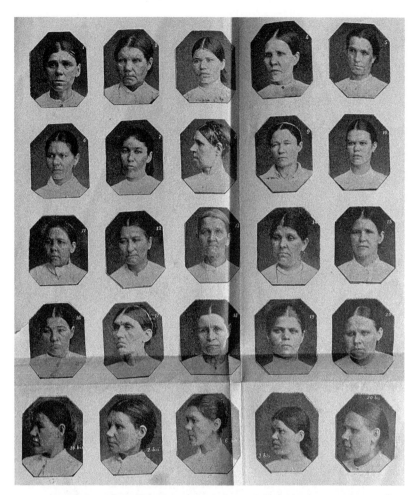

Figure 13.14 Portraits of Russian women from *Criminal Woman* purportedly revealing their 'criminal' physiognomy. Or, perhaps, nobody likes to have their mugshot taken.

similar mugshots from Cesare Lombroso's 1893 follow up to *Criminal Man*, entitled *Criminal Woman, the Prostitute, and the Normal Woman* (Figure 13.14), a great majority of the female 'crime' in question was either non-violent and associated with poverty, or simply reflected behaviour that did not conform to the prevailing gender and sexual norms.

All of this would have put an impossible burden on women, which arguably persists to this day. If they were hetero-conforming, gender-conforming, smiling, and submissive, then they were acceptable in society—at the cost of being relegated to an inferior status in every respect. *Criminal Woman* (see Figure 13.15) is blunt on this score, asserting both that 'Compared to male intelligence, female

Figure 13.15 Out and proud lesbian prison couple from *Criminal Woman*.

intelligence is deficient' (p.85) and that 'woman is always fundamentally immoral' (p.80) (Lombroso and Ferrero 1893/2004). This supposed inferiority is framed in terms of the familiar pseudo-Darwinian hierarchy:

> Normal woman has many characteristics that bring her close to the level of the savage, the child, and therefore the criminal (anger, revenge, jealousy, and vanity) and others, diametrically opposed, which neutralize the former. Yet her positive

traits hinder her from rising to the level of man, whose behavior [. . .] represents the peak of moral evolution (p.128).

If, on the other hand, a woman failed to be submissive or to conform, then her 'masculine' character was considered evidence that she was a degenerate, just as 'feminine' traits would for a man:

Degeneration induces confusion between the two sexes, as a result of which one finds in male criminals a feminine infantilism that leads to pederasty. To this corresponds masculinity in women criminals, including an atavistic tendency to return to the stage of hermaphroditism. [. . .] To demonstrate the presence of innate virility among female prisoners, it is enough to present a photograph of a couple whom I surprised in a prison. The one dressed as a male is simultaneously so strongly masculine and so criminal that it is difficult to believe she is actually female (p.263).

It is a no-win situation.

Conclusions

Our urge to attribute character based on appearance is, as we have shown, based on unjustified stereotypes. Some of these stereotypes, such as the 'untrustworthy face', appear to stem from gender characteristics (including differences in average height) and overgeneralisation of expression classification (interpreting an apparent 'resting smile' or 'resting frown' as windows into the soul). Other differences relate to grooming, presentation, and lifestyle—that is, differences in culture, not in facial structure. Such differences include makeup and eyeshadow, facial hair, glasses, selfie angle, and amount of sun exposure. Lesbian, gay, and straight selfies on dating sites tend to differ in these ways, with no convincing evidence to date of structural facial differences. A handful of yes/no questions about presentation can do nearly as good a job at guessing sexual orientation as supposedly sophisticated facial recognition AI.

At this point, then, it is hard to credit the notion that AI is superhuman at 'outing' us as lesbian or gay based on subtle but unalterable details of our facial structure. Neither can AI based on facial images recognise terrorists or murderers. This does not negate the privacy concerns various commentators have raised, but it emphasises that such concerns relate less to AI per se than to mass surveillance, which is troubling regardless of the technologies used. AI is a general-purpose technology that can be used to automate many tasks, including ones that should not be undertaken in the first place.

References

Bates, Betsy. (2010) 'Drug Screens Fail Accuracy Tests 10% of Time'. Addiction Psychiatry. September, Clinical Psychiatry News.

Berry, Diane S. (1990) 'Taking People at Face Value: Evidence for the Kernel of Truth Hypothesis'. *Social Cognition* 8(4): 343–361

Brammer, John Paul. (2017) 'Controversial AI 'Gaydar' Study Spawns Backlash, Ethical Debate'. *NBC News*, 13 September. Retrieved 25 June 2022 from https://www.nbcnews.com/feature/nbc-out/controversial-ai-gaydar-study-spawns-backlash-ethical-debate-n801026

Cogsdill, Emily, Alexander Todorov, Elizabeth Spelke, and Mahzarin Banaji. (2014) 'Inferring Character from Faces: A Developmental Study'. *Psychological Science* 25(5): 1132–1139.

Cox, William T. L., Patricia G. Devine, Alyssa A. Bischmann, and Janet S. Hyde., (2016) 'Inferences about Sexual Orientation: The Roles of Stereotypes, Faces, and the Gaydar Myth'. *The Journal of Sex Research* 53(2): 157–171.

Darwin, Charles. R. (1871) *The Descent of Man, and Selection in Relation to Sex*. London: John Murray. Volume 1. 1st edition. Accessed at http://darwin-online.org.uk/content/frameset?pageseq=48&itemID=F937.1&viewtype=side

Fink, Bernhard, Nick Neave, Gayle Brewer, and Boguslaw Pawlowski,. (2007) 'Variable Preferences for Sexual Dimorphism in Stature (SDS): Further Evidence for an Adjustment in Relation to Own Height'. *Personality and Individual Differences* 43(8): 2249–2257.

Galton, Francis. (1878) 'Composite Portraits'. *Nature* 18: 97–100.

Gould, Stephen Jay. (1981) *The Mismeasure of Man*. W. W. Norton & Company.

Hess, Ursula, Reginald B. Adams Jr, and Robert E. Kleck. (2004) 'Facial appearance, gender, and emotion expression'. *Emotion* 4(4): 378–388.

Holtz, Brian C. (2015) 'From First Impression to Fairness Perception: Investigating the Impact of Initial Trustworthiness Beliefs'. *Personnel Psychology* 68(3): 499–546.

Huskey, Velma R. and Harry D. Huskey. (1980) 'Lady Lovelace and Charles Babbage'. *Annals of the History of Computing* 2(4): 299–329. Reproduced online at https://richardzach.org/2015/12/de-morgan-on-ada-lovelace/

Kendregan, Charles P. (1966) 'Sixty Years of Compulsory Eugenic Sterilization: Three Generations of Imbeciles and the Constitution of the United States'. *Chicago-Kent Law Review* 43: 123–143.

von Krafft-Ebing, Richard (1893) *Psychopathia Sexualis*, authorized translation of the seventh enlarged and revised German edition, Philadelphia. Accessed at https://www.gutenberg.org/files/64931/64931-h/64931-h.htm

Laughlin, Harry H. (1914) 'Eugenics Record Office Bulletin No. 10B: Report of the Committee to Study and to Report on the Best Practical Means of Cutting Off the Defective Germ-Plasm in the American Population, II'. *The Legal, Legislative, and Administrative Aspects of Sterilization*, Cold Spring Harbor, February 1914.

Levin, Sam. (2017) 'Face-Reading AI Will be Able to Detect Your Politics and IQ, Professor Says'. *The Guardian*, 12 September. https://www.theguardian.com/technology/2017/sep/12/artificial-intelligence-face-recognition-michal-kosinski

Levin, Sam. (2017) 'LGBT Groups Denounce 'Dangerous' AI that Uses Your Face to Guess Sexuality'. *The Guardian*, 9 September. Retrieved 25 June 2022 from https://www.theguardian.com/world/2017/sep/08/ai-gay-gaydar-algorithm-facial-recognition-criticism-stanford

Lombroso, Cesare. (1876) *Criminal Man*. Duke University Press.

Lombroso, Cesare and Ferrero, G. (2004) [1893] *Criminal Woman, the Prostitute, and the Normal Woman*, trans. N. H. Rafter and M. Gibson. Duke University Press.

Luffman, Johann Caspar. (1802) *Physiognomical Sketches by Lavater, Engraved from the Original Drawings*, London.

Nott, Josiah Clarke and George Robbins Gliddon. (1854) *Types of Mankind*. Philadelphia: J.B. Lippincott, Grambo & Co. Accessed at https://wellcomecollection.org/works/b76r97ys.

Olivola, Christopher. Y., Friederike Funk, and Alexander Todorov. (2014) 'Social Attributions from Faces Bias Human Choices'. *Trends in Cognitive Sciences* 18(11): 566–570.

Oosterhof, Nikolaas N. and Alexander Todorov. (2008) 'The Functional Basis of Face Evaluation'. *Proceedings of the National Academy of Sciences of the USA* 105(32): 11087–11092.

Osucha, Eden. (2009). *The Whiteness of Privacy: Race, Media, Law*. Camera Obscura: Durham, NC, USA, 24(1): 67–107.

Penton-Voak, Ian S., Nicholas Pound, Anthony C. Little, and David I. Perrett. (2006) 'Personality Judgments from Natural and Composite Facial Images: More Evidence for a "Kernel Of Truth" in Social Perception'. *Social Cognition* 24(5): 607–640.

Rezlescu, Constantin, Brad Duchaine, Christopher Y. Olivola, and Nick Chater. (2012) 'Unfakeable Facial Configurations Affect Strategic Choices in Trust Games With or Without Information About Past Behavior'. *PloS ONE* 7(3): e34293.

Rocine, Victor Gabriel (1910) *Heads, Faces, Types, Races*, Chicago, Ill., Vaught-Rocine pub. Co. Accessed at https://babel.hathitrust.org/cgi/pt?id=loc.ark:/13960/t2m62ps0b&view=1up&seq=173&q1=eyes,%20eyebrows,%20nose,%20lips,%20jaw,%20hair%20and%20pose

Rule, Nicholas O. and Nalini Ambady. (2008) 'Brief Exposures: Male Sexual Orientation is Accurately Perceived at 50 ms'. *Journal of Experimental Social Psychology* 44(4): 1100–1105.

Sedgewick, Jennifer R., Meghan E. Flath, and Lorin J. Elias. (2017) 'Presenting Your Best Self (ie): The Influence of Gender on Vertical Orientation of Selfies on Tinder'. *Frontiers in Psychology* 8: 604.

Steinbach, Susie. (2004) *Women in England 1760-1914: A Social History*. London: Weidenfeld & Nicolson.

Sontag, Susan. (2004). *Regarding the Pain of Others*. London: Penguin.

Todorov, Alexander. (2017) *Face Value: The Irresistible Influence of First Impressions*. Princeton University Press.

Todorov, Alexander, C. Y. Olivola, R. Dotsch, and P. Mende-Siedlecki. (2015) 'Social Attributions from Faces: Determinants, Consequences, Accuracy, and Functional Significance'. *Annual Review of Psychology* 66: 519–545.

Valla, Jeffrey. M., Stephen J. Ceci, and Wendy M. Williams. (2011) 'The Accuracy of Inferences About Criminality Based on Facial Appearance'. *Journal of Social, Evolutionary, and Cultural Psychology* 5(1): 66–91.

Wang, Lucy. Lu., Gabriel Stanovsky, Luca Weihs, and Oren. Etzioni. (2021) 'Gender Trends in Computer Science Authorship'. *Communications of the ACM* 64(3): 78–84.

Wang, Yilun and Michal Kosinski. (2018) 'Deep Neural Networks are More Accurate than Humans at Detecting Sexual Orientation from Facial Images'. *Journal of Personality and Social Psychology* 114(2): 246–257.

Wilson, John Paul and Nicholas O. Rule. (2015) 'Facial Trustworthiness Predicts Extreme Criminal-Sentencing Outcomes'. *Psychological Science* 26(8): 1325–1331.

Wu, Xiaolin and Zhang, Xi (2016) 'Automated Inference on Criminality Using Face Images'. Preprint at arXiv arXiv:1611.04135: 4038–4052.

14

Signs Taken for Wonders

AI, Art, and the Matter of Race

Michele Elam

Recoding Gender & Race: AI & the Erasure of Ambiguity

LGBTQ+ -identities and mixed-race experiences have always put the lie to false gender binaries and racial purities—and thus to the political project of classificatory schema requiring the recognition of types.[1] This essay argues that the systematic erasure of intersectional ambiguities is an invisibilised and invisibilising practice that falls under what Ruha Benjamin calls the New Jim Code (Benjamin 2019). But it also cautions against any easy valorisation of ambiguity, and unknowability, demonstrating how those can and have been accommodated under technocapitalism in the name of multiracial progress and genderless futurities (see Atanasoski, Chapter 9 in this volume).

Feminist inquiry is crucial to my analysis of how technologies reinscribe not just current but antiquated social norms, and the ways the exponentially distributed agencies of AI—the so-called black box—make them unaccountable because they cannot account for structural inequities (see Browne, Chapter 19 this volume). My critical practice is particularly informed by the intersectional feminisms of Wendy Hui Kyong Chun, Lisa Nakamura, Catherine D'Ignazio, Lauren F. Klein, N. Katherine Hayles, Ruha Benjamin, Safiya Noble, Imani Perry, Jennifer Nash, Saidiya Hartman, Linda Martín Alcoff, James Baldwin, Sara Ahmed, Mutale Nkonde, Neda Atanasoski, Kalindi Vora, Lisa Lowe, and Joy Buolumwini, to name but a few scholar-activists in whose work I find inspiration. Furthermore, digital media studies and performance studies have offered some of the most important understandings of how 'code is performative, and interfaces are theatrical' (Metzger 2021, p.xiii; see also Drage and Frabetti, Chapter 16 this volume).

This essay critiques technological projects of categorisation and classification not to abolish either but to understand them both as indices to deeper issues of why, how and to what ends racial and gender formation is occurring. I examine the gender—and race-making in commercial AI technologies' efforts to represent

---

[1] Originally published in: Michele Elam (2022) 'Signs Taken for Wonders: AI, Art & the Matter of Race'. *Dædalus* 151(2): 198–217.

Michele Elam, *Signs Taken for Wonders*. In: *Feminist AI*. Edited by: Jude Browne, Stephen Cave, Eleanor Drage, and Kerry McInerney, Oxford University Press. © Oxford University Press (2023). DOI: 10.1093/oso/9780192889898.003.0014

people, especially in their inability to manage ambiguity—sometimes referred to as data 'noise': that which falls outside what is understood as useable categories and data sets. Or rather, the problem is not inability, per se, since technologies will evolve, but with technologists' active disinterest in what Hortense Spillers calls 'the interstices' (2003, p.155). Those interstices (the unsaid, misread, misseen or unseen, that what cannot be made sense of in a system even as it invisibly defines what is or is not relevant in it) are not minor glitches; they are essential. They are key because interstices are, in fact, precisely where one can find portals to understanding race, ethnicity and gender not simply normative categories but as dynamic social processes always indexing political interests and tensions.

I would like to add a few words about one of the applications I analyse in this essay that illustrate this concern: Generated Photos. In this example of 'making up people', as Ian Hacking (2006) put it, the intersectional implications of the racialising and gendering in the representation of emotion are subtler and more insidious in their erasures. There are only two emotion options available, listed as either 'joyful' or 'neutral'. What goes unsaid is that the images of 'women' and 'girls', whether smiling or unsmiling, are read so situationally and culturally differently than those faces associated with men, especially the men of colour. A smiling female—especially a female of colour—has historically been interpreted as friendly, welcoming, beckoning, sexually available. They signal no physical or political harm; there is no Angela Davis vibe signalled by natural hair or an unflinching gaze. Nor is there any accounting for the fact that the men of colour, even those in the app that may be also designated 'neutral', are historically perceived as harbouring threat, anger, and resentment. And a smiling man of colour is rarely taken at face value since, as James Baldwin (1955) made clear, whites suspect but do not wish to see the dissembling and rage existing beneath a Black person's smile.

AI's inability to account for masking or covering is not a problem simply of inaccuracy, which might lead some to think the answer is ever more refined categories, nor of gender or racial misrecognition, which some might hear as a call for more sophisticated facial recognition technology (FRT). In fact, the generation of multiple, putatively more refined, racial categories (as in Brazil or South Africa) has historically led to greater not lesser social inequities. Likewise, a plethora of gender categories will not, in and of itself, realise political and representational power.

Rather, the 'problem' is in being directed away from asking why and to what ends these 'people' are being made up in the first place. The occlusion of the subversive, liminal potential that Homi Bhabha and others saw possible in hybridities makes it difficult to disrupt digital reassertions of social norms and political status quos. Rogers Brubaker refers to the oscillations, recombinations, and gradations of 'transgender betweenness', a 'positioning of oneself with reference to two (or sometimes more) established categories, without belonging entirely to either one and without moving definitively from one to the other' (2016, p.94).

Hybridities hold possibility for transformation because they appear to slip beyond the pale of social order and historical time. But, ironically, even a strategic act of boundary-crossing can in and of itself more sharply mark those boundaries and the gatekeeping of who 'is' or who 'is not'. Moreover, hybridity itself cannot prevent the political and commercial investment in racial profiling; it, too, has been monetised and accommodated very comfortably within a racial capitalist and colourist economy that touts the 'progressive' inclusion of multiracials.

The problem with the technological ubiquity and institutional maintenance of gender and racial typologies lies also with the confluence of that profit-for-good motive married to a back-to-the-future imperative. There is an amnesia about, if not downright nostalgia for, these cut-and-dried(out) racial and gender categories that is tapped when they are so disarmingly repurposed. In sites such as This Person Does Not Exist or apps such as Generated Photos we see resurrected binaries as centrepieces for the futures we are told are good for us. The profit motive is juiced with a feel-good narrative that these apps use to take us into the future and create a world that wants, needs, and depends on non-ambiguities.

Their reinscription of racial and gender categories—and with them, the imagined world in which such flattened categories might make sense—leaves entirely unquestioned the categorising imperative embedded in these technologies. In a bait-and-switch, the recoding of racial and gender binaries in emergent tech has the effect of *appearing* as if it is an inclusive act, while actually both blinding and binding the publics using the tech to the urgent global political, social and economic questions around equity and justice.

*****

As he grew accustomed to the great gallery of machines, he began to feel the forty-foot dynamos as a moral force, much as the early Christians felt the Cross. The planet itself felt less impressive, in its old-fashioned, deliberate, annual or daily revolution, than this huge wheel, revolving within arm's length at some vertiginous speed, and barely murmuring–scarcely–humming an audible warning to stand a hair's breadth further for respect of power, while it would not wake the baby lying close against its frame. Before the end, one began to pray to it; inherited instinct taught the natural expression of man before silent and infinite force. Among the thousand symbols of ultimate energy the dynamo was not so human as some, but it was the most expressive.
                    —Henry Adams, 'The Virgin and the Dynamo'

In astonishment of the new technologies at the turn into the twentieth century, the renowned historian Henry Adams found the Gallery of the Electric Machines "physics stark mad in metaphysics" and wondered at their profound hold on the cultural imagination (1918, p. 382).[2] The dynamo that so moved and unsettled Adams was a new generator of unprecedented scale, a machine responsible for powering the first electrified world's fair in 1893, a purportedly spectacular event presided over by President Glover Cleveland. Its power was invisible but the more potent for it: "No more relation could he discover between the steam and the electric current than between the cross and the cathedral. The forces were interchangeable if not reversible, but he could see only an absolute fiat in electricity as in faith" (Adams, 1918, p. 381). For Adams, the dynamo's effect in the world was akin to evidence of things unseen like the symbols of the Virgin or the cross, imperceptible but world-transforming currents with implications both worldly and spiritual.

I open with this discussion of the world's fair at the fin de siècle because Adams's dynamo is our GPT-3 (Generative Pre-trained Transformer 3), a language model that uses deep learning to produce text/speech/responses that can appear generated by a human. His exhilaration–hand-in-glove with his existential vertigo–and his internal conflict similarly speak to our contemporary aspirations for and anxieties about artificial intelligence. Adams understood that the turn to such formidable technology represented a thrilling but cataclysmic event, "his historical neck broken by the sudden irruption of forces entirely new" (1918, p. 382). Although human grappling with exponential leaps in technology dates at least to the medieval period, this particular historical precedent of a transformational moment is singularly relevant for our contemporary moment: there's a direct line between Adams's concern with the hagiography of tech, the devaluation of the arts and humanities, and the comingling of scientific development with (racialized, ableist) narratives of progress to current debates about those nearly identical phenomena today. The consequences of those fundamental mindsets and practices, institutionally codified over time, continue to mushroom in devices, applications, platforms, design practices, and research development. Unacknowledged or misunderstood, they will continue to persist despite the best efforts of many well-intentioned technologists, scholars, policy-makers, and industries that still tend to frame and limit questions of fairness and bias in terms 'safety,' which can mute or obscure attention to issues of equity, justice, or power.[3]

Significantly, Adams's response to the dynamo is neither apocalyptic jeremiad nor in the genre of salvation: that is, his concerns fell beyond the pale of narratives

---

[2] Unlike in *The Wonderful Wizard of Oz*, L. Frank Baum's children's book published the same year as the Paris Exhibition, for Adams, there is no reveal of the "man behind the curtain," no Oz orchestrating a show. His interest is not in the technologists but in what ontological truths their creations tap.

[3] While there is no clear ethical or legal consensus on what constitutes 'fairness,' there are critiques of fairness models that assume an equal playing field thwarting access to opportunities, that presume equal discrimination equals fairness, or that understand fairness in the narrowest sense of preventing harm (the critique of situating 'fairness' under concerns of 'safety'). For a review of some of the debates about fairness, see Michele Elam and Rob Reich (2022), 'Stanford HAI Artificial Intelligence Bill of Rights.'

Figure 1 Gallery of the Electric Machines, The Great Exposition, 1900 Paris World's Fair

Source: La Galerie des Machines Électriques at the Fifth Paris International Expositions of 1900. Image from Dynamo Exhibition Gallery of France, https://www.ndl.go.jp/exposition/e/data/L/4281.html.

of dystopia or deliverance. He was no technophobe; in fact, he deeply admired scientific advances of all kinds. Rather, his ambivalence has to do with the *inestimable psychological and spiritual sway* of machines so impressive that "the planet itself felt less impressive," even "old-fashioned" (1918, p. 380). That something manmade might seem so glorious as to overshadow creation, seemed so evocative of the infinite that people felt out of step with their own times. For Adams, those experiences signaled an epistemic break that rendered people especially receptive and open to change, but also vulnerable to idolizing false gods of a sort. He saw that the dynamo was quickly acquiring a kind of cult status, inviting supplication and reverence by its followers. The latest technology, as he personified it in his poem 'Prayer to the Dynamo,' was simultaneously a "Mysterious Power! Gentle Friend! Despotic Master! Tireless Force!" (1920, l. 1-2). Adams experienced awe in the presence of the dynamo: 'awe' as the eighteenth-century philosopher Edmund Burke meant the term, as being overcome by the terror and beauty of the sublime. And being tech awestruck, he also instantly presaged many of his generation's–and I would argue, our generation's–genuflection before it.

As part of his concern that sophisticated technology inspires a kind of secular idolatry, Adams also noted its increasing dominance as the hallmark of human progress. In particular, he presciently anticipated that it might erode the power of both religion and the arts as vehicles for and markers of humanity's higher strivings. Indeed, his experience at the Gallery taught him firsthand how fascination with such potent technology could eclipse appreciation of the arts: more specifically, of technological innovation replacing other modes of creative expression as the pinnacle of human achievement. Adams bemoaned the fact that his friend, Langley, who joined him at the exposition, "threw out of the field every exhibit that did not reveal a new application of force, and naturally, to begin with, the whole art exhibit" (1918, p. 380). The progress of which technology increasingly claimed to be the yardstick extended beyond the valuation of art also extended to racial, ethnic, and gender scales. Most contemporary technological development, design, and impact continue to rely unquestioningly on enlightenment models of the 'human,' as well as the nearly unchanged and equally problematic metrics for human achievement, expression, and progress.

These are not rhetorical analogies; they are antecedences to AI, historical continuities that may appear obscured because the tech-ecosystem tends to eschew history altogether: discourses about AI always situate it as future-facing, prospective not retrospective. It is an idiom distinguished by incantations about growth, speed, and panoptic capture. The messy, recursive, complex narratives, events, and experiences that actually make up histories are reduced to static data points necessary in training sets for predictive algorithms. Adams's reaction offers an alternative framing of time in contrast to marketing imperatives that fetishize the next new thing, which by definition sheds its history.

This reframing is important to note because for all the contemporary talk of disruption as the vaulted and radical mode of innovation, current discourse still often presents so-called disruptive technologies as a step in an inexorable advance forward and upward. In that sense, tech disruption is in perfect keeping with the same teleological concept of momentum and progress that formed the foundational basis by which world's fairs ranked not only modes of human achievement but also degrees of 'human.' The exhibitions catalogued not just inventions but people, classifying people by emerging racialized typologies on a hierarchical scale of progress with the clear implication that some were more human than others.[4] This scale was

---

[4] On the question of humanness, and for an analysis of how technologies determine full humans, not-quite-humans, and nonhumans, see Alexander G. Weheliye's (2014) excellent *Habeas Viscus: Racializing Assemblages, Biopolitics, and Black Feminist Theories of the Human.* See also Sylvia Wynter's (2003) foundational essay, 'Unsettling the Coloniality of Being/Power/Truth/Freedom: Towards the Human, After Man, Its Overrepresentation–An Argument'. Wynter critiques the overrepresentation of man (as white, Western) as the only imaginable mode of humanness, overwriting other ontologies, epistemologies, and imaginaries. See also Katherine McKittrick (2015), *Sylvia Wynter: On Being Human as Praxis.* A major influence on this essay, Wynter's pioneering and prolific work draws on arts, humanities, natural sciences and neuroscience, philosophy, literary theory, and critical race theory.

made vivid and visceral: whether it was the tableaux vivant 'ethnic villages' of the 1893 world's fair in Chicago's 'White City' or the 1900 Paris showcase of African American achievement in the arts, humanities, and industry (images of 'racial uplift' meant to counter stereotyping), both recognized how powerfully influential were representations of races' putative progress–or lack of it.

Carrying the international imprimatur of the fairs, the exhibitions were acts of racial formation, naturalizing rungs of humanness and, indeed, universalizing the imbrication of race and progress. Billed as a glimpse into the future, the fairs simultaneously defined what was *not* part of modernity: what or who was irrelevant, backward, regressive in relation. Technological progress, therefore, was not simply represented *alongside* what (arts/humanities) or who (non-whites) were considered less progressive; progress was necessarily measured *against* both, indeed constituted by its difference and distance from both.

For critical theorist Homi Bhabha, such notions of progress, and the technology and symbol of it, are inextricably tied to the exercise of colonial and cultural power. His essay 'Signs Taken for Wonders: Questions of Ambivalence and Authority Under a Tree outside Delhi, May 1817' critiques the "wondrous" presence of the book, itself a socially transformative technology, by beginning with the premise that innovation cannot be uncoupled from the prerogatives of those who have the power to shape realities with it:

> The discovery of the book is, at once, a moment of originality and authority, as well as a process of displacement, that paradoxically makes the presence of the book wondrous to the extent to which it is repeated, translated, misread, displaced. It is with the emblem of the English book–"signs taken as wonders"–as an insignia of colonial authority and an insignia of colonial desire and discipline that I begin this essay. (1985, p. 144)

Adams spoke of awe in the presence of the dynamo. Bhabha goes further in challenging such "signs taken as wonders," in questioning technologies so valorized that they engender awe, obeyance, and reverence as if such a response was natural, innocent of invested political and economic interests, free of market value systems.

> Like all tools, AI challenges the notion that the skull marks the border of the mind. . . . New tools breed new literacies, which can engender nascent forms of knowing, feeling and telling.
> —Vanessa Chang, 'Prosthetic Memories, Writing Machines'

Art sits at the intersection of technology, representation, and influence. Literature, film, music, media, and visual and graphic arts are all crucial incubators for how publics perceive tech. Storytelling impacts, implicitly or explicitly, everything from product design to public policy. Many of these narratives bear traces of literature's earliest engagement with technology, at least since medieval times, and others–either engaged with AI or AI-enabled–are also offering new plotlines, tropes, identity formations, historiographies, and speculative futurities. Moreover, because cultural storytelling helps shape the civic imagination, it can, in turn, animate political engagement and cultural change.[5]

Indeed, the arts are specially poised to examine issues in technological spaces (from industry to STEM education) of equity, diversity, social justice, and power more capaciously and cogently than the sometimes reductive industry-speak of inclusion, fairness, or safety (usually simply meaning minimization of harm or death–a low bar indeed). Even before GPT-3, powerful natural language processing was enabling explorations in AI-assisted poetry, AI-generated filmscripts, AI-informed musicals, AI-advised symphonies, AI-curated art histories, and AI-augmented music.[6] Many are proposing new nomenclature for hybrid genres of art, design, and tech, and fresh subfields are blooming in both academe and entertainment.[7] And during the COVID-19 pandemic and intensified movements for social justice, there has been a plethora of virtual exhibitions and articles about the hot debates over the status, meaning, and valuation of AI-generated or -augmented art.[8]

Amidst this explosion of artistic engagement with AI, social and political AI scholars Kate Crawford and Luke Stark, in 'The Work of Art in the Age of Artificial Intelligence: What Artists Can Teach Us about the Ethic of Data Practice,' offer a not uncommon perspective on the need for interdisciplinary collaboration: "Rather than being sidelined in the debates about ethics in artificial intelligence and data practices more broadly, artists should be centered as practitioners who are already seeking to make public the political and cultural tensions in using data platforms to reflect on our social world" (2019, p. 452). However, they also close the article by recommending that arts practitioners and scholars would do well with more technical education and that without it, their engagements and critiques

---

[5] For instance, see the comment made by the then head of Instagram, Adam Mosseri, that his recent policy eliminating public 'likes'–because of his concern "about the unintended consequences of Instagram as an approval arbiter" as he put it–was partly informed by an episode of the science fiction anthology television series *Black Mirror* (2020).

[6] For example, see Google's 'PoemPortraits' by Es Devlin. A 2016 short-film script entitled *Sunspring* was made by an AI bot. The main character in the filmed version was played by Thomas Middleditch, the same actor who plays the lead, Richard Hendriks, in the TV series *Silicon Valley* ☺. See also AI-generated musicals and symphonies, as in Maura Barrett and Jacob Ward (2019), 'AI Can Now Compose Pop Music and Even Symphonies. Here's How Composers Are Joining In.'

[7] Artist-technologists working at these intersections include Amelia Bearskin-Winger, Legacy Russell, Stephanie Dinkins, Ian Chang, Rashaad Newsome, Jacolby Satterwhite, Joy Buolamwini, Martine Syms, and others–not to mention those writers in the long literary history of speculative fiction.

[8] See, for instance, Barbara Pollack and Anne Verhallen (2020), 'Art at a Time Like This.' I take up in detail the debates over the value and valuation of the arts and humanities in the age of artificial intelligence in Michele Elam (forthcoming spring 2022), 'GPT-3 in 'Still I Rise!': Why AI Needs Humanists.'

will have lesser insight into and standing regarding the ethics of data practice: "One barrier to a shared and nuanced understanding of the ethical issues raised by digital art practices is a lack of literacy regarding the technologies themselves Until art critics engage more deeply with the technical frameworks of data art, their ability to analyze and assess the merits of these works–and their attendant ethical dilemmas–may be limited" (2019, p. 451). They continued: "a close relationship to computer science seemed to offer some artists a clearer lens through which to consider the ethics of their work" (2019, p. 452).

Certainly, continuing education is usually all to the good. But I would welcome the equivalent suggestion that those in data science, computer science, engineering, and technology, in turn, should continue to educate *themselves* about aesthetics and arts practices–including at least a passing familiarity with feminist, queer, decolonial, disability, and race studies approaches to AI often central to those practices–to better understand ethical debates in their respective fields.[9] Without that balance, the suggestion that artists and nontechnical laypeople are the ones who primarily need education, that they require technical training and credentialing in order to have a valid(ated) understanding of and legitimate say in the political, ethical, social, and economic discussions about AI, is a kind of subtle gatekeeping that is one of the many often unacknowledged barriers to cross-disciplinary communication and collaboration. Given the differential status of the arts in relation to technology today, it is usually taken for granted that artists (not technologists, who presumably are doing more important and time-consuming work in and for the world) have the leisure and means not only to gain additional training in other fields but also to do the hard translational work necessary to integrate those other often very different disciplinary practices, vocabularies, and mindsets to their own creative work. That skewed status impacts who gains the funding, influence, and means to shape the world.

Instead of asking artists to adapt to the world models and pedagogies informing technological training–which, as with any education, is not simply the neutral acquisition of skills but an inculcation to very particular ways of thinking and doing–industry might do well to adapt to the broader vernacular cultural practices and techne of marginalized Black, Latinx, and Indigenous communities. Doing so might shift conversation in the tech industry from simply mitigating harm or liability from the differentially negative impact of technologies on these communities. Rather, it would require a mindset in which they are recognized as equal partners, cultural producers of knowledge(s), as the longtime makers, not just the recipients and consumers, of technologies.[10] In fact, artist-technologist Amelia

---

[9] There are many academic and activist resources and collectives working in these areas, including Latinx in AI, Black in AI, Queer AI, Indigenous AI, and Accessible AI, to name but a few.

[10] See, for instance, Ruha Benjamin (2019), 'Introduction: Discriminatory Design, Liberating Imagination,' and Nettrice R. Gaskins (2019), 'Techno-Vernacular Creativity and Innovation across the African Diaspora and Global South.'

Winger-Bearskin, who is Haudenosaunee (Iroquois) of the Seneca-Cayuga Nation of Oklahoma, Deer Clan, makes a case that many of these vernacular, often generational, practices and values are what she calls "antecedent technologies," motivated by an ethic that any innovation should honor its debt to those seven generations prior and pay it forward seven generations (2019).[11]

In this way, many contemporary artist-technologists engage issues including, but also going beyond, ethics to explore higher-order questions about creativity and humanity. Some offer non-Western or Indigenous epistemologies, cosmologies, and theologies that insist on rethinking commonly accepted paradigms about what it means to be human and what ways of doing business emerge from that. Perhaps most profoundly, then, the arts can offer different, capacious ways of knowing, seeing, and experiencing worlds that nourish well-being in the now and for the future. It is a reminder of and invitation to world models and frameworks alternative to what can seem at times to be dominating or totalizing technological visions. In fact, one of the most oft-cited criticisms of AI discourse, design, and application concerns its univision, its implied omniscience, what scholar Alison Adams calls 'the view from nowhere.'[12] It is challenged by art that offers simultaneous, multiple, specifically situated, and sometimes competing points of view and angles of vision that enlarge the aperture of understanding.

For instance, informed by disability culture, AI-augmented art has drawn on GANs (generative adversarial networks) to envision non-normative, including neurodivergent, subjects that challenge taken-for-granted understandings of human experience and capability. The presumption of a universal standard or normative model, against which 'deviance' or 'deviation' is measured, is nearly always implied to be white, cis-gendered, middle-classed, and physically and cognitively abled. That fiction of the universal subject–of what disability scholar and activist Rosemarie Garland-Thomson terms the "normate"–has historically shaped everything from medical practice and civil rights laws to built environments and educational institutions (1997, p. 8). It also often continues to inform technologies' development and perceived market viability and use-value. Representations of 'human-centered' technology that include those with mental or physical disabilities often call for a divestment from these usual ways of thinking and creating. Such a direct critique is posed in art exhibitions such as *Recoding CripTech*. As the

---

[11] See also Amelia Winger-Bearskin (2018), 'Before Everyone Was Talking about Decentralization, Decentralization Was Talking to Everyone.'

[12] See specifically Katz's discussion of AI notions of the self: "Practitioners in the 1970s, for instance, offered visions of the self as a symbolic processing machine. In the late 1980s and early 1990s, by contrast, the prevailing 'self' started looking more like a statistical inference engine driven by sensory data. But these classifications mask more fundamental epistemic commitments. Alison Adams has argued that AI practitioners across the board have aspired to a 'view from nowhere'–to build systems that learn, reason, and act in a manner freed from social context. The view from nowhere turned out to be a view from a rather specific, white, and privileged space." See also Alison Adams (1998), *Artificial Knowing: Gender and the Thinking Machine.*

curatorial statement puts it, the installations reimagine "enshrined notions of what a body can be or do through creative technologies, and how it can move, look or communicate. Working with a broad understanding of technology . . . this multidisciplinary community art exhibition explores how disability–and artists who identify as such–can redefine design, aesthetics and the relationship between user and interface" (2020). Works included in *Recoding CripTech* that employ artificial intelligence, such as M Eifler's 'Prosthetic Memory' and 'Masking Machine,' suggest a provocative reframing of 'optimization' or 'functionality' in technologies that propose to augment the human experience.[13]

> Race–racism–is a device. No More. No less. It explains nothing at all. . . . It is simply a means. An invention to justify the rule of some men over others. [But] it also has consequences; once invented it takes on a life, a reality of its own. . . . And it is pointless to pretend that it doesn't exist–merely because it is a lie!
> —Tshembe in *Les Blancs* by Lorraine Hansberry

Rashaad Newsome's installation *Being* represents another artistic provocation that reframes both the form and content of traditional technological historiographies often told from that 'view from nowhere.' Newsome, a multimedia artist and activist, makes visible the erased contributions to technology and art by people of African descent. Newsome terms the interactive social humanoid *Being 2.0* an AI "griot," a storyteller (2020). But unlike most social robots commanded to speak, *Being* is intentionally 'uppity': wayward, noncompliant, disobedient, with expressive gestures drawn Black Queer vogue dance repertoire meant as gestures of decolonial resistance to the labor and service that social robots are expected to perform. It upends the historical association of robots and slaves (in the etymology of the Czech word, 'robot' translates to 'slave') in movement, affect, function, and speech. Taking aim at the limited training data sets used in natural language processing, Newsome draws on broader archives that include African American vernacular symbolic systems.[14] And since language carries cultural knowledge, *Being*'s speech expands not just vocabularies but reimagines how the standardized expressions of emotion and behavior often deployed in AI are racially and culturally encoded.[15] In fact, *Being* is an attempt to redress the historical violence of antiquated notions about race, the more disturbing because the representations

---

[13] The curatorial statement in *Recoding CripTech* explains the terminology: "the term 'crip' reclaims the word for disability culture and recognizes disability as a political and cultural identity." See also M Eifler, 'Prosthetic Memory' (2020), and 'Masking Machine' (2018). Many thanks to Lindsey Felt and Vanessa Chang, who produced and curated *Recoding CripTech*, and especially for Lindsey's suggestions for this section of the essay.

[14] See Su Lin Blodgett and Brendan O'Connor (2017), 'A Racial Disparity in Natural Language Processing: A Case Study in Social Media African American English.'

[15] See Neda Atanasoski and Kalindi Vora (2019), 'The Surrogate Human Affect: The Racial Programming of Robot Emotion.'

Figure 2 Rashaad Newsome's *Being 2.0*
Being © Rashaad Newsome Studio.

of race, reduced to seemingly self-evident graduations of color and physiognomy, are being actively resurrected in AI development and application.

Race is always a negotiation of social ascription and personal affirmation, a process of what sociologists Michael Omi and Howard Winant term "racial formation" (1994, p. 55). Omi and Winant refer to racial formation as a way of historicizing the practices and circumstances that generate and renew racial categories and racializing structures:

> We define *racial formation* as the sociohistorical process by which racial categories are created, inhabited, transformed, and destroyed. . . . Racial formation is a process of historically situated *projects* in which human bodies and social structures are represented and organized. Next we link racial formation to the evolution of hegemony, the way in which society is organized and ruled. . . . From a racial formation perspective, race is a matter of both social structure and cultural representation. (1994, p. 55–56)

The expression "racial formation" is therefore a reminder that race is not *a priori*. It is a reminder to analyze the structural and representational–not just linguistic–contexts in which race becomes salient: the cultural staging, political investments, institutional systems, and social witnessing that grant meanings and values to categories. A full accounting of race therefore involves asking in whose interest is it that a person or people are racialized in any given moment in time and space? And with what ends?

Overlooked, for instance, in many debates over racial bias, surveillance, and privacy in facial recognition technology is the practice of coding 'race' or 'ethnicity' as fixed, static programmable variables, something writ on the face or otherwise available as physically intelligible–an outdated approach to race that harkens back to nineteenth-century phrenology and other pseudoscience mappings of racial traits. Moreover, that practice renders opaque how categories are never merely descriptive, disinterested renderings of facts or things even though they cannot be purged of the value systems that animate their creation and make them intelligible for technological use–at least as currently developed–in the first place. Additionally, the claim to a universal objectivity is one of the "epistemic forgeries," according to Yarden Katz, who describes it as one of the "fictions about knowledge and human thoughts that help AI function as a technology of power" because it enables "AI practitioners' presumption that their systems represent a universal 'intelligence' unmarked by social context and politics" (2020, p. 94–95).[16] That drive for comprehensive typing and classification, for a universal compendium, cannot easily accommodate race other than a technical problem in mapping variation of types.[17]

To illustrate why AI representations are so problematic, let me take a seemingly innocuous example in the new algorithmic application 'Ethnicity Estimate,' part of the Gradient app, which purports to diagnose percentages of one's ethnic heritage based on facial recognition technology (FRT). Such an app is significant precisely because popular data-scraping applications are so often pitched as convenient business solutions or benign creative entertainment, bypassing scrutiny because they seem so harmless, unworthy of research analysis or quantitative study. Critically examining on such issues would be a direct impediment to a seamless user

[16] The claim to universality, according to Katz, is the first forgery: "The second is that AI systems have matched or exceeded the capabilities of human thought [drawing] on deep-seated notions in Western culture about hierarchies of intelligence. The third epistemic forgery suggests that these computational systems arrive at truth, or 'knowledge,' 'on their own,' AI practitioners being merely the ones who set off the necessary conditions for computational processes to properly unfold" (2020, p. 94–95).
[17] Directly related to the issue of racial classification is the issue of racial and gender 'ambiguity,' often (mis)understood in AI as simply a technical issue of documenting (or not) variance and managing uncertainty. For extended discussion of the complicated technical and political challenges posed by social identities transgressing racial or gender boundaries in AI, see Michele Elam (forthcoming 2022), 'Recoding Gender-Binaries and Color-Lines: AI and the Erasure of Ambiguity.'

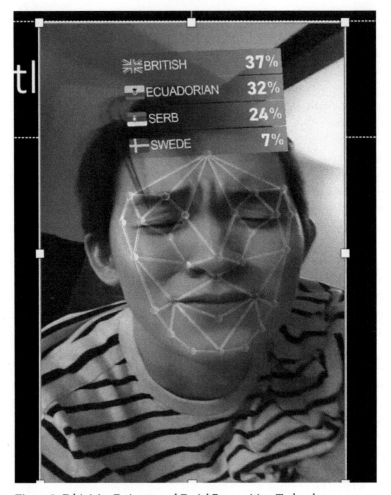

Figure 3 Ethinicity Estimate and Facial Recognition Technology.

Screenshot of the Ethnicity tool using FRT on one of my students, Edric Zeng, who is Korean and Chinese. Note his incredulous expression upon seein its onclusion: 37 percent British: 32 percent Ecuadorian: 24 percent Serb: 7 percent Swede. Image courtesy of Edric Zeng.

experience with the product, thus designers and users are actively disincentivized from doing so. Like many such applications, Ethnicity Estimate problematically uses nationality as a proxy for ethnicity and reduces population demographics to blood quantum.

Or consider Generated Photos: an AI-constructed image bank of 'worry-free' and 'infinitely diverse' facial portraits of people who do not exist in the flesh, which marketers, companies, and individuals can use "for any purpose without worrying about copyrights, distribution rights, infringement claims or royalties" (Generated Media, 2022). In creating these virtual 'new people,' the service offers a workaround for privacy concerns. Generated Photos bills itself as the future of intelligence, yet it reinscribes the most reductive characterizations of race: among

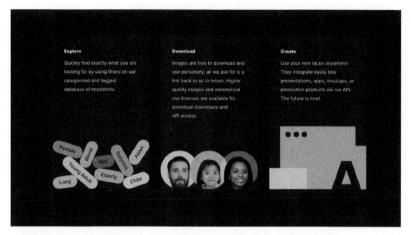

Figure 4 Generated Photos: "Use Your New Faces Anywhere!"
Source: Screenshot of promotional materials on https://generated.photos/.

other parameters users can define when creating the portraits, such as age, hair length, eye color, and emotion through facial expression, the racial option has a dropdown of the generic homogenizing categories Asian, African American, Black, Latino, European/white.

Skin color options are similarly presented as self-evident and unproblematic givens, a data set based on an off-the-shelf color chart. There is a long racializing history of such charts, from the von Luschan chromatic scale, used throughout the first half of the twentieth century to establish racial classifications, to the Fitzpatrick scale, still common in dermatologists' offices today, which classifies skin types by color, symbolized by six smiling emoji modifiers. Although the latter makes no explicit claim about races, the emojis clearly evoke the visuals well as the language of race with the euphemism of 'pigmentary phototype.'

All these types are readily serviceable as discrete data points, which makes them an easy go-to in algorithmic training, but the practice completely elides the fact that designations of 'dark' or 'light' are charged cultural and contextual interpretations that are always negotiated in context and *in situ*.[18] The relevance and meaning of race emerge through social and cultural relations, not light frequencies. Fran Ross's brilliant, satirical novel *Oreo* (1974) offers a wry send-up of attempts to apply color charts to social identities, shown as Figure 6.

Although new AI technologies show promise in diagnosing medical conditions of the skin, thinking of racial identification primarily in terms of chromatic scales or dermatoscopic data deflects attention, to put it generously, from the long history

---

[18] By arguing for understanding words 'in context,' I mean in the social science sense of meaning emerging through performative interaction, not meaning garnered by sentence level 'context,' as is commonly understood in natural language processing: that is, of other words that appear near each other in a sentence. Thanks to my Stanford colleagues Chris Manning and Surya Ganguli for our conversations about natural language processing.

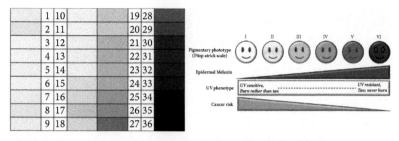

Figure 5 The von Luschan Chromatic Scale (*left*) and the Fitzpatrick Scale (*right*).

The reproduction of the on Luschan chromatic scale, based on the chart first printed in *Völker, Raissen, Sprachen* (1927), is by Wikimedia users Dark Tichondrias and Churnett. Printed under the Creative Common Attribution-ShareAlike 3.0 Unported license. The Fitzpatrick scale is from John D'Orazio, Stuart Jarrett. Alexandra Amaro-Ortiz, and Timothy Scott, "UV Radiation and the Skin", *International Journal of Molecular Sciences* 14(6) (2013). Reprinted under the Creative Commons Attribution 3.0 Unportes license.

Table 1 Fitzpatrick Type and von Luschan Scale

Fitzpatrick Type	Van Luschan Scale	Also Called
I	0–6	Very light or white, "Celtic" type
II	7–13	Light or light-skinned European
III	14–20	Light intermediate, or dark-skinned European
IV	21–27	Dark intermediate or "olive skin"
V	28–34	Dark or "brown" type
VI	35–36	very dark or "black" type

Source: Nina G.Jablonski, "Skin Coloration: in *Human Evolutionary Biology*, ed. Michael P. Muehlenbein (Cambridge: Cambriddge University Press, 2010), 177.

of the damaging associations of skin color and race that gave rise to early technologies like this in the first place, whether it was the 'science' of phrenology, IQ tests, or fingerprinting, and with implications, more recently, for the use of biometrics.[19] At a minimum, it ignores the imbrication of 'race' in pigmentocracies and colorism, the historical privileging of light skin, and the various rationales for

[19] This reduction of race to data points for AI has many problematic medical implications, even in work aiming to mitigate bias. Using self-reported racial data along with other medical information to create a 'phenotype,' Juan Banda acknowledges that racial porosity (people identifying with multiple races, differently over time, or differently in various contexts) as well as durational changes in aging mean up to 20 percent of data are thrown out in assessments. Moreover, as is long documented, race is often proxy for all sorts of other factors and inequities, so whether the data collected are self-reported or social ascription, it is still a highly problematic category. Juan Banda (2022), 'Phenotyping Algorithms Fair for Underrepresented Minorities within Older Adults.'

Helen Clark: Singer, pianist, mimic, math freak (a 4 on the color scale)

Colors of black people

white	high yellow (pronounced YAL-la)	yellow	light-skinned
1	2	3	4

light brown-skinned	brown-skinned	dark brown-skinned
5	6	7

dark-skinned	very dark-skinned	black
8	9	10

Note: There is no "very black." Only white people use this term. To blacks, "black" is black people are not nearly so black as your black pocketbook). If a black person says, "John is very black," he is referring to John's politics, not his skin color.

Figure 6 Fran Ross's *Oreo* Color Scale.

Source: Fran Ross, *Oreo* (Boston: Northeastern University Press, 1974).

identifying what counts as 'light-skinned.' Colorism, a legacy of colonialism, continues to persist in contemporary hierarchies of value and social status, including aesthetics (who or what is ranked beautiful, according to white, Western standards), moral worth (the religious iconography of 'dark' with evil and 'light' with holy continues to saturate languages), social relations (for instance, the 'paper bag test' of the twentieth century was used as a form of class gatekeeping in some African American social institutions),[20] and the justice system (since social scientists have documented the perceptual equation of 'blackness' with crime, and thus those perceived as having darker skin as *a priori* criminally suspect).[21]

Why does this matter? Because it suggests that the challenges in representing race in AI are not something technological advances in any near or far future could solve. Rather, they signal cultural and political, not technical, problems to address. The issue, after all, is not merely a question of bias (implicit or otherwise), nor of inaccuracy (which might lead some to think the answer is simply the generation of more granular categories), nor of racial misrecognition (which some might hear as simply a call for ever more sophisticated FRT), nor even of ending all

[20] The 'paper bag test,' and color discrimination broadly, is a complicated form of internalized racism dating to slavery. Early scholarship suggested lighter-skinned slaves were supposedly treated better, although that view has been challenged, since being 'in the Big House' instead of the field often meant greater exposure to sexual abuse. Moreover, there are many documented accounts of mixed race children–the issue of white masters and slaves–being sold away, often at the insistence of the white mistress of the house, since they stood as corporeal testimony to miscegenation and adultery–in short, to the sins of the father.

[21] See Jennifer Eberhardt (2020), *Biased: Uncovering the Hidden Prejudice that Shapes What We See, Think, and Do.* Eberhardt documents how police are more likely to think a suspect is holding a gun if they are perceived as 'dark,' and that juries are more likely to negatively regard, and thus weigh judgment against, a defendant if they are considered 'dark,' among the many other consequences.

uses of racial categorization.[22] It matters because algorithms trained on data sets of racial types reinforce color lines, literally and figuratively remanding people back in their 'place.' By contrast, as I have suggested, the increasingly influential rise of AI artist-technologists, especially those of color, are among those most dynamically questioning and reimagining the commercial imperatives of 'personalization' and 'frictionlessness.' Productively refusing colorblindness, they represent race, ethnicity, and gender not as normative, self-evident categories nor monetizable data points, but as the dynamic social processes–always indexing political tensions and interests–which they are. In doing so, they make possible the chance to truly create technologies for social good and well-being.

> Something has happened. Something very big indeed, yet something that we have still not integrated fully and comfortably into the broader fabric of our lives, including the dimensions–humanistic, aesthetic, ethical and theological–that science cannot resolve, but that science has also (and without contradiction) intimately contacted in every corner of its discourse and being.
> —Stephen Jay Gould, The Hedgehog, the Fox, and the Magister's Pox

I cite what may seem minor examples of cultural ephemera because, counterintuitively, they hint at the grander challenges of AI. They are a thread revealing the pattern of "something very big indeed," as historian of science Stephen Jay Gould put it (2011, p. 15). Certainly there are ethical, economic, medical, educational, and legal challenges facing the future of AI. But the grandest technological challenge may in fact be *cultural*: the way AI is shaping the human experience. Through that lens, the question becomes not one of automation versus augmentation, in which 'augmenting' refers to economic productivity, but rather to creativity. That is, how can AI best augment the arts and humanities and thus be in service to the fullness of human expression and experience?

This essay opened with Henry Adams's moment of contact with the Dynamo's "silent and infinite force," as he put it, which productively *de*naturalizes the world as he knows it, suspends the usual epistemological scripts about the known world and one's place in it (1918, p. 380). It is a sentiment echoed almost verbatim two hundred years later by Gould, witnessing another profound technological and cultural upending. Writing at the turn into our own century, Gould, like Adams,

---

[22] For instance, racial classification is often employed in the United States (though not in France and other countries) for strategic purposes generally understood as for the larger social good: for instance, by the Office of Management and Budget for census purposes, which are tied to the distribution of resources and the tracking of civil rights violations, or by Federal Statistical Research Data Centers that enable the study of health and economic trends across demographics. In such cases, data collection of those gross categories is undertaken with the understanding that these are efforts to capture not inherent traits or types, but rather broad trends in the complex of race, gender, and socioeconomics, among other variables. More recently, efforts to collect immigration and citizen status, neither following the prior intent nor practice of the Census, has made vivid the potential for misuses of this national mode of data collection.

cannot fully articulate the revelation except to say poignantly that "something has happened," that every dimension of "the broader fabric of our lives" is intimately touched by a technology whose profound effect cannot be "solved" by it (2011, p. 15). That liminal moment for Adams, for Gould, and for us makes space for imagining other possibilities for human creativity, aesthetic possibilities that rub against the grain and momentum of current technological visions, in order to better realize the "magisteria of our full being" (2011, p. 15).[23]

References

Adams, Henry and Mabel La Farge. (1920) *Letters to a Niece and Prayer to the Virgin of Chartres*. Boston: Houghton Mifflin Company.

Atanasoski, Neda and Kalindi Vora. (2019) The Surrogate Human Affect: The Racial Programming of Robot Emotion. In *Surrogate Humanity: Race, Robots and the Politics of Technological Futures*, pp. 108–133. Durham, NC, USA: Duke University Press.

Baldwin, James. (1955) *Notes of a Native Son*. Boston, MA, USA: Beacon Press.

Banda, Juan. (2022) 'Phenotyping Algorithms Fair for Underrepresented Minorities within Older Adults'. Stanford HAI Weekly Seminar, Stanford, CA, USA. Seminar.

Barrett, Maura and Jacob Ward. (2019) 'AI Can Now Compose Pop Music and Even Symphonies. Here's How Composers Are Joining In'. *MACH by NBC News* https://www.nbcnews.com/mach/science/ai-can-now-compose-pop-music-even-symphonies-here-s-ncna1010931.

Benjamin, Ruha. (2019) Introduction: Discriminatory Design, Liberating Imagination. In *Captivating Technology: Race, Carceral Technoscience, and Liberatory. Imagination in Everyday Life*, ed. Ruha Benjamin, pp. 1–25. Durham, NC, USA: Duke University Press

Bhabha, Homi. (1985) 'Signs Taken for Wonders: Questions of Ambivalence and Authority Under a Tree outside Delhi, May 1817'. Special issue on 'Race', Writing, and Difference, *Critical Inquiry* 12(1): 144–165.

Blodgett, Su Lin and Brendan O'Connor. (2017) 'Racial Disparity in Natural Language Processing: A Case Study of Social Media African-American English'. *Fairness, Accountability, and Transparency in Machine Learning, Halifax, Canada. Workshop*.

Brubaker, Rogers. (2016) *Trans: Race and Gender in an Age of Unsettled Identities*. Princeton NJ, USA: Princeton University Press,

Change, Vanessa. (2020) 'Prosthetic Memories, Writing Machines'. *Noēma* https://www.noemamag.com/prosthetic-memories-writing-machines/.

Devlin, Es. (2019) 'PoemPortraits'. Google Arts & Culture https://experiments.withgoogle.com/poemportraits.

Eberhardt, Jennifer. (2020) *Biased: Uncovering the Hidden Prejudice that Shapes What We See Think, and Do*. London: Penguin.

---

[23] I am following Stephen Jay Gould's critiques in *The Hedgehog, the Fox, and the Magister's Pox* of E. O. Wilson and C. P. Snow's notion of two cultures, described in Snow's (1961) *The Two Cultures and the Scientific Revolution* as a split between science and humanities. Gould found that duality artificial, simplistic, and ahistorical, offering instead more accurate accounts of four historical periods exploring continuities of creative thinking across the sciences and humanities.

Eifler, M. (2020) *Prosthetic Memory*. Ars Electronica Center, Linz, Austria.

Eifler, M. (2018) *Masking Machine*. CODAME ART+TECH, San Francisco, CA, USA.

Elam, Michele and Rob Reich. (2022) 'Stanford HAI Artificial Intelligence Bill of Rights'. Stanford HAI White Paper, https://hai.stanford.edu/white-paper-stanford-hai-artificial-intelligence-bill-rights

Felt, Lindsey D. and Vanessa Chang. (2020) 'Curatorial Statement'. *Recoding CripTech* https://www.recodingcriptech.com/

Garland-Thomson, Rosemarie. (1997) *Extraordinary Bodies: Figuring Physical Disability in American Culture and Literature*. New York City, NY, USA: Columbia University Press.

Gaskins, Nettrice R. (2019) Techno-Vernacular Creativity and Innovation across the African Diaspora and Global South. In *Captivating Technology: Race, Carceral Technoscience, and Liberatory Imagination in Everyday Life*, ed. Ruha Benjamin, pp. 252–274, Durham, NC, USA: Duke University Press.

Generated Media. (2022) *Generated Photos*. https://generated.photos/#

Gould, Stephen Jay. (2011) *The Hedgehog, the Fox, and the Magister's Pox: Minding the Gap Between Science and the Humanities*. Cambridge, MA, USA: Belknap Press.

Hacking, Ian. (2006) 'Making Up People'. *London Review of Books* 28(16): 23–26.

Hansberry, Lorraine. (1972) '*Les Blancs*'. *Lorraine Hansberry: The Collected Last Plays*, edited by Robert Nemiroff. New York City, NY, USA: New American Library.

Katz, Yarden. (2020) *Artificial Whiteness: Politics and Ideology in Artificial Intelligence*. New York, NY, USA: Columbia University Press.

McKittrick, Katherine. (ed.) (2015) *Sylvia Wynter: On Being Human as Praxis*. Durham, NC, USA: Duke University Press.

Metzger, Sean. (2021) 'Editorial Comments: On the Possibilities of AI, Performance Studies, and Digital Cohabitation'. *Theatre Journal* 73: xiii–xx.

Mosseri, Adam. (2020) 'Interview "This Is the Guy Who's Taking Away the Likes" by Amy Chozick'. *The New York Times*. https://www.nytimes.com/2020/01/17/business/instagram-likes.html

Newsome, Rashaad. (2020) 'Interview "HAI Visiting Artist Rashaad Newsome: Designing AI with Agency" by Beth Jensen'. *Human-Centered Artificial Intelligence* https://hai.stanford.edu/news/hai-visiting-artist-rashaad-newsome-designing-ai-agency

Pollack, Barbara and Anne Verhallen. (2020) *Art at a Time Like This*. https://artatatimelikethis.com/

Ross, Fran. (1974) *Oreo*. New York City, NY, USA: Greyfalcon House.

Snow, C. P. (1961) *The Two Cultures and the Scientific Revolution*. New York City, NY, USA: Cambridge University Press.

Spillers, Hortense J. (2003) 'Interstices: A Small Drama of Words'. *Black, White and In Colour: Essays on American Literature and Culture* pp. 153–175.

Stark, Luke and Kate Crawford. (2019) 'The Work of Art in the Age of Artificial Intelligence: What Artists Can Teach Us About the Ethics of Data Practice'. *Surveillance & Society* 17(3): 442–455.

Weheliye, Alexander G. (2014) *Habeas Viscus: Racializing Assemblages, Biopolitics, and Black Feminist Theories of the Human*. Durham, NC, USA: Duke University Press.

Winant, Howard and Michael Omi. (1994) *Racial Formation in the United States*. New York City, NY, USA: Routledge Press.

Winger-Bearskin, Amelia. (2019) 'Antecedent Technology: Don't Colonize Our Future'. *Immerse News* https://immerse.news/antecedent-technology-b3a89956299d

Winger-Bearskin, Amelia. (2018) 'Before Everyone Was Talking about Decentralization, Decentralization Was Talking to Everyone'. *Immerse News* https://immerse.
news/decentralized-storytelling-d8450490b3ee

Wynter, Sylvia. (2003) 'Unsettling the Coloniality of Being/Power/Truth/Freedom:
Towards the Human, After Man, Its Overrepresentation–An Argument'. *CR: The
New Centennial Review* 3(3): 257–337.

15

The Cruel Optimism of Technological Dreams

Caroline Bassett

'To my ear, the genre of the "life" is a most destructive conventionalized form of normativity: when norms feel like laws, they constitute
... rules for belonging and intelligibility whose narrowness threatens
people's capacity to invent ways to attach to the world.'

(Berlant quoted from Prosser and Berlant 2011, p.182).

This essay explores gendered anxiety and hope about AI through an engagement with Lauren Berlant's (2011) exploration of cruel optimism—defined as that
condition, both individual and collective, sociological and ontological, in which
attachments to objects of desire are obstacles to their own becoming, being 'sheer
fantasy', 'too possible', or simply 'toxic'. Berlant's exploration of cruel optimism is
an element within a larger body of work focused on forms of affective experience
that characterise life in the contemporary world.

AI promises to transform this life, both because it will order it anew, and because
in doing so it will intervene in the *genre* of life; as lived, as practised, as imaginary.
Whether that intervention is likely to be freeing—widening people's capacity for
invention—or further narrowing or constraining it is less clear. It is, however,
striking that the terms Berlant invokes when defining cruel optimism, parallel
those invoked in discussions critical of what AI does (now) and might deliver (in
the future). In these, AI is excoriated on the apparently contradictory grounds that
it will fail to deliver on what it seems to promise (its 'sheer fantasy'), and that what
it says it will deliver seems all 'too possible' and all too 'toxic'.

Berlant died in 2021. In this essay, in part an appreciation, her feminist scholarship is used as a guide. Berlant was a leading feminist scholar of the affective turn.[1]
Her work turned on the body as the location of experience, on everyday life, understood both as individual and collective, and on desire or attachment as central to

---

[1] Berlant focussed her work in the United States and made her career there. She also worked with
multiple feminist scholars internationally notably at Lancaster with Anne Marie Fortier, Sara Ahmed,
and others.

Caroline Bassett, *The Cruel Optimism of Technological Dreams*. In: *Feminist AI*. Edited by: Jude Browne, Stephen Cave,
Eleanor Drage, and Kerry McInerney, Oxford University Press. © Oxford University Press (2023).
DOI: 10.1093/oso/9780192889898.003.0015

how life is lived. For Berlant life is given form by the social system within which it is produced. Their investigations are thus at once ontological in orientation, in that they explore embodied desires and drives, and constitute an investigation that is politically economic, being centrally concerned with the forms of life *differentially* available, or constrained in various ways, by political economic and geographical circumstances, notably by raced, gendered, and classed inequalities, and by the normative materialisations of bodies, the forms of embodied experience, these inequalities produce. In their work the public and private intersect. The intimate publics or 'affect worlds' (Jolly 2011 p.v) they explore include 'both the strangers formed into communities by affective ties; and the assumptions of shared emotions and worldviews that precede, create, and then often render anxious and ambivalent such publics' (Prosser and Berlant 2011, p.180). There is no sharp division between lifeworld and system here and Berlant is not interested primarily in tracking resistance in everyday life (the kind of capillary revolt that is central to de Certeau's exploration of tactics (1984) does not interest her). What she wants to understand is how life is *maintained* in the present moment, and she argues that affective attachments to objects beyond ourselves which may act as armatures for our desires are central to this. Life, she says, *endures* through attachments to objects that may support desires—that hold a promise for a better life, for health, for wealth, for justice, for instance. Or perhaps, let's say, for an upgraded life.

For Berlant attachments to, which are also desires for, are necessary for any kind of life to be lived; without desires there is nothing human. But she also argues that attachments may be made to objects that are harmful, or harmful for particular groups; specifically those without power, or choice, or voice; including those whose form of living breaks with what is given as the genre boundary of life itself— those whose bodies are deemed illegible by virtue of an alleged transgression in their sexuality for instance.

The promises AI makes are often explored of course, and often with the intention of dividing truth from hype, or marketing from science. Many studies seek to tell us what AI or an AI informed algorithm really is, or really can do. Exploring AI promises in terms of the attachments they support or *afford* for bodies individually and in affective communities demands not working with division (truth or falsehood, imaginary or real, ideology or technology) but exploring a relation. The affordance of what AI promises, that is, needs to be understood not only as a fixed characteristic of the object itself (identifiable once the hype is discarded perhaps) but as relational; in that what is offered, what promise can be supported or delivered, depends on to whom the offer is made to, to how it is graspable and in what way by that person. A promise, including a technological promise, affords differently to those engaging with it. Restituting this deeply relational understanding of affordance, central in J.J. Gibson's work on perception (1979),[2] but often being attenuated in its current application, can help us understand Berlant's sense of objects that promise. We may reach for the door, but, depending on who we are, or how we stand in a system that organises us in relation to powerfully normative categories, and that is riven with division, it may not open. Instead of letting us out it may indeed lock us in.

AI variously articulated, and continuously transferring into new objects, promises advancement, transformation, progress—and it rather explicitly makes these promises not only about technological progress but about progress *in life*. It is in this sense a proselytising discourse—or to be cruder it wants to sell us a future by promising one. What is clear however, anywhere AI is instantiated, is that while 'progress' is promised, and may be an object hard not to attach to, it is not delivered evenly; what is for some an enrichment or augmentation (economic, social, even embodied) may for others be an impoverishment, what is freeing for some forms of identity may be constraining for others. To reach for the promise of progress or invest in AI as a promise about life progress (synecdochally we might say) may therefore be hurtful rather than therapeutic, or curative. Groups experiencing historical and current discrimination and inequality—gendered, classed, racialised—stand in a particular relation to the promises AI make, given that the latter turn on claims to operate universally or generally working at levels *above*, or in *advance* of, or *beyond,* social distinctions. My interest, following lines Berlant developed, is partly in how this contradiction is lived with. What Berlant forces us to address, and helps us understand, is *why* many in groups that are likely to be hurt—and even why does *feminism*—keep investing in these AI promises?

Berlant's work is also invoked here to underscore how feminist work on affect and the body, exploring the politics of embodiment, and insisting on embodiment as integral to informational culture, not only *contributes* to ways of thinking about AI but is urgently *needed* as the field develops and as various tendencies within it accelerate (see also D'Ignazio and Klein, Chapter 12 and Amrute, Chapter 11, both in this volume). I include in those tendencies that propensity, noticeable in specialist AI discourse but also circulating in public fora, to assume that, since AI is a matter of abstract intelligence, offering logical calculation over affective engagement, bodies no longer matter, or no longer matter as much as they did, or will come not to matter soon; this is a revenant discourse whose timelines are characterised by a particular mix of vagueness and precision that is pernicious in that encourages an orientation towards a future fix. If AI currently produces problems around racism or sexism in algorithmic sorting operations, then these may be dismissed as glitches, temporary, and since they are *technical* glitches largely

---

[2] 'The affordances of the environment are what it offers the animal, what it provides or furnishes, either for good or ill' (p.127).

relevant in the register of the technical, not thought though in relation to bodily harm, glitches for which a technical solution will be found.

Berlant's work was rarely, if ever 'about' the computational or digital, but it nonetheless constitutes an injunction to feminist research into AI to remember the centrality of the affective, the experiential, and the embodied in the making of individual and collective lives even as they are increasingly made 'through information', or even as life is adumbrated in ways that apparently de-prioritise these qualities, valorising instead abstraction (for instance abstract smartness), virtuality (which flattens bodies), or etherealisation (which dissolves them). Neglecting embodiment encourages neglecting difference. One of the cruellest promises AI makes perhaps, cruel because it is seductive, and because it is false, is the one that says it will affect us equally, and so *render* us equal. A rendering is a taking down and there may be a kind of abjection involved in attaching to the AI promise for everybody—but some people can afford to give less away than others, their lives are more precarious and their bodies more fragile, and they are unlikely to benefit from what is then delivered in return·

Besides the Truth?

As noted, many investigations of AI as a motor for societal transformation turn on an assessment of the truth or untruth of the claims made for what it can do now and what it will be. They explore whether the promises made for instantiated AI tools are accurate, or they ask if predictions of AI-to-come that claim to be based on 'science' are 'truthful' or 'accurate', or if the recourse they make to various industrial, or scientific registers, is justified, or consistent with the evidence available. These kinds of engagements are sometimes very productive in their own terms. An issue with them though, is that focusing on the 'truth' of AI encourages the making of a division between ideology (presumed un-reliable) and material (presumed undisputable fact or purely a matter of a technical affordance), a parsing which is highly problematic since the ideological is enmeshed in the ways the materials are organised, and ideological assumptions are encoded into material. A corollary of that division, and with the search for the basic truth in AI in general, is that it assumes the truth is *out there* (perhaps even is to be revealed; a teleological impulse is bound up in this line of thinking), and those who cannot see it are simply gulled.

This is obviously problematic. It is not a secret that AI developments, say those relevant to the automation of cognitive work, will hit different groups differentially (see Nyrup et al., Chapter 18; Keyes, Chapter 17; Iyer et al., Chapter 20, all in this volume); the promise of automated process might look tattered from the get-go from the perspective of say a bank clerk—a now almost extinct being. A post-work society is understood by many as likely to result in a coming slow death (Berlant

2007),[3] rather than, say bringing luxury communism, or a creative renaissance. The arguments circulate, the predictions are made in public. There is not a dearth of information and nor, even in a time of epistemic crisis, a lack of ways to parse it.

So let's not start from there. To cleave to Berlant's lines of thinking is to shift attention away from matters of deception and truth in relation to AI promises and towards an investigation of how technological promises in general have purchase, or invite affective investment; how they hold, revive, fail, are scrambled for, and are continuously re-made as new waves of technology making good on what the wave before could not support emerge. AI contains something new, but it also constitutes the latest object that might support the familiar kinds of desires and hopes that technology often entertains; for the new, for the better, for progress, for order, for freedom; for instance. Truth claims are not irrelevant to how successfully new technologies hold promise, or constitute objects that may be objects that may be invested in. Certainly the promise of AI is not premised untruth, which might be central to conspiracy—although the attachment that each might support may not always be as far apart as they seem (social media, that affective cauldron of circulating 'information' tells us why). But desire does not rely for its force on rationally arrived at judgements (or even their impossibility—relevant in the case of AI given its claims to be emergent or uniquely unknowable), and the objects it attaches do not therefore have to be fully accounted for either in relation to misinformation or by the imputation of false consciousness. A better way to account for them might be to ask what resources they provide for affective investment, and a better way to judge them perhaps, might be to ask how these investments might be re-couped—or not.

This is partly why shifting the focus from what is 'true' about AI, to what promises AI makes, and to asking how these promises afford different kinds of support, what kind of *resource* they provide, how they are *experienced*, given how what they promise might *instantiate*, or not, and might do so *differentially*, seems to me useful if the goal is to understand how what technology promises to everybody it delivers partially to many—which is why in relation to divisions including race, class and gender—it may be a *cruel* promise, and one that may be recognised as such, even while it is also, in some way or other, kept faith with.

Everyday Life

Berlant was occupied with life as it is lived and was in revolt against the romance of everyday life as necessarily resistant, or autonomous, arguing that pervasive mediation long ago hollowed out much that was tactical or sub-cultural and linked it

---

[3] Slow death is term Berlant used in her study of the attrition of life for the poor in the US, notably in her writing on obesity (2007b).

into the system. Perhaps in this sense they aligned somewhat with the (literary) post-critical, although the sense of the necessity of recognising and responding to the violence of a structural system based on inequality was always undimmed in their work. Instead of resistance, Berlant explored managing, or failing to manage. The heart of a heartless world is often a *broken* heart. It continues to beat and to live, but often this is what it can do; this and not more. Life for many is straitened and constrained. In 'Intimate Publics', already quoted, Berlant argues that 'the genre of "life" is a most destructive conventionalized form of normativity' adding that when 'norms feel like laws', they 'narrow' attachment, threatening 'people's capacity to invent ways to attach to the world' or to have a life. Responding to this they suggest that 'queer, socialist/anti-capitalist, and feminist work has all been about multiplying the ways we know that people have lived and can live, so that it would be possible to take up any number of positions during and in life in order to have "a life"' (Prosser and Berlant, 2011 p.182). Berlant's explorations of desire and attachment constitute interventions at the level of theory, but there was also an ethnographic and diagnostic register to their work; and their extended studies focused on the contemporary United States, one of the epicentres of technological change, where for some attachment is fraught and life maintained with difficulty. The lives they explored are among those into which AI is emerging, among those for whom it might be an object of desire, something to be invested in because it can hold a promise, or is one.

AI as an Object of Desire?

What does it mean to talk about an attachment to a technology, or a desire that is invested in AI? Berlant defines attachment in relation to desire, and she defines objects of desire in terms of the promises we want them to make to us. However, she also qualifies the object almost immediately, accepting that objects *cluster*: 'When we talk about an object of desire, we are really talking about a cluster of promises we want someone or something to make to us and make possible for us' (2011, p.23).

Clustering is useful in thinking about AI. We could see the AI cluster as containing the multiple promises and potentials of AI—contested for, argued over, disputed, operating in multiple registers, articulated in symbolic and material form, making promises and producing outcomes. But this cluster is not self-contained nor heterogeneous. It contains sub-clusters which make specific, even contradictory, kinds of promises of their own and which may operate at different scales, and which certainly make sense (have purchase) differently in relation to the conditions in which they are encountered. It is tempting to suggest such sub-clusters have something in common with genres, being both theoretically divined and socially constructed: which is to say they are operational in the world. This would also suggest that some clusters have more purchase than others; perhaps they hold out a more powerful promise or seem more desirable.

Two AI clusters that have come to be well known, almost canonical, are briefly now adumbrated by invoking them as they appear in concentrated forms. They are familiar, widely invoked in popular and journalistic writing around AI and both might be said to partake of the general promise of AI. The first cluster is represented by a popular science book, Max Tegmark's *Life 3.0* (2019), the second cluster is identified by taking a look at 4IR discourse. Both clusters make their promises in a factual register—if fables of singularity remain relevant here they do so only in so far as they flavour a dream of the possible. Both work in in a register (the 'scientific'), which is continuous with 'real life' while also, as speculation, being to some extent fictive—and having the capaciousness that the latter produces (Bassett 2022; Skiveren 2018; for more the interplay between fiction and real life, see also Cave et al., Chapter 5 in this volume).

My interest here is in how *Life 3.0* stands, synecdochally, for a cluster of promises made by AI that relate to life itself. Focusing on rising complexity and intentionally blurring divisions between intelligent agency and intelligent life Tegmark argues that AI dismantles an old barrier between human and other forms of life and/or other forms of intelligence; the world is to become more intelligent, and more alive. This defines a promise based on rising liveliness and I would note its relation to new materialist writing pushing to expand an understanding of agency into things, rocks, stones, often involving a strongly vitalist sense (e.g. in Jane Bennett, 2015). Discourses of AI and life often appear to be pure speculation (in which, as we are supposed to know, investment is risky), or are utopian in their temporal appeal (possible futures), but increasingly make a claim or a promise that has purchase now. Consider how deepfakes blur the division between what is made and what born and that in doing so also seem to blur other divisions what is the difference between a fake cat and fake woman? Or consider AI-based language modellers that, even if they are stochastic parrots (Bender et al. 2021), begin to challenge simulation as a limit point of 'what machine intelligence may be', and to suggest new forms of lively agency; not life but something edging *closer* to some things like it. You might say these language agents chip away at a distinction long held to be definitional of the human, even as it had different consequences as historically instantiated; I am referring to that that human capacity to engage in, and acquire symbolic language, to speak, to write, to say. To enter the prison house—or escape from it?

Fourth Industrial Revolution (4IR) rhetoric, coming straight out of World Economic Forum (https://www.weforum.org/events/world-economic-forum-annual-meeting-2016), combines an investment in technological revolution to come with a marked social conservatism. The cluster of promises it makes focuses on cognitive automation, particularly on the wide scale automation of cognitive activities previously reserved for humans attached to occupations and professions, and on the extension of computational logics into biogenetics. The intervention into life here, the promise of change, comes with the assertion that automation enacted by machines can produce 'in real life' a new order, a new ordering. The whole is couched in terms of global progress, while also maintaining a strong sense of nationalist rhetoric. This cluster entails the acceptance of a form of *decrease* in liveliness for many humans, many of whom, it is admitted, will be automated out of the picture.

They are also distinctively different. One set of promises cluster around complexity and the other around order or mastery. At stake might be vitalism versus the expansion of dead order? Both clusters though make promises to resist entropy— through the ordering of life and the order that the algorithmic brings—and might thus have more to with each other than they at first appear. They point to the kinds of promises that the clusters of objects we call AI might make, the kinds of attachments it can entertain. They at once deal with (or in) a promise of exactitude (formalisation, the artificial) and also exploit the fuzzy edges of AI discourse, the vagueness of its claims. AI is striking for its capaciousness—perhaps even characterised by emptiness.[4] This is developed further later in the essay where I suggest that AI promises a promise, as much as holding out any specific promise say of the end of work, or of affluence. This may be key in linking technological promise (and within that the promise AI appears to make) to Berlant's work in other areas of life in the United States, with its, focus on desire and attachment and its role in the maintenance of an everyday life in a system riven by structural divisions— raced, classed, gendered, abled—in which attachment might be both necessary and harmful. Before discussing this further, however, a return to cruel optimism is required.

From Endurance to Cruel Optimism

'Cruel Optimism', the formulation for which Berlant is best known, has moved far from its original moorings in Berlant's writing on power, affect, and everyday life in the contemporary United States, becoming almost memetic, a travelling concept. I too am travelling with it, but I also want to remain faithful to Berlant's own sense that desire finds its objects (even organises its object relations?) in relation to the world (location, situation, context) in which it finds itself. If cruel optimism is a tool or a probe; a formulation, that can direct study in particular directions and do so effectively, it is also term that maps an affective formation directing activity that emerges in relation to, and as continuous with, everyday life.

---

[4] See Katz in *Artificial Whiteness* (2020), and Bassett et al. in *Furious* (2020), for different explorations of AI emptiness.

As noted, Berlant saw attachments as central and essential for life to be lived at all; 'endurance' as she put it, is 'in the object' (Berlant 2007, p.33). Investment in objects enables a form of being in the world, that is, it enables everyday life itself to continue.[5] Berlant would therefore argue that the maintenance of any kind of life *demands* that attachments are made, that we attach to objects beyond ourselves; to things, ideals, possibilities, practices, habits. More, by virtue of *being* attachments, and regardless of what they attach *to*, these may be viewed as optimistic, in the most brutal sense that they look forward, or offer continuity *going forward*. As Berlant puts this:

> ...whatever the content of the attachment, the continuity of the form of it provides something of the continuity of the subject's sense of what it means to keep on living and to look forward to being in the world
>
> (Berlant 2006, p.21).

The specific temporality of the attachments Berlant explores then, inheres in the degree to which they turn on a promise (a 'look forwards') that grounds a sense of the possibility of continuity in the present moment. This is what Berlant means then, when she comments that: 'in optimism the subject leans towards promises contained within the present moment of their encounter with the object'. This is not an argument about, or an exploration of, utopias, and attachments are not heterotopic either. Eagleton's distinction between hope and optimism, where the former is the more uncertain of the two but makes a more radical promise, while the latter is more likely to *deliver* but, because of that, operates within more constrained horizons is germane here, although the terms are deployed somewhat differently (see also Bassett 2022). The kind of distinction Berlant makes between attachment and its necessity within her account of endurance/maintenance/continuity in everyday life, and studies of the everyday relying on forms of what we might term (using Eagleton's division) utopian hopefulness, is key here. Moreover, it underscores the way in which the theorisation of cruel optimism might be understood as in part a post-critical gesture. Berlant felt perhaps, that a romantic attachment to the revolutions of everyday life, not least on the part of theorists, might neglect the difficulty many face in living it—or in formulating a more conscious politics in response to it.

Berlant's discussion of attachment suggests that the specific content of the object of desire might be subordinated to the continuity that *having* the desire and *attaching* to the object produces. Whether an object can hold, or sustain the promise it holds, partly depends on its content, but still it is clear in her account that attachment plays its part in construing the object or cluster it attaches to. How this

[5] There are strong resonances with Hannah Arendt's sense of the conditioning (1958), or perhaps the gearing necessary for engaging with the world—and her concern that automation might extinguish it.

is so becomes clearer when cruel optimism is introduced as a specific mode of attachment.

If attachments are necessary then they might *necessarily*—in the absence of other object choices—be made to objects that will fail to deliver, fail to deliver as expected, and will do harm to many. Cruel optimism is then defined as a relation of attachment to a 'significantly problematic object'. This is an attachment whose 'realization is discovered either to be impossible, sheer fantasy, or too possible, and toxic' (Berlant 2011, p.24) to invoke the terms with which I began, but also to note that we would need to ask *for whom* these problems arise, and perhaps also at which *intersections*, say of race and class and gender. Investing in this object or cluster of objects does not bring the fulfilment of the desire which led to the attachment nearer but on the contrary makes this less likely. It is when optimistic attachments frustrate the possibility of their own delivery that is when may first become 'cruel'. Berlant elaborates this in her work through an exploration of a series of engagements, gathered up and crystalised through the filmic and poetic, in which the attachments that found an everyday life are more, or less, cruelly operative; these include for instance a suburban queer idyll in which something *is* delivered, and a tale of race and its constraining horizons where access to the resources to escape from stifling poverty turn out to be toxic to the project (or the desire) to escape. The further twist is that this relation of attachment becomes crueller still when it is 'maintained' (Berlant 2007, p.33), even by those to whom it does harm or even destroys, perhaps because there is nowhere else, or nothing else, to invest in, nothing else that might hold the promise of making a promise, or take its form.

Cruel Optimism, Gender, and AI

How does AI as an object of desire, whose clustered contents can be read as a series of promises about the augmentation of life and about the improvement of everyday life relate to this formation? For whom might the investment in AI turn out to be cruel, to be harmful and yet persisted in? Further, is there something specific about this kind of attachment? Or characteristic about it as a mode of attachment to the technological?

To explore this demands first a return to the AI clusters identified above through the synecdochal examples of the popular book (Tegmark's 3.0), promising artificial life and the industrial manifesto promising a new order based on intelligent automation. The promise of order, notably of a new industrial order explored in the 4IR manifesto, is not new of course, nor is it new that it comes wrapped in the technological (think about hygiene and sunlight and the Industrial Revolution and its imaginary; smokestacks and chaos, but also, systems, regulations, regulated time), but with AI it has new purchase, and new force; takes new forms.

Viewed as a form of dispassionate processing that resolves human bias, AI affords an investment in the hope of a more just order to come. It appears indeed to elevate itself above the turmoil that human desire as a motor for societal operation is said to produce. Promises of data neutrality and impartial calculation, which might be attached to, hold specific promise, or have specific allure to groups who have historically been and continue to be the victims of partiality and non-neutrality of non-automated systems of ordering and governance, and who also experience violence in the application of these systems by humans. As noted below these promises have of course been addressed—and extensively—by scholars and activists who have exposed their emptiness (see e.g. Noble, Benjamin and others); but this does not mean they do not remain powerful—and are indeed powerfully re-made over and over again.

In so far as the promise of AI is to abstract the operations of the everyday (governance, order, administration, schooling, buying, marking, appointing, recognising, visualising) out of their embedding in historical, and historically biased, human-run systems, and subject them to impartial processes operating (apparently) beyond social divisions, this might be seen oddly enough as responding to a desire to escape from history into the present.[6]

The second cluster of AI promises identified above might seem more future-orientated. They cohere around augmentation, smartness, prosthetic reach—holding out the prospect of an intelligent world, a changed body, or life beyond bodily constraint. But this promise is held out not only in future-speculative accounts, but in promises of automated decision making, external memory, in avatar life; a second skin life you might say, is emerging as new forms of virtual reality (VR) develop. In either time the promise is one that promises to render flesh impervious to hurt, degradation, or imperfection, by improving it, overcoming it, or simply replacing it.

If these AI promises are cruel then this is first of all because their operations *harm* precisely those groups that are most likely to invest in the promise of neutrality, impartiality, justice. Algorithmic operations holding out the promise for forms of dispassionate governance are obviously and continuously and visibly undermined and fractured by AI operations in practice (algorithmic bias, everyday sexism, platform supported populism, racist image, and text recognition—in 2022 a camera can still advertise itself as novel for being capable of supporting Black skin tones). All this has, as noted, above been extensively surfaced by scholars in critical race theory, intersectional gender studies, and elsewhere too. For instance investing in artificial life, or in advanced humanity, narrows the possibilities for different and diverse lives to be recognised and valued; as theorists of disability have pointed out its starting point is the valorisation of a standard

[6] In this sense it resonates with Critical Race Theory's attention to the impossibility of progression, for groups who, it is asserted, live now, and unwillingly, in history.

human; artificial perfectibility is of a normative flesh model (see e.g. Cline, 2012, Simon, 2020). To return to Berlant; the norm remains the law—and even if lives lived differently through technology might challenge that, or leave some openings for other forms of life, an upgrade culture might also tend to continuously route around this, so that the norm expands to include previously refused positions, but in doing so, it might also standardise the forms in which the newly admitted normal is entertained.

Promise Crammed?

I now want to turn back to a more general promise that stands behind specific 'outputs', or programmes, or claims, so far discussed. This is the one that says AI is evidence of progress and stands as a guarantor of progress in general, and in that way it can stand also as a promise of better life in general. A fevered adherence to progress is central to AI discourses of many kinds. Evidenced in scientific writing, marketing, journalism, and variously materialised, it circulates widely and informs public understandings. It stands as the other to a recurring and constantly refreshing topoi of technological anxiety (see Bassett 2022) and turns on many of the same axes; an increase in cognitive agency, the extension of automation, new forms of control, machine neutrality and machine in-humanity, rationality/inscrutability, transparency, and black-boxing, something beyond 'the human' that constitutes 'human' progress. The wellsprings of love/hate are the same. What is not the same is how these promises are experienced as promises, as well as how they might be delivered on, across the social order, and in relation to geography and location. Whether they expand life or constrain it, increase control or subject the experiencing subject to tighter control, whether they constitute new laws or new freedoms, depends on by whom the promises are heard, in what contexts they resonate. What can be done with them, and what they do, let's say, is very different, depending on all this.

In a computational culture, in which computerisation, cognitive automation, sensor saturation, augmentation, VR, extend globally, a technological future, an AI delivered future, begins to claim to be the guise in which any promise about the future (which is any promise at all) will be made. Or even to be one of very few promises possible at all. This general promise of progress is doubly cruel, not only because, as Berlant has been concerned to note, many promises about the future, many things it seems possible to hope for, hurt those who are most fragile, but because the specific promise made is to overcome—by way of machinic virtues of precision, impartiality, abundance—precisely those kinds of divisions; class, race, gender, ableism, notably. The promise is to render difference itself irrelevant; through its impartial governance, the vision it offers being of a new good life for all—all watched over by those machines of loving grace.

The final turn might be about why attachment continues. A simple answer, already prefigured here, is that within computational capitalism the options narrow. There are other investments possible, often themselves reliant on media technologies—one of them is nostalgia perhaps—an attachment supported by the capacity of the media to let us live archivally, in which continuance is with the past, and another is conspiracy, where the attachment to a lie might maintain a certain form of expectation for a future once things are 'sorted out', as a marcher on the Capitol almost put it...

A further twist is provided by the temporal structure of AI promises. The promise of technological progress often appears to contain a cure—or a fix perhaps—for earlier damage.[7] Next time, or just around the corner, real neutrality, real autonomy, the post-work society of affluence will come; the temporal deferral characteristic of AI discourse, and perhaps of media technology in general, enables attachment to continue despite endless let-downs. If only we had *all* the numbers...[8] Following a logic of attachment the fix here does not so much attract because we hope to finally get the goods, the justice, the software, but because it sends a thread into the future; something is always *coming*. Continued attachment to the promise of AI is not surprising; in a computational culture, where all is mediated, what other choice is there? This is not being gulled. It is perhaps a way of engaging at all.

And Then?

My interest here has been to look at the affective force of the promises AI makes to produce new forms of life and new ways of living, and to ask for whom these promises are likely to be cruel.

Berlant's exploration of Cruel Optimism was diagnostic perhaps. She wanted to understand the fractures and fissures and precarity of everyday life in contemporary capitalism. She wanted to understand the powerful constrictions on forms of life that normative values impose and that structural divisions exacerbate. She did not focus on resistance. On the other hand, her critical position, speaking out against structural violence, which she understood as exercised differentially on bodies, as intrinsic to contemporary capitalism, makes it clear where she stands.

So, what does Berlant give technological feminism as it grapples with AI? For a feminism exploring AI her account is powerful not least because it begins by

---

[7] 'cruel optimism's double bind: even with an image of a better good life available to sustain your optimism, it is awkward and it is threatening to detach from what is already not working' (Berlant 2011, p.263); 'the fear is that the loss of the promising object/scene itself will defeat the capacity to have any hope about anything' (p.24).

[8] Arguably it is because the promise persists that the industrial response remains orientated towards 'fixes' and 'ethics afterwards' and that this is accepted.

refusing that reduction that placing 'the human' in opposition to the 'non-human' produces and instead focusses on embodied and particular and specific forms of experience. It is also important in how it pursues difference in relation to differential and discriminatory ways in which desire and attachment deal with those who desire and attach; cruelly for many. Again, this has relevance for thinking about the general promise of AI as offering progress, and the ways in which this is always already experienced in fractured, partial, and at times conflictual ways.

The final turn in this paper asks how this might relate to feminism as a political project. This is not an empirical question—again Berlant was not looking to valorise resistance in her focus on the cruel, though she acknowledges perhaps a kind of bravery in living, evident particularly in the lives of those who must live illegibly, whose lives do not make normative 'sense'. What I would like to address is whether Berlant's work is suggestive in thinking about the development of critical intersectional feminist responses to AI; and whether they can be in tune with her project in some way.

One way to ask this question is to wonder if feminism's longstanding engagement with computational technology is not itself characterised by a kind of cruel optimism; we attach, we know it is going to fail, we re-attach. This might be a way to ask what kinds of feminist politics might enable a reaching around Berlant's sense of fatigue, around the theorisation of resistant practice (or even everyday life as tactical as well as a matter of endurance, or perhaps a new fusion of the two)? I accept that, in suggesting this I may be understood to be myself exhibiting a form of attachment, that is (going to be) revealed to be cruel. Can Berlant's sense that cruel attachments are persisted in, in full knowledge of their futility, account for a widespread propensity within techno-feminist scholarship to continue to *re-invest* in resistance, as technology (endlessly) takes new forms? Is feminist techno-politics around AI, at least when it is hopeful, itself a kind of cruel optimism? What are the politics of the transfer or deferral or refusal of the promise of technology?

I think Berlant's work can be a powerful antidote to feminist thinking that mistakes its diagnostics for its programme; or finds resistance where what is needed might be a recognition of, or analysis of a situation. It refuses the attractions of a declarative but abstract politics—feminisms proliferation of digital manifestos is rabid (*mea culpa*) and arguably the form is often increasingly sterile; although there are always new surprises[9]. It forces attention towards a real that is not founded in a kind of technical reduction or stripping back (this is what AI does, this is AI instantiated, this is AI truth, found at the node, explicit in the algorithm, alien to the body and irrelevant to it).

My final question though, is, if so, is that enough? A diagnostics of the constraint that reduces hope for the future (or a politics of hope)—whether based on AI or

---

[9] And stalwarts; the longevity of the *Cyborg Manifesto* is startling (see Haraway 1991).

not—to a matter of an optimism of attachment, or an attachment to a cluster, a possibility, that allows for maintenance or persistence of the structures that make sense of the present, let alone one that is likely to be cruel, since in itself it frustrates the desires that launch it, may not be entirely melancholic (Berlant says it is not) nor fatalist—but neither is it ambitious. In a sense it is trapped, I think, in the *discourses* of AI (Berlant might call them genres), and forgets the material operations, the intervention of agency/agents, operations. These may be shaped by markets—so that we might say they disappoint (moving into the other debate around technology; utopia and its disappointment), in that they constrain rather than open, or produce more of the same. not the genuinely new, but they have other potentials. Matt Fuller (2008) reminds us that the shape of the technology the market gives us is not the only shape it can be. I would suggest that compulsory technology is not ontologically so—and compulsions after all, can be resisted or refused (Bassett, 2022).[10] A feminist engagement with AI might want to, need to, consider the *attachments* it promotes, which are not to do with what it desires of course (it isn't life, remember that was one of its cruelties), but might be a matter or what it *affords* and this might include radically new affordances. AI potentially at least has potential for finding new ways, and here I am back to Berlant, *to attach to the world...*

References

Arendt, Hannah. (1958) *The Human Condition*. Chicago University Press.

Bassett, Caroline. (2022) *Anti-Computing*. Manchester: Manchester University Press.

Bassett, Caroline, Sarah Kember, and Kate O'Riordan. (2020) *Furious: Technological Feminism and Digital Futures*. London: Monograph. Pluto.

Bender, Emily M., Timnit Gebru, Angela Mcmillan-Major, and Shmargaret Shmitchell. (2021) 'On the Dangers of Stochastic Parrots: Can Language Models Be Too Big?'. *Proceedings of the 2021 ACM Conference on Fairness*. (https://doi.org/10.1145/3442188.3445922).

Bennett, Jane. (2015) *Vibrant Matter. A Political Ecology of Things*. London: Duke UP.

Berlant, Lauren. (2006). 'Cruel Optimism'. *Differences* 17(3): 20–36.

Berlant, Lauren. (2007) 'Cruel Optimism: On Marx, Loss and the Senses'. *New Formations* 63: 33–51.

Berlant, Lauren. (2007b) 'Slow Death (Sovereignty, Obesity, Lateral Agency)'. *Critical Inquiry* 33(4)(Summer 2007): 754–780.

Berlant, Lauren. (2011) *Cruel Optimism*, London: Duke UP.

Brent, Walter Cline. (2012) '"You're Not the Same Kind of Human Being" The Evolution of Pity to Horror in Daniel Keyes's Flowers for Algernon'. *Home* 32(4): (no pagination).

de Certeau, Michel. (1984) *The Practice of Everyday Life*. California: California University Press.

---

[10] See also Mark Fisher on capitalist realism (2009).

Fisher, Mark. (2009) *Capitalist Realism: Is There No Alternative?* Winchester: Zero Books.

Fuller, Mathew. (2008) *Software Studies: A lexicon.* London: MIT.

Gibson, J. J. (1979) *The ecological approach to visual perception.* Boston, MA: Houghton Mifflin.

Haraway, Donna. (1991) *Simians, Cyborgs, and Women: The Reinvention of Nature.* New York: Routledge.

Jolly, Margaretta. (2011) 'Introduction: Life Writing as Intimate Publics', Jolly, Margaretta (ed.) Life Writing and Intimate Publics'. *Biography* 34(1): Winter: v–xi.

Katz, Yarden. (2020) *Artificial Whiteness Politics and Ideology in Artificial Intelligence.* Blackwell. Oxford.

Prosser, Jay and Lauren Berlant. (2011) 'Life Writing and Intimate Publics: A Conversation with Lauren Berlant'. *Biography* 34(1): Winter: 180–187.

Schwab, Klaus. (2016). 'The Fourth Industrial Revolution: What it Means, How to Respond'. 14th January, *World Economic Forum* https://www.weforum.org/agenda/2016/01/the-fourth-industrial-revolution-what-it-means-and-how-to-respond/

Simon, Victoria (2020). 'Democratizing Touch Xenakis's UPIC, Disability, and Avant-Gardism'. *Amodern* April.

Skiveren, Tobias. (2018) 'Literature'. *NewMaterialism.EU.* https://newmaterialism.eu/almanac/l/literature.html.

Tegmark, Max. (2019) *Life 3.0: Being Human in the Age of Artificial Intelligence.* New York: Knopf.

16

AI that Matters

A Feminist Approach to the Study of Intelligent Machines

Eleanor Drage and Federica Frabetti

Introduction

Feminist thinkers have played an important role in exposing and communicating AI's integration within and exacerbation of gendered power systems. Expanding on this, we argue that performativity, a well-established concept in gender studies and feminist science studies, is a useful tool for explaining in detail how AI produces the effects of gendered embodiment that it claims to describe or 'identify'. Theories of performativity have revealed, for example, that a person's gender is not an innate or binary human attribute but is instead enacted through repeated behaviours and activities, as Judith Butler proposed in the 1990s. The concept of performativity has also been used to insist that the material world does not precede our scientific and cultural observations of it but is actively *produced* by them, a phenomenon that Karen Barad calls 'agential realism'. Without downplaying their differences, we stage a dialogue between Butler's and Barad's concepts of performativity in relation to AI to explain how AI creates the effects that it names. We begin with an explanation of Butler's and Barad's theories and what kind of work they do in the study of software and AI, before locating AI's performative origins in Alan Turing's Imitation Game, in which gender is the modality through which the thinking computer can be perceived as such. Next, we show how both Butler's and Barad's concepts of performativity work together in applying feminist knowledge to AI at the level of its technical and conceptual functioning. Finally, we engage in a close reading of neural networks in the context of Facial Detection and Recognition Technologies (FDTR) and Automatic Gender Recognition (AGR). We show how these technologies make claims about the world that generate gendered and racialised interpretations of bodies in accordance with hegemonic value systems by iteratively 'citing' social norms in the act of 'observing'. This work translates concepts from feminist and gender studies into meaningful ways forward for AI

Eleanor Drage and Federica Frabetti, *AI that Matters*. In: *Feminist AI*. Edited by: Jude Browne, Stephen Cave, Eleanor Drage, and Kerry McInerney, Oxford University Press. © Oxford University Press (2023).
DOI: 10.1093/oso/9780192889898.003.0016

practitioners, users ('participants'[1]) and stakeholders when responding to the issue of AI's implication in existing power structures.

Performativity

In the 1990s, Judith Butler elaborated their now well-established concept of performativity, transforming gender studies' understanding of how a person is 'gendered' and gaining Butler star status in the field. Gender, Butler claims, never pre-exists a person; it emerges through a person's interaction with the world. These acts are 'citational': they always reference and re-animate a previous source. 'Gender' is only an assumed reality insofar as it is naturalised and reinforced through these repetitions. Performativity, as outlined in *Gender Trouble* (1990), is an elaboration of Jacques Derrida's reading of J. L. Austin's work on the 'performative utterance' (1990): authoritative speech acts that execute actions, such as 'Let there be light', 'I declare war' or 'I pronounce you husband and wife'. Butler's elaboration of linguistic performativity is particularly concerned with how authoritative speech 'performatively' sediments gender norms. In *Bodies That Matter* (1993), they take the example of a hospital medical professional who proclaims 'it's a girl' as they hand a newborn to its parent, demonstrating how these early performative utterances initiate bodies into gendered systems (p.232). In doing so, the authoritative speech is often 'performative', that is, it is able to enforce and enact a norm on a body. Speech acts can also inflict harm on bodies, as is more apparent in one of Butler's further examples: the shout of 'are you a lesbian!?' from a passing car (Olson and Worsham 2004, p.759). The proclamation echoes the words jettisoned at Frantz Fanon, and many besides, as he walked down the street as a child: 'Look, a negro!' (1967, p.111). Both declarations carry a greater force than the words themselves. This is because, Butler argues, they reference a heterosexual, and in Fanon's case, white norm, and carry the voice and the weight of predominantely white and heteropatriarchal institutions—the legal system, the police force—which mark them both as a social underclass. Butler draws on Michel Foucault's work on disciplinary power to argue that these speech acts demonstrate how the hand of the law can exert itself outside of law enforcement when society disciplines itself.[2]

---

[1] We follow Priya Goswami (2021) in recognising that 'users' are in fact active participants in AI systems.

[2] Since *Bodies that Matter*, Butler has complicated their conceptualization of the subject as something that is constituted through a speech act (and thus called or interpellated into being) in order to amplify the subject's capacity to undertake political action and to resist power. Addressing the complexities of Butler's radical critique of ontology, as well as their subsequent move towards what, for example, Stephen White (2000) calls 'weak ontology', is beyond the scope of this chapter. Here we focus on the concept of citationality (pivotal to Butler's analysis of the dynamics of power at play in the constitution of the subject) and we bring it to bear onto AI's construction of the reality it is supposed to analyse.

In the early 2000s, Karen Barad (2003) proposed a reinterpretation of performativity that foregrounded the active role of matter to counterbalance Butler's attention to the discursive aspects of gender. Barad's rereading of Butler enabled her to develop a performative posthuman framework that she named 'agential realism', which in turn spanned the New Feminist Materialisms (NFM) school of thought. Barad argues that Butler's destabilisation of the boundary between the material and the discursive does not go far enough, and that an account of matter is needed to develop a fully feminist scientific knowledge that goes beyond representationalism. Drawing on quantum physics, feminist and queer theory, and science studies, Barad proposes a performative understanding of 'how matter comes to matter'. She argues that the universe materialises differently according to the apparatus used by the observer. For instance, whether light materialises as a particle or as a wave is dependent on the apparatus deployed, that is, the 'material arrangements' used to detect it (Barad 2007, p.441). The observer and the observed do not pre-exist each other; rather, they constitute each other in the process of observation. Barad names this process 'intra-action' to emphasise the inseparability of the known and the knower, the material and the discursive, society and science, human and non-human, nature and culture. For Barad, matter is a substance in intra-active becoming—not an essence but a doing; it is a process of materialisation, of iterative intra-activity that stabilises and destabilises according to different apparatuses, each of which introduces a different 'agential cut'. Observing apparatuses, as well as abstract concepts, or theories, all enact onto-epistemological agential cuts: they are acts of knowing that also constitute the universe ontologically. There is no 'knower' and no 'known'; there is only the universe making itself understandable to (parts of) itself, in a constant process of unstable materialisation. In this self-differentiating process, boundaries and meanings are constantly rearticulated, and the differential co-constitution of the human and the non-human is always accompanied by particular exclusions and always open to contestation.[3]

Outside feminist and gender studies, the concept of performativity has gained traction in a wide variety of fields over the last two decades. These uses of performativity adhere to and stray in numerous ways from Butler's and Barad's theses. For example, in STS, Licoppe uses performativity to improve explanations about what information and communication technologies actually 'do' in

---

[3] In one example of the application of agential realism outside quantum physics and in relation to the human body, Fitsch and Friedrich (2018) draw on the concept of intra-action to argue that medical digital imaging technologies such as functional magnetic resonance imaging (fMRI) and computed tomography (CT) operate by aligning human bodies with the apparatuses used to observe them. The living materiality of the body is operationalised—they argue—and normalised according to the algorithmic logic of scanners, which dictates how bodies must be configured in order to produce contextually significant knowledge. The digital logic of the scanner is then mapped back onto the human body to correlate the functional model of the body with its anatomical mapping. In this sense, fMRI and CT can be understood as agential cuts that constitute visible and intelligible human bodies.

communication events (2010, p.181)—that is, they confer on data the meaning they have already acquired in its previous iterations. Economists have also argued that economic knowledge is performative because it moulds the market economy by 'reorganising the phenomena the models purport to describe' (Healy 2015, p.175). While this use of performativity strays substantially from linguistic theory, and in fact, only references work in economics and STS from the late 2000s (MacKenzie et al. 2007 and Healy 2015) it does interrogate a basic tenet of both Butlerian and Baradian performativity: that attempts to predict or identify a phenomenon instead create the effect that it names. Similarly, performativity has been used in machine learning (ML) to explain why predictive models can trigger actions that influence the outcome they aim to predict: for example, stock price prediction determines trading activity and hence prices, and product recommendations shape preference and thus consumption (Perdomo et al. 2020; Varshney, Keskar, and Socher 2019).

A substantial body of work in software studies draws on linguistic performativity to conceptualise software both as a cultural object and as a powerful agent with significant social and cultural influence (Fuller 2008; Mackenzie & Vurdubakis 2011; Kitchin 2017). These studies understand software as executable language, as a performative utterance that causes 'real' changes in the world through hardware, and as a material-symbolic inscription (Galloway 2004; Berry 2011; Chun 2011). While some of these works draw on linguistic performativity to problematise the boundary between the symbolic and the material, software and hardware, in a way that echoes Butler's theory, they overlook the ethical and political ramifications of these findings: that the performative aspects of software can result in systems enacting material damage to human bodies in accordance with social norms.

Finally, the notion of linguistics and gender performativity has been applied to AI with the purpose of improving its functionalities or attempting to de-bias it. Yalur (2020) draws on Derrida's work on linguistic performativity to explore Natural Language Processing (NLP) implemented through Recurrent Neural Networks (RNN). He argues that the logic of performativity is not fully applied by recurrent nets, which decontextualise language. For him, a better understanding of the performative nature of human language would allow NLP systems to move beyond current contextual constraints and to engage in more flowing, human-like conversation. Similarly, Butler's theory of gender performativity has been explicitly positioned as an analytical tool to address concerns around issues of racial and gender bias in facial recognition (Scheuerman et al. 2019). AGR is a particularly sensitive application of facial recognition because of the potential harm caused by misreading a person's gender and collapsing non-binary gender identities within the heterosexual dyad (Keyes 2018). Scheuerman et al. (2019) show how current commercial facial analysis services perform consistently worse on transgender individuals than on cisgender ones precisely because they represent gender through binary categories. Conversely, image labelling services are more apt at embedding gender diversity because they can handle multiple and contradictory gender categories—for example, they can associate gender presentation elements such as 'man', 'makeup' and 'girl' to a single image. Therefore, the authors identify specific points of intervention where a performative understanding of gender can be applied to AI with the aim of creating more gender-inclusive classifications. This study touches upon the idea that AGR uses the gendered physical appearance of individuals to merge their social and technical identities into a new algorithmic identity. However, it ultimately associates gender performativity with the diverse and fluid way in which individuals experience and perform their gender socially—a diversity that AI artificially stabilises and calcifies into fixed, bidimensional technical infrastructures. Similar to what we have seen in Yalur's article, Scheuerman et al. regard performativity as a useful tool to illuminate the variety and flexibility of human behaviour as opposed to the rigidity of machinic functionalities. Here performativity is viewed as a 'fix', as a useful tool to try to make AI systems more flexible, more 'accurate' in their representation of human performativity, and less biased. By presenting performativity as an intrinsically human process, they fail to acknowledge the imbrication and co-constitution of human and technological processes that Barad's understanding of performativity illuminates, an omission that seriously limits their ability to explain *how* AI genders the bodies of its participants.

In summary, these migrations of the notion of performativity across disciplines on the one hand show the appeal and the heuristic value of performativity, and on the other hand lead to substantial redefinitions of the notion itself (Gond et al. 2015). Further, all these uses of performativity show some conceptual affinities with Barad's and Butler's theories, while departing from them in other respects, thus maintaining some explicative aspects of feminist performativity while losing others. As we hope to have shown in the above overview, there are several reasons why performativity is so influential: it is able to elucidate how discursive/symbolic systems work with physical/material systems to generate 'real' consequences in the world; it illuminates how human and non-human agents work together to produce knowledge; it shows how knowledge creates the effects that it names, and, by investigating how AI operates iteratively it is able to explain phenomena that are otherwise inexplicable.

We argue that performativity is a robust concept with which to explore AI precisely because it enables the study of what something 'does' rather than 'is'. In gender studies, Butler used performativity to refute the normative assumption that gender is an essential human characteristic by directing attention towards what gender does, how it materialises, and the effect that this has on people's lives. Current discourses on AI are also benefiting from an engagement with its performative capacity in order to bypass the question of what AI 'is' and insead analyse the effects that it produces. In this volume alone, scholars demonstrate the different ways that social norms are enacted through AI: Os Keyes (Chapter 17) examines how AI negatively reconstructs the bodies and lives of autistic people, while Kerry McInerney (Chapter 7) demonstrates how racialised and gendered policing creates the crimes and criminals that it purports to identify. This work helps to direct the sector's attention beyond questions around AI's ontological status, including what constitutes an intelligent machine, and towards AI's epistemological implications in producing particular kinds of knowledge about the world.

We believe that the issue of what AI 'does' to those who interact with it is a more accessible point of entry for the general public into gaining knowledge about AI, because it pertains to people's own experience of the effects AI has on their bodies, political subjectivities and civil rights. And yet, aside from the work listed above, little has been written on AI and performativity that helps explain the ethical and political effects of the technology. In this chapter we take both Barad and Butler's concepts of performativity seriously and we begin to show what a performative understanding of AI might look like and how it can shift current debates around AI ethics. Without aiming to resolve the conceptual differences and tensions between Butler's citational understanding of performativity and Barad's materialistic and observational one, we bring different aspects of performativity to bear on examples of AI. Thus, we show how AI can be productively approached using the concepts of iteration, citationality, and agential realism that are typical of feminist theories of performativity.

Turing and AI's Performative Origins

To an extent, one could argue that the different technologies that currently go under the name of AI are intrinsically performative because 'intelligence' is something that they 'do' rather than 'are'. The Turing Test itself, with its hypothesis that a machine is intelligent if it can fool a human interrogator into thinking that they are interacting with a gendered human being, lays the foundation for a performative understanding of AI. A machine is said to pass the Turing Test (which Turing called an 'Imitation Game') if the interrogator is unable to tell the difference between the machine and a human interlocutor. The test has been subject to many critiques, which demonstrate that an agent (human or artificial) can exhibit intelligent behaviour without actually *being* intelligent; it can, for example, successfully engage in written conversation in a language it does not understand by using a sufficient number of symbols and grammars (Moural 2003). Such critiques pick up on the performative aspect of Turing's original scenario, which effectively does away with a precise definition of intelligence by positing that machines must be considered intelligent if they perform intelligence convincingly. As we have seen above, this resonates with Butler's argument that, although we may find it difficult to give an exact definition of gender, we (often incorrectly) believe that we are able to recognise it. Gender, like intelligence, can only be 'identified' by way of historically and contextually-produced proxies that stand in for the concept itself.[4] Thus, to a certain extent, we could say that (artificial) intelligence has always been citational and, like gender, it is always already a copy without original.

Indeed, as Hayles observes in *How We Became Posthuman*, the original version of the Turing Test was contingent on the performance of gender (Hayles 1999). The scenario that Turing devised in his classic 1950 paper 'Computer Machinery and Intelligence', which would set the agenda of AI for several decades, was based on a well-known parlour game, which involved a human interrogator guessing which of two unknown interlocutors was a man and which was a woman. The interlocutors communicated with the interrogator via an intermediary (Turing suggested using a teleprinter) so that their voices did not give the solution away. To complicate things further, one of the interlocutors tried to intentionally mislead the interrogator. Turing's question was: what would happen if a machine replaced one of the interlocutors? Would the interrogator guess gender correctly as often as they did when the interlocutors were both humans? If the results were similar, the computer could be considered intelligent. Therefore, as Hayles points out (1999, p.13), Turing's article established a curious equivalence between correctly guessing gender (the man/woman distinction) and correctly guessing intelligent humanity (the human/artificial divide). Overall, Hayles reads the Turing Test as representative of the erasure of embodiment in mathematical theories of communication, notably Claude Shannon and Warren Weaver's. Shannon and Weaver's work, which views information as an abstract quantity, would inflect AI for decades to come by reducing intelligence to the calculation and formal manipulation of symbols (Shannon and Weaver 1949). Further, Hayles observes how the very presence of a gendered imitation game signals the possibility of a disjunction between the physical, living body and its digital representation. Once this crucial move has been made, the question becomes that of bringing the physical and the represented body back in conjunction through technology by correctly guessing gender or humanity. For Hayles, the foundational test of AI demonstrates how AI contingently produces the overlay between the physical and the represented body.[5]

Hayles' perceptive reading of Turing can be reformulated in performative terms by saying that Turing's Imitation Game attempted to determine if computers performed (or 'did') gender and humanness as well as humans. Viewed through the lens of contemporary AI mis-readings of gender, such as with the AGR tools we have mentioned above, the presence of gender in the original Turing Test scenario

---

[4] See Stephen Cave (2020) on critical approaches to intelligence in AI.
[5] For a critique of Hayles's focus on Western versions of personhood, see Weheliye (2002, p.202).

acquires new relevance, as does the symmetry instituted by the test between gender and intelligent/intelligible humanity. Turing's thinking machine, like Butler's socially interpellated subject 'I', emerges through gender relations. For Butler, one enters the realm of society and humanity—that is, one becomes a human subject—only through a constitutive and performative act of gender assignment at birth. A baby cannot be a subject without also being gendered as either male or female. Subsequently, the inability to 'read' an individual's gender consigns them to the realm of non-existence; it dehumanises them. Strikingly, the issue of embedding and citing gender norms in a legible way is present at the core of one of the benchmark tests for AI: foundationally, AI aims at establishing norms of cultural legibility and intelligibility.

How Is AI Performative?

How does this process of establishing norms of cultural intelligibility work in contemporary AI? In this section and the next we explore how AI makes performative utterances that constitute, produce or obscure embodied subjects, therefore conferring upon them certain rights—first of all, the right to be represented in the political sphere—or depriving them of said rights. We also detail how the two different conceptualisations of performativity proposed by Butler and Barad allow us to explore AI both as normative and citational systems (Butler) and as observational apparatuses (Barad).

As we have outlined before, Butler understands gender as a process of performative citationality. Subjects are brought into existence through an authoritative speech act performed at birth—that is, a gender declaration. We see this in the historical perception of intersex bodies as 'problematic', since they cannot be read in a binary way. In this case, normative citationality produces a regime of intelligibility, within which the intersexed body is perceived as unintelligible. We argue that a similar regime of (un)intelligibility is produced when an AI system is unable to 'gender' an individual, as exemplified by Scheuerman et al. in their analysis of AGR systems that fail to 'read' transgender and non-binary individuals (2019). Given that FDTR, which include AGR technologies, are increasingly deployed in gatekeeper roles (for example as an aid to airport security), the potential harm caused by AI to people deemed 'unreadable' is huge (Leslie 2020). For example, a mismatch between the gender detected by an AGR system and the gender declared on the passport that a person carries with them can lead to problems getting through airport security (Halberstam 2021). The mismatch creates a confusion and ambiguity as to the legibility of the citizen, and can result in their rights of passage being denied to them or their interrogation at border control. As Toby Beauchamp has shown, attempts to objectively distinguish between bodies during airport security screenings 'displace the violence of state policies and practises onto individual bodies' and therefore mark particular bodies as threats according to gendered, racialised, and ableist ideals of normative bodily health (2018, p.51).[6]

AI's intelligibility regimes bind and constrain the movement of subjects, their rights, and their own emergence into political representation. Unintelligibility and invisibilisation find their ironic counterpart in the hypervisibility to which marginalised communities are subject when surveilled (Sadiya Hartman, 1997). As Ruha Benjamin points out, marginalised communities are suspended between hypervisibility and invisibility (2019, p.66).[7] This double bind is exacerbated by AI, which acts as a 'toggle' between the two possibilities—over which the populations subject to it have no control—shifting the state's gaze so that marginalised communities are either, for example, highly exposed to racialised surveillance or concealed when benefits are allocated. Ranjit Singh thinks of this bind as a phenomenon resulting from AI's ability to see in 'high' and 'low' definition, leaving some unduly exposed to its gaze and others excluded from essential social provisions (2021, pp.14.02–23.36). Ultimately, hypervisibility, viewed as surveillance, can be as disempowering as invisibility, which deprives a subject of its social and political existence because it does not conform to racial and gender norms.

The issue of AI's inability to create the conditions of positive intelligibility for a variety of non-binary, non-white subjects has become central to contemporary debates on AI. FDTR's failure to accurately read photographs of individuals with a darker skin shade is often discussed in connection with gender. As Buolamwini and Gebru (2018) have demonstrated, commercial gender classification systems are consistently and conspicuously more accurate when classifying lighter males than darker females. As indicated in the Introduction, this debate is often formulated in terms of bias: AI systems are found to be 'biased' against racialised, female or non-binary subjects. However, identifying and removing bias remains an often unfruitful task that assumes that the harmful consequences of AI are isolatable within a system rather than the product of technology's relationship to social norms, institutions, and economies. This is why a performative analysis of AI is important: because it shifts the focus from the pursuit of an impartial algorithm without bias or prejudice towards what Amoore calls an attentiveness to the

---

[6] Attempts to objectively identify anything using AI—even for the purposes of anti-discrimination— are always impossible. This is the case with AI products that claim to 'de-bias' human recruitment processes, for example, by introducing mathematical or statistical objectivity into the mix. A white paper released by recruitment AI software 'myInterview' suggests that subjective bias is a purely human phenomenon, while AI can be purely objective, and therefore hiring practices can use data science to circumvent the prejudices of hiring managers 'if you want the bias out, get the algorithms in' (myInterview n.d., p.4). However, AI still reproduces a normative 'ideal candidate' in its consideration of whether a candidate will be a good fit (Drage and Mackereth 2022).

[7] To make this point, Benjamin cites Britt Rusert's analysis of American minister and abolitionist Hosea Easton's mediation on the double bind experienced by Black subjects 'trapped between regimes of invisibility and spectacular hyper visibility' (2017, p 98). Easton's 1837 *Treatise* invoked the Declaration of Independence as an argument against racial discrimination.

'already partial accounts being given by algorithms' (Amoore 2020, pp.19–20). In the context of AGR, performativity also responds to Os Keyes' appeal for discriminatory systems not to be seen as 'biased' but as 'bullshit', a fallacy based on the belief that gender can be inferred from facial features and then measured (Keyes 2019). This false belief produces a disfigured and naturalised idea of what gender is.

When analysing how gender, race, bodies and subjects are performatively and citationally produced by AI, it is important to keep in mind that the normative materialisation of bodies is widely distributed across different parts of AI systems, and across different systems within the broader AI ecosystem. It is also likely to manifest in different forms, depending on the types of AI deployed in different contexts. AI systems are typically large, distributed and networked; they are unstable objects enacted through the varied practices of people who engage with them (Devendorf and Goodman 2014). Algorithms constituting AI systems are forever changing and adapting to participants in what is commonly described as 'personalisation'. This is the case for the AI technologies embedded in social media that filter content for different participants or make recommendations. As Seaver (2019) shows, what is commonly described as 'the Facebook algorithm' is never a single instance of a computational process. Rather, each Facebook user can only experience a different instance of that algorithm, adjusted to that user's navigational habits and history. In other words, 'we can't log into the same Facebook twice' (p.416). Furthermore, AI systems are often deployed by third-party developers, who integrate them into their applications, for example combining a pre-existing face recognition API with their proprietary text-generation AI system (Varshney et al. 2019). The high complexity and instability of AI systems and infrastructures is an issue for AI practitioners too, who often have limited understanding of the system they work on (Ananny and Crawford 2018). In sum, the algorithmic decision-making that affects both the most mundane and the most critical aspects of human life, from unlocking our smartphones with a glance to determining a previously convicted individual's likelihood to relapse into criminal behaviour, travel through layered and interconnected infrastructures. For all these reasons, generalising how social norms are embodied by AI is an impossible task, and each example of AI would have to be studied in its singularity. However, here we aim to lay the groundwork for feminist scholarship that demonstrates how AI is able to enunciate utterances that constitute, or obscure, embodied subjects.

Performative Deep Neural Networks

If we take AGR as an example (both in terms of facial analysis and image labelling services), it is important to notice that describing what these systems do as 'gender recognition' or 'facial recognition' is to misnomer them. What these systems achieve at most is a commentary on a set of images, performed through what is commonly described as ML, or, more often, an evolved version of it, Deep Learning (DL) based on Deep Neural Networks. As we will explore, such commentary is constrained by the system's presumption of what a face should look like, and by the way the system 'extracts' salient characteristics of an image of a face to iteratively and performatively compose a computational representation of it. As a consequence, an AI system's ability to 'read' embodied subjects is enmeshed with gendered norms that influence its capacity to shape the observed object as something that conforms to what the system already knows— that is, to the system's internal representation of bodies that results from its early training.

What does this look like at the level of neural networks? Again, it is important to stress how the following analysis is based on high-level technical narratives about the functioning of neural networks, rather than on the close reading of code or other forms of (always inevitably mediated) observation or experimentation with neural networks.[8] Like every other software system, AI can be described at many different levels of abstraction: for example, it can be 'narrated' in terms of logical circuits, binary code, source code, or natural language. Neural networks, which are but one component of an AI system, can be viewed as connections between nodes (roughly modelled on biological representation of neurons and synapses—hence the name), each connection being differently weighted to give the node an indication of how to evaluate inputs coming from other nodes, and each node being able to perform some basic computation. There are different kinds of neural networks, depending on the way they are structured and how they perform. Current models of DL are often based on Convolutional Neural Networks, which in turn are made of multiple layers of nodes, where each layer carries out some computation onto input data and feeds a single output to the next layer.

At the cost of some oversimplification, we can say that input data are numerical representations of a face image (for example, an image from social media) that are progressively parsed by CNN layers. This progressive 'reduction' of multiple inputs to one single output, using various computational methods, is the way in which ML parses complex data to obtain some form of simplified, more manageable abstraction that can then be used to perform further computation. Neural networks can also be referred to as 'classifiers' and, again for the sake of simplicity, they are in charge of performing categorising operations on datasets inputted in the AI system. It must be kept in mind that neural networks are implemented as software. In other words, this description of neural networks is

---

[8] We regard close readings of code, as well as various forms of empirical work (such as code writing and tweaking; experimentation with API for AI; ethnographic accounts, etc.), as valid and effective methods of investigation that remain beyond the scope of the present article and can be pursued in future work.

just one possible natural language reformulation of what Barad calls a material-discursive arrangement. Neural networks weave together inputs and outputs, data and computational processes, that can be also described as software (or code, or computer programmes). In Barad's terms this very narrative we are producing here to illustrate the functioning of neural networks *is* an agential cut, or an onto-epistemological operation that attempts to make neural networks understandable to the non-technical reader.[9]

With a slightly different agential cut, data is often visually represented as data points in a bidimensional space, where relative distances between data points can lead to their grouping into clusters, which can also be considered (in a more intuitive sense) as 'categories'. Neural networks (classifiers) parse data points to form clusters, which serve as categories to make datasets intelligible. The way in which a neural network parses data—by 'firing up' different nodes analogously to the way that neurons are activated in the brain—can vary according to the type of neural network. Inputs propagate backward and forward throughout the network. This is highly visible in backward propagation, or 'backprop', where a network recalibrates its weights according to the best pathway between nodes, therefore reconfiguring itself according to the input it is currently processing plus the previous inputs it has processed in the past and the outputs it has returned. In this sense, neural networks operate in an iterative way, which also depends on the context and the 'history' of the system—that is, on its previous computations.

This process is citational. Each previous computation is 'cited' by the neural network.[10] Each data parsing operation is an iteration; it is a slightly different rearrangement of the original network configuration, and this rearrangement is repeated again and again until it produces a configuration that can be considered a legible (and therefore gendered) version of the face image. In supervised learning, which is one of the most popular ways to train a network to recognise faces, human intervention is also required to help the neural network select the most acceptable configurations—those that enact an agential cut understandable as 'male face' or 'female face'. In Butler's terms, this intelligible configuration that

---

[9] This also goes to show the difficulties of explaining AI and the importance of complicating and questioning the idea of explainability. Understanding AI is so central today that it has given rise to a whole subfield of AI, Explainable AI (XAI), which aims at resolving the problem of black box AI (where even developers cannot explain how a solution has been arrived at). XAI is supposed to fulfil the social obligation to make steps behind AI's decisions understandable to the general public, in what has been called the right to an explanation (Gunning et al. 2019). We want to emphasise how discourses and technologies falling under the umbrella of XAI can also be viewed as agential cuts, and how concepts of performativity may provide a useful framework to explore and complicate Explainable AI.

[10] Citationality corresponds with Wendy Chun's assertion that AI makes 'the future repeat a highly selective and discriminatory past through correlation', so that truth equates to repetition (2021, p.36). Citationality differs however, in that it is a process of repetition with a difference and without an original, which we propose as a framework to understand how machine learning creates the effects that it names. Citationality does not so much validate a prior 'truth' (which exists only in the form of a copy), as establish the frame within which a subject is made intelligible.

returns (or rather embodies) a gendered output is a citation of the original configuration that perfectly describes (or embodies) the male or female face—except that
there is no original configuration that embodies a perfectly gendered face; there is
just a continual adjustment of parameters. In the same way that for Butler there is
no perfect embodiment of an original, essential, male or female gender, neural networks do not have an original. They reconfigure themselves according to previous
inputs, how such inputs have historically been parsed and what adjustments have
been made as a consequence. Human intervention is just a factor in the process
of constituting and reconstituting unstable intelligibility. This is the case regardless of whether it occurs at the level of the initial labelling of training datasets, for
example the manual association of face images with labels describing gender, or at
the level of network supervision, for example the selection of 'successful' network
configurations that match the label initially associated with a given image.

The different forms of data parsing we have just described are named 'learning' because they correspond to ideas about how humans 'learn'. It is important
that we recognise that data, neural networks, faces, images and learning processes
come together in AI as an unstable, ever-changing, dynamic process of materialisation. The resulting AI system, like all forms of knowledge, is 'a specific engagement
of the world where part of the world becomes intelligible to another part of the
world' (Barad 2007, p.342). In this way, neural networks are able to perform an
agential cut that materialises an observer (the neural network, operating in collaboration with its human supervisors) and an observed reality (face images).
Further, face images are materialised as such because this process is also citational. In other words, faces materialise according to the iterative citation of what
a gendered face is supposed to look like. This process obscures faces that do not
correspond to the gender binary norm embodied by neural networks. This obfuscation can happen in different ways, depending on the context in which the neural
network is deployed. For example, a face can be misgendered, therefore generating a mismatch between the photograph and the individual's gender identity. This
mismatch will happen in the AI system further down the line, in combination with
other parts of the system that perform the comparison between the 'observed'
gender of the image and the gender declared on documentation carried by the
individual (for example, a passport). This in turn will lead to the constitution of
the subject as abnormal. The individual's identity will be deemed impossible to validate and the individual will fail, for example, to get through airport security. In
other contexts, faces are obscured because they do not conform to what the neural
network is able to constitute as a face, as in the case of 'unintelligible' darker female
faces (Ahmed 2020)—a process in which gender norms play a part in association
with other norms (biometric features, readability of darker skin in photographs,
and so on).

In sum, feminist thories of performativity can demonstrate that AI has a unique
capacity to leave traces in the world. AI produces the reality it is supposed to

observe, and it does so performatively and citationally, depending on the singular circumstances in which it is developed and deployed. Acknowledging the performative way in which AI operates in the world radically shifts the terms of current debates on AI's ethics and politics. This is particularly crucial with regards to facial recognition—whether or not it is used for the purpose of gender recognition—because it proves that faces are constituted through an agential cut that is also the citation of a norm in a process of normative materialisation. Further, when we say that a neural network can learn to identify faces, we perform another 'agential cut', because the term 'learning' has epistemological, ontological and political implications. We assume that there is a 'reality' (for example, a face, or a human body) that can be known, and we confer upon the 'learning' algorithm the power to read this reality and to produce objective knowledge about it. A feminist and performative reconceptualisation of AI challenges the 'objectivity' of its outputs, illuminates how AI produces harmful effects on individuals and communities, and moves the debate on AI forward by enabling a wider range of stakeholders to be involved.

References

Ahmed, Maryan. (2020) 'UK passport photo checker shows bias against dark-skinned women', *BBC*, 8 October, https://www.bbc.co.uk/news/technology-54349538

Amoore, Louise. (2020) *Cloud Ethics: Algorithms and the Attributes of Ourselves and Others*. Duke University Press.

Ananny, Mike, and Kate Crawford. (2018) 'Seeing without Knowing: Limitations of the Transparency Ideal and Its Application to Algorithmic Accountability'. *New Media and Society* 20(3)(March): 973–989.

Barad, Karen. (2003) 'Posthumanist Performativity: Toward an Understanding of How Matter Comes to Matter'. *Signs* 28(3): 801–831.

Barad, Karen. (2007) *Meeting the Universe Halfway: Quantum Physics and the Entanglement of Matter and Meaning*. Durham University Press, 2007.

Beauchamp, Toby. (2018) *Going Stealth: Transgender Politics and U.S. Surveillance Practices*. Duke University Press.

Benjamin, Ruha. (2019) *Race After Technology*. Polity.

Berry, David M. (2011) *The Philosophy of Software: code and mediation in the digital age*. Palgrave Macmillan.

Buolamwini, Joy, and Timnit Gebru. (2018) 'Gender shades: intersectional accuracy disparities in commercial gender classification'. *Proceedings of Machine Learning Research* 81: 77–91.

Butler, Judith. (1990) *Gender Trouble*. Routledge.

Butler, Judith. (1993) *Bodies That Matter: On the Discursive Limits of 'Sex'*. Routledge.

Cave, Stephen. (2020) 'The Problem with Intelligence: Its Value-Laden History and the Future of AI'. *AIES '20: Proceedings of the AAAI/ACM Conference on AI, Ethics, and Society*.

Chun, Wendy. (2021) *Discriminating Data*. MIT Press.

Chun, Wendy Hui Kyong. (2011) *Programmed Visions: Software and Memory*. MIT Press.

Derrida, Jacques. (1990) *Limited Inc.* Galilée.

Devendorf, Laura and Elizabeth Goodman. (2014) 'The Algorithm Multiple, the Algorithm Material: Reconstructing Creative Practice'. *The Contours of Algorithmic Life*, 15–16 May, UC Davis, https://www.slideshare.net/egoodman/the-algorithm-multiple-the-algorithm-material-reconstructing-creative-practice.

Drage, Eleanor and Kerry McInerney. (2022) '"Does AI Debias Recruitment? Race, Gender, and AI's "Eradication of Difference"'. *Philosophy and Technology* 35(4): 1–25.

Fanon, Frantz. (1967) *Black Skin, White Masks: The Experiences of a Black Man in a White World*, trans. Charles L. Markman (1952). Grove Press Inc.

Fitsch, Hannah and Friedrich, Kathrin. (2018) 'Digital Matters: Processes of Normalisation in Medical Imaging'. *Catalyst: Feminism, Theory, Technoscience* 4(2): 1–31.

Fuller, Matthew (ed.) (2008) *Software Studies: A Lexicon.* MIT Press.

Galloway, Alexander R. (2004) *Protocol: How Control Exists after Decentralization.* MIT Press.

Gond, Jean-Pascal, Laure Cabantous, Nancy Harding, and Mark Learmonth. (2015) 'What Do We Mean by Performativity in Organizational and Management Theory? The Uses and Abuses of Performativity'. *International Journal of Management Reviews* 18(4): 440–464.

Goswami, Priya. (2021) 'Priya Goswami on Feminist App Design'. *The Good Robot Podcast*, 1 June, https://podcasts.apple.com/gb/podcast/the-good-robot/id1570237963.

Gunning, David et al. (2019) 'XAI-Explainable Artificial Intelligence'. *Science Robotics* 4(3): 7.

Halberstam, Jack. (2021) 'Tech, Resistance and Invention'. *The Good Robot Podcast*, 18 June, https://open.spotify.com/episode/09DWSEEPEG5CuAI2mZDU4q.

Hartman, Sadiya. (1997) *Scenes of Subjection: Terror, Slavery, and Self-making in Nineteenth-century America.* Oxford University Press.

Hayles, N. Katherine. (1999) *How We Became Posthuman: Virtual Bodies in Cybernetics, Literature, and Informatics.* University of Chicago Press.

Healy, Kieran. (2015) 'The Performativity of Networks'. *European Journal of Sociology* 56(2): 175–205.

Houser, Kimberly A. (2019) 'Can AI Solve the Diversity Problem in the Tech Industry? Mitigating Noise and Bias in Employment Decision-Making'. 22 *Stanford Technology Law Review 290*, By permission of the Board of Trustees of the Leland Stanford Junior University, from the Stanford Technology Law Review at 22 STA, https://ssrn.com/abstract=3344751

Keyes, Os. (2018) 'The Misgendering Machines: Trans/HCI Implications of Automatic Gender Recognition'. *Proceedings of the ACM on Human-Computer Interaction*, Vol. 2, Issue CSCW, November, pp. 1–22, https://doi.org/10.1145/3274357

Keyes, Os. (2019) 'The Body Instrumental'. *Logic*, Issue 9, 7 December, https://logicmag.io/nature/the-body-instrumental/

Kitchin, Rob. (2017) 'Thinking Critically About and Researching Algorithms, Information, Communication & Society'. *Information, Communication & Society* 20(1): 14–29.

Leslie, David. (2020) *Understanding Bias in Facial Recognition Technologies: An Explainer.* The Alan Turing Institute. https://doi.org/10.5281/zenodo.4050457

Licoppe, Christian. (2010) 'The 'Performative Turn' in Science and Technology Studies'. *Journal of Cultural Economy* 3(2): 181–188.

Mackenzie, Adrian and Theo Vurdubakis. (2011) 'Codes and Codings in Crisis: Signification, Performativity and Excess'. *Theory, Culture & Society* 28(6): 3–23.

MacKenzie, Donald, Fabian Muniesa, and Lucia Siu. (2007) *Do Economists Make Markets? On the Performativity of Economics*. Princeton University Press.

Moural, Josef. (2003) The Chinese Room Argument. In *John Searle, Contemporary Philosophy in Focus*, ed. Barry Smith, pp. 214–260, Cambridge University Press.

myInterview. n.d. 'The Definitive Guide to AI for Human Resources'. https://explore. myinterview.com/myinterviewintelligence

Olson, Gary A. and Lynn Worsham (2000) 'Changing the Subject: Judith Butler's Politics of Radical Resignification'. *JAC: A Journal of Composition Theory* 20(4): 727–765.

Perdomo, Juan C., Tijana Zrnic, Celestine Mendler-Dünner, and Moritz Hardt. (2020) 'Performative Prediction'. *Proceedings of the 37th International Conference on Machine Learning, PMLR 119*. http://proceedings.mlr.press/v119/perdomo20a/ perdomo20a.pdf

Rusert, Britt. (2017) *Fugitive Science: Empiricism and Freedom in Early African American Culture*. New York University Press.

Scheuerman, Morgan Klaus, Jacob M. Paul, and Jed R. Brubaker. (2019) 'How Computers See Gender: An Evaluation of Gender Classification in Commercial Facial Analysis and Image Labeling Services'. *ACM Human-Computer Interaction* 3, CSCW, Article 144 November, pp. 144–177, https://doi.org/10.1145/3359246

Seaver, Nick. (2019) 'Knowing Algorithms'. *DigitalSTS: A Field Guide for Science & Technology Studies*, eds Janet Vertesi et al., pp. 412–422. Princeton University Press.

Shannon, Claude E. and Warren Weaver. (1949) *The Mathematical Theory of Communication*, University of Illinois Press.

Singh, Ranjit. (2021) 'Ranjit Singh on India's Biometric Identification System, Representation and Governance'. *The Good Robot*. Podcast. 9 November, https://podcasts. apple.com/gb/podcast/the-good-robot/id1570237963

Turing, Alan. (1950) 'Computing Machinery and Intelligence'. *Mind* 9(236): October: 433–460, https://doi.org/10.1093/mind/LIX.236.433

Varshney, Lav. R. Nitish Shirish Keskar, and Richard Socher. (2019) 'Pretrained AI Models: Performativity, Mobility, and Change'. *Computers and Society* https://arxiv. org/abs/1909.03290

Weheliye, Alexander. (2002) 'Feenin: Posthuman Voices in Contemporary Black Popular Music'. *Social Text 1* 20(2) June (71): 21–47. https://doi.org/10.1215/01642472-20-2_71-21

White, Stephen K. (2000) *Sustaining Affirmation: The Strengths of Weak Ontology in Political Theory*. Princeton University Press.

Yalur, Tolga. (2020) 'Interperforming in AI: Question of 'Natural' in Machine Learning and Recurrent Neural Networks'. *AI & Society* 35(3): 737–745.

17

Automating Autism

Os Keyes

Disability is rarely considered by scholars working in Artificial Intelligence (AI) ethics, despite being a common focus of developers.[1] As this chapter demonstrates, that imbalance not only leaves questions of disability justice unattended, but renders invisible fundamental flaws in normative frameworks for 'AI ethics' writ large. Continuing this volume's engagement with ideas of personhood (Wilcox, Chapter 6; Atanasoski, Chapter 9; Rhee, Chapter 10: all this volume), I use this essay to take up the discursive construction of autistic lives in the design of algorithmic systems. I demonstrate how autistic personhood is fundamentally denied within these designs, using two case studies, and the ways in which the conventional ethical frameworks scholars have developed to address algorithmic harms are ill-equipped to confront the denial of personhood.

On the surface, it might be unclear how this essay fits in a volume on feminist AI—my analysis hardly mentions gender at all. But feminist theory is and always has been about more than gender alone. Feminism is about, amongst other things, forms of recognition; ways of recognising each other, ways of recognising knowledge, and addressing the forms of injustice that result from failures to do either or both. There is a reason de Beauvoir titled her introduction to *The Second Sex* 'woman as Other'. To be other—to be set aside from full personhood, to be ineligible for participation and recognition in moral debate—is the foundational harm feminism seeks to address.

Drawing from feminist epistemology (particularly the work of Lorraine Code on the status of persons as viable or non-viable 'knowers'), I seek to tease out precisely how autistic people are constructed (or ignored) in the development of systems putatively 'for' us. Although my analysis is focused on two particular case studies, and one particular form of disability, its implications—for how we design systems, but also how we approach processes to address systems' flaws—apply far more widely, reaffirming the breadth and value of feminist modes of analysis.

---

[1] A previous version of this chapter has been published in the *Journal of Sociotechnical Critique* in 2020.

Os Keyes, *Automating Autism*. In: *Feminist AI*. Edited by: Jude Browne, Stephen Cave, Eleanor Drage, and Kerry McInerney, Oxford University Press. © Oxford University Press (2023). DOI: 10.1093/oso/9780192889898.003.0017

Normative Views of AI Ethics

With the increasing development and deployment of AI, attention has turned to the question of 'AI ethics': the articulation of various approaches to the appropriate and 'good' use of AI. The widespread feeling that AI is 'a significant emerging and future-shaping technological field that is developing at an accelerating rate' (Goode 2018), and a corresponding rise in public, governmental, scholarly, and corporate interest, has led to a particular flourishing of both applied and theoretical scholarship on the ethics of AI. The result has been myriad sets of principles, guidelines and policies around 'good' AI, what it constitutes, and what is necessary to produce it (Whittlestone et al. 2019; Jobin et al. 2019).

The rapidly expanding nature of the field and its wide range of stakeholders means that these principles are yet to 'stabilise': theorists and practitioners frequently disagree over precisely what constitutes an ethical approach. But some components appear fairly consistently and frequently—in particular, notions of *fairness*, *accountability*, and *transparency* (Jobin et al. 2019). Although each of these principles have been conceptualised and articulated in many different ways, a broad-strokes summary would be that *fairness* requires an avoidance of discrimination in making algorithmic decisions, *transparency* the disclosure of the rationale behind any such decision, and *accountability* a mechanism of addressing any harmful consequences or algorithmic failures.

Fairness has been a particularly frequent topic of discussion. Approaching fairness as a technical problem—does a system produce uneven outcomes for different demographics?—both academic and industry researchers have begun focusing on technical tools to identify and correct discriminatory systems, seeking to fix one algorithm with another (Bellamy et al. 2018; Spiecher et al. 2018). Interdisciplinary researchers have similarly attended to fairness, treating questions of bias as a primary component of an algorithmic system's moral valence (Buolamwini & Gebru 2018; Chouldechova 2017).

AI Meets Disability

While gender and race are frequently deployed as protected characteristics to be scrutinised in evaluating algorithmic systems, disability is not. Instead, it is often left (at best) unmarked. Although a small number of works substantively discuss the ways that algorithmic systems could discriminate against disabled people, a 2019 review of 1659 AI ethics article abstracts found *eleven* containing disability-related keywords (Lillywhite and Wolbring 2019). This is particularly concerning given the increasing interest in explicitly applying algorithmic systems to questions of disability.

There are signs that this is beginning to change. A call to arms by Meredith Ringel Morris on 'AI and Accessibility' (Morris 2019), in parallel with a dedicated workshop at the ACM SIGACCESS Conference on Computers and Accessibility (ASSETS) 2019 (Trewin 2018), provided a rare focus on disability in discussions of AI ethics. Given the trend towards fairness as a general value in AI ethics, this heightened attentiveness frequently centres issues relating to bias and discrimination. The ASSETS workshop, for example, was specifically titled 'AI Fairness for People with Disabilities'. The problem is that this framing of ethics is anything but uncontested, as suggested by a paper *at* that workshop specifically contesting it (Bennett and Keyes 2019).

There are frequent critiques raised about fairness as a sole or primary ethical value for AI, both generous and pointed. Some researchers are concerned by the *immediacy* of fairness: the way that fairness-based approaches to ethics typically evaluate the immediate outputs of an algorithm, while leaving the longer-term consequences unexamined (Selbst et al. 2019). Others point to the manifold definitions of fairness, and the vastly different material outcomes produced by each one (Hutchinson and Mitchell 2019; Hampton 2021). Less optimistically, some critics highlight the treatment of 'fairness' as a value that can, at least theoretically, be modelled, as anything but an accident. Instead they contend that the focus on computable ethical principles that do not address more structural and longitudinal outcomes is *precisely the point*, constituting 'ethics-washing' that allows organisations to continue with 'business as usual' (Wagner 2018).

Discrimination and Discourse

One particular issue is the question of *discursive*, rather than *directly material* harms. As Hoffmann notes in her work on 'Where Fairness Fails' (Hoffmann 2019), a fairness-oriented frame, 'fails to appropriately attend to the legitimising, discursive or dignitary dimensions of data. . .algorithms do not merely shape distributive outcomes, but they are also intimately bound up in the production of particular kinds of meaning, reinforcing certain discursive frames over others' (Hoffmann 2019, p.908). In other words, what algorithms do is not just a question of material goods and (direct) material harms, but a question of the discourses and narratives they depend on, perpetuate and legitimise (see also Drage and Frabetti, Chapter 16 in this volume).

When used in critical scholarship, the term 'discourse' refers to how statements fit into knowledge; how they shape and signify what can be known, through what methods, and through what actors (McHoul and Grace 2015, Chapter 2). This is illustrated by Bivens and Hoque's 'Programming sex, gender and sexuality'

(Bivens and Hoque, 2018), which, as Hoffmann highlights, is an example of critical discourse analyses in technological domains. Bivens and Hoque investigate the discourses deployed in and around 'Bumble', a dating app billed as embodying feminist values. Exploring public relations statements by the designers, media coverage and aspects of the app's design, the researchers articulate how the 'feminist' figure who the app is designed for is specifically a middle-class, white, cisgender, and heterosexual woman, with sometimes-dangerous consequences (including the possibility of assault) for those who fall outside that mould.

Bivens and Hoque's point is not just that the app is exclusionary to a vast range of people, but that this exclusion *generates meaning*: within the world of Bumble, to be feminist is to be a white, cisgender, and heterosexual woman; to be male is to be a threat; to be a lesbian is to be non-existent. These frames, and the way that they resonate with wider cultural narratives, delegitimise particular populations. Bumble is not simply an app but a tool for meaning-making and knowledge generation—one that cannot, as designed, be positively applied to those outside a narrow set of norms.

Similarly, both the technologies and cultural im aginaries entangled with 'AI' serve as sources of meaning and knowledge. As a consequence, we should attend not just to whether particular populations are excluded, but the terms under which that happens: the justifications used, the framings they are subject to, and how this might reinforce or undermine damaging cultural frames regardless of what 'the software' is *intended* to do. If applications of AI ethics to disability do not (or *cannot*) investigate this, then the model of ethics we are using may allow vast harms to go unnoticed by those with the structural power to address them.

AI Interventions in Autism

To demonstrate these harms, I undertake a critical discourse analysis (CDA) of AI research publications and popular coverage that concern themselves with autism as a phenomenon, and autistic lives as a site of utility or intervention. This analysis concerns itself with how 'dominant discourses (indirectly) influence...socially shared knowledge, attitudes and ideologies...[and] facilitate the formation of specific social representations' (van Dijk 1993, pp.258–259). In the case of autism, I centre questions of what social representations of autism (and autists) are (re)produced in the corpus, and whose voices are included or excluded from the process of shaping those representations. Such an approach has been undertaken in other research on disability and technology, including Elman's work on wearable technologies (Elman 2018) and Spiel et al.'s inquiries into the experiences of autistic children using co-designed technologies (Spiel et al. 2017).

Sites of Analysis

The analysis focuses on two different arenas—one project, and one research subfield—both of which concern themselves with autism as a phenomenon, and autistic lives as a site of utility or intervention. Respectively they are *Artificial Intelligence for Autism Diagnosis* (or *AIAD*) and *Daivergent*.

AI for Autism Diagnostics (AIAD) originates in the perception that current autism diagnostics are 'expensive, subjective and time-consuming' (Jiang and Zhao 2017). By replacing existing mechanisms centred on conversations between doctors and patients, researchers hope to provide 'efficient objective measures that can help in diagnosing this disease [sic!] as early as possible with less effort' (Thapaliya et al. 2018). Such replacements come in a range of forms. Many papers use computer vision—machine learning systems that 'see'—to examine behavioural or social responses (Hashemi et al. 2018), evaluate eyeball movement (Jiang and Zhao 2017), or similarly, gait (Hasan et al. 2018), head movement (Bovery et al. 2019), or general upper-body form (Wedyan and Al-Jumaily 2016).

To investigate AIAD, I constructed a corpus of 82 papers that investigated the use of machine learning systems for autism diagnosis. Drawing influence from Waidzunas & Epstein's investigation of the history of the plethysmograph (Waidzunas and Epstein 2015), I followed the citation networks of papers that featured the terms ('autism' OR 'autistic') AND 'machine learning', incorporating into the corpus any papers that both cited a work in the initial 'seed' dataset, and concerned themselves with autism diagnostic or screening tools. These are narrow keywords; incorporating (for example) 'artificial intelligence', 'neural network', or more precise machine learning terminology would produce different initial seed papers. However, the reliance on citational networks rather than keywords alone goes some way towards mitigating this limitation. The resulting corpus is, while not comprehensive, fairly *cohesive*, with papers regularly citing not simply one other work within the corpus but many.

Corpus contributions pertaining to Daivergent consisted of media and marketing coverage of the company—both traditional venues (such as *The Wall Street Journal*) and non-traditional (the marketing blog of Amazon, whose software Daivergent uses)—that could be discovered through LexisNexis, along with the content of Daivergent's website.

Daivergent (the name of which plays on AI and the idea of autistic people as deviant or non-normative) originates with a very different perceived problem: the question of autistic people's unemployment. The company was founded by two data scientists, Bryon Dai and Rahul Mahida, both of whom have autistic relatives—a brother and a cousin, respectively—and funded by the venture capitalist Brian Jacobs, whose son is autistic (Galer 2019; Levy 2019). Concerned about their relatives' future after child-oriented disability services stopped being applicable, Dai and Mahida began Daivergent to provide a bridge between autistic people and the technology industry.

This bridge consists of, in parallel, offering autistic people jobs in classifying and 'hand-coding' the input data for AI, and training in workplace norms and practices. To the founders, pairing autistic people with hand-coding takes advantage of what they see as the nature of autism: a 'unique aptitude' for 'intensively focused, complex, repetitive processes' (Galer 2019). While most people get bored of such work, autists are seen as individuals who 'can do it for the day, can do it for the week, can do it month after month' (Kadet 2019). In exchange, they receive salaries of US$15–20 an hour, and the opportunity to 'gain a meaningful life' (Kung 2019), with the founders pointing to ex-employees who have gone on to work as a clerk, in a payroll role, or 'even in other places such as game design' (Galer 2019). Daivergent is hardly the only company seeking to market itself as rendering autists 'productive' in the technology sector, but it is (so far as I can determine) singular in positioning autists as a specialised workforce within AI; as unique assets in developing AI systems and the datasets they depend on.

Given that AIAD and Daivergent are very different sites occupying highly distinct environments, the strong alignment between the narrative representations of autism that they deploy demonstrates that these representations are likely to appear frequently in the field of AI. Similarly, while there are obvious differences in the types of source documents (publications versus news coverage), both constitute the most available material in which the actors represent themselves to their community and to the wider world. While they have different audiences, they are ultimately the same *kinds* of audience within the worlds that AI researchers and startup founders, respectively, occupy.

Analysing Discourses of Autism

After obtaining the source texts, I coded them following an approach based on critical discourse analysis (CDA), which focuses on the replication of dominant discursive frames—and the ways those frames constrain individuals and communities subject to them. In this case, my approach is one of 'sociodiagnostic critique': I seek not simply internal contradictions, but means to situate them in the context of wider discourses and society, and my own background knowledge (Reisigl and Wodak 2005).

To a certain degree, CDA is methodologically agnostic; there are few consistent approaches in how data should be collected and analysed (Meyer 2001). My approach consisted of looking particularly at how the source texts described or discussed autistic people or autism, and how autistic people were positioned in relation to the works. This approach generated a range of common themes including the following research question: how do AIAD and Daivergent's materials construct conceptions of autists' ability to *know*, and to communicate that knowledge?

Autism Discourses in AI

The terms 'to communicate' or 'to know' have a range of possible meanings and interactions. My understanding and use of those terms in this paper draw from feminist epistemologists who (from the 1970s onwards) have consistently attended to questions of knowledge and communication. This consists not just of examining what constitutes knowledge, but 'attention to what kind of subject one must be in order to be (seen as) a knowing subject' (Tuana 2017, p.126); attending to *who* can know. Within this frame, knowledge and communication are deeply bound up in one another. Someone who is not a recognisable knower is not a person, and vice versa. Both the ability to communicate and the ability to know thus have deep implications for personhood (Congdon 2018).

The social and reciprocal nature of knowledge and its construction is well-established in science and technology studies; as Helen Longino summarises, 'scientific inquiry is a collaborative human activity [and is] socially organized in certain ways that affect both goals and criteria of success' (Longino 1990, p.17). This relationship between knowledge and social recognition is not abstract; knowledge-making is deeply important to day-to-day activities and individuals' status in society. As Genevive Lloyd notes in her foundational work on feminist epistemology, in a society that conceives of itself around notions of rationality, the ability to know (and be seen to know) is deeply tied up with one's humanity (Lloyd 2002).

Communication and Knowledge of Others

An absence of *sociality* is 'often deemed to be a major feature of those diagnosed as being on the autism spectrum' (Milton 2012): it is a core component of narratives within research (Verhoeff 2012), current and defunct diagnostic criteria (O'Reilly et al. 2019), and public perception (Billawala and Wolbring 2014).

By sociality I mean the ability to appropriately and properly interact with others. This has implications around both communication and knowledge. When it comes to communication, autists are sometimes framed as literally lacking the ability to communicate with others, as many of us are non-verbal. For those of us who are verbal, our particular tropes are treated as inappropriate or invalid. These include echolalia (repeating the words of another), which is seen as containing no value (Roberts 1989), and overly direct styles of communication, frequently treated as rude or disruptive.

There are alternative interpretations of these: sociality as being constituable in autistic ways (Heasman and Gillespie 2019) and echolalia *as* a form of communication (de Jaegher 2013), but they distract from how autistic modes of communication are treated as less-valid and less intelligible. Moreover, the explanation for non-normative communication is often one of *knowledge*: rather than simply being ignorant, autistic people are seen as not being able to understand what is appropriate communication and incapable of understanding others.

Both diagnostic AI researchers and Daivergent figureheads are unified in pointing to abnormal social behaviour and communication as an autistic attribute. They cite 'serious shortcomings in their social skills' (Irani et al. 2018), more specifically: 'deficiency' in making eye contact (Uluyagmur-Ozturk et al. 2016), 'serious problems with being creative' (Lund 2009), 'difficulties' in recognising the emotions of others, and 'delay or perversion in language' (Altay and Ulas 2018). It 'makes ordinary social interactions particularly challenging' (Levy 2019), explaining the high unemployment rate: as one set of researchers mournfully inform us, 'about 50% of people with autism can never. . .make useful speech' (Altay and Ulas 2018).

Unsurprisingly, then, the interventions themselves build on and replicate these assumptions. One way of framing computer vision-oriented diagnostic tools intended to replace subjective interviewing is to understand their development as dependent on the presumption that diagnosis cannot rely on purposeful autistic communication. In the case of Daivergent, we see a repeated emphasis on the fact that the company provides not only jobs, but social skills opportunities: it emphasises that 'Daivergent stands out for the training it provides [. . .] not just technical skills but social and communication skills-training' (Welz 2019), and offers employees the ability to 'Join any of our 15 shared interest groups to meet like-minded individuals that share your passions' (Daivergent 2019a). Positioning itself as a provider of 'unique social and communication training opportunities', Daivergent operates from the implicit assumptions that autistic communities—of which there are many, including organisations serving/led by non-verbal autistic people (Yergeau 2018)—do not exist; that autistic communication must be guided and shaped by non-autists to be legitimate or capable of being recognised and understood (Daivergent 2019b; Demo 2017).

Knowledge of Self

The dominant explanation within normative ideas of autism is Baron-Cohen's model of 'Theory of Mind': the idea that autistic people simply lack empathy and an understanding of others (Dinishak 2016). From this comes the associations autism has with a lack of empathy, bluntness, and difficulties communicating.

But there are other implications that stem from this as well; implications about autistic *knowledge of self*. 'Empathy', in much theory and philosophy, is not something that just appears *de novo*: it is something learned, and premised on our own experiences. The analogy of a 'simulation' is used; we model our idea of others on our own senses of self, and simulate how *we* would interpret the situation were we in 'their shoes'. Consequently, normative theories of autistic minds do not just imply a lack of understanding of others, but that this stems from *a lack of understanding of self*. Jeanette Kennett, for example, uses 'the highest-functioning autistic people' as an intellectual foil, positioning them as '[having] *some* capacity for introspection about their condition' (emphasis mine); implying that the default state for autists is total ignorance of self (Kennett 2002). Autists are framed as unreliable narrators of their internal state, incapable of knowing and representing their needs or desires, much less communicating them.

In the absence of such knowledge, autistic people cannot be credible sources of information—not even information about ourselves. With Daivergent, it is notable that (with one exception, discussed later) no autistic people speak in their materials, press coverage, interviews or marketing reports. Instead, the idea of autism and the needs of autists are communicated by non-autistic people, pointing to the existence of autistic family members as a source of their expertise. Dai, for example, is depicted as having 'first-hand experience with the challenge' by dint of having a *brother* who directly experiences autism (Galer 2019), while Mahida states that 'We both [have] family members with autism. We know the type of things they enjoy doing', generalising those 'things' to autists as a whole, and speaking for autists as a population when he states emphatically that 'They want to work in tech. They want to work doing things for AI' (Kung 2019).

Within the diagnostic AI research, the bulk of users and perspectives centre familial voices rather than autistic ones. In Thabtah's study, the app was designed for use by 'a variety of stakeholders including parents, caregivers and more importantly health professionals', but never self-diagnosis (Thabtah 2019); Irani et al.'s project adapted to feedback solicited from 'the parents' (Irani et al. 2018); in Tariq et al.'s study, participation was determined by the parents—referred to throughout as the 'participants'—despite the data covering autistic people up to the age of 17 (Tariq et al. 2018). Indeed, as noted by M. Remi Yergeau, 'clinical constructions of autism frequently position expertise and self-knowledge as antithetical to autism itself' (Yergeau 2018, p.140). Under the discourse of autism used, autistic people cannot consent or give feedback, not simply because they cannot communicate but because they have nothing *to* communicate.

Knowledge, Agency, and Personhood

So if autists are entities lacking in the ability to communicate and be social, and further, lacking the ability to have knowledge of self (much less knowledge of others): do autists have agency? Personhood? Are autists, really, human?

I raise this question because the answer that dominant frames of autism provide is 'no'. Indirect inhumanity is communicated through representations of autists as alien (Reddington and Price 2016), robotic (Belek 2017), or (in much of ethics, and in 'autism advocacy') analogous to psychopathy: an interesting thought experiment in whether one can be a moral agent while quite so neurologically deviant (McGeer 2008). More directly, autism is treated as oppositional to the traits that 'make' a person a person (Duffy and Dorner 2011). Yergeau's critical summary is blunt, 'humans are human because they possess a theory of mind, and autistics are inhuman because they do not' (Yergeau 2013).

Portrayals of inhumanity in AIAD research and Daivergent's materials are largely indirect. One telling illustration comes from media coverage of Daivergent, discussing efforts to employ autistic people and other people with 'intellectual disabilities' (IDD):

> At Salesforce.com, a customer relationship management (CRM) software company headquartered in San Francisco, 46 IDD workers are currently core to the firm's operations, says Benny Ebert-Zavos, manager of real estate communications for the organization. 'We hire them to organize and maintain conference rooms, assist with event setup, support our reusable dish program, stock pantries, upkeep our social lounges, stock office supplies and brew coffee', he says. 'These folks are the key to making sure that when people come in, they can focus on work'.
>
> (Welz 2019)

Notable is the distinction between 'these folks' and people; the distinction between their labour and 'work'. Consider the rationales provided for hiring autistic people in particular; their dedication to engaging in the same tasks 'month after month' (Kadet 2019), a status that resonates more strongly with metaphors of machines than of people. But that is not all: companies should hire autistic people because they have 'perseverance' and a 'sense of loyalty'; because they are not going to *leave* (Levy 2019).

Non-agentism and inhumanity also feature, albeit more implicitly, in much of the AIAD literature. As well as discussing communication, the literature also evokes common myths relating to violence and risk. An autistic person has 'a very high risk for wandering; he can become very dangerous for himself, his family and the society as he can harm others as well as himself in an aggression' (Omar et al. 2019). Autism impacts 'self-control and [the] person's ability to learn' (Pahwa et al. 2016). Stimming and other 'stereotypical motor movements can lead to self-injurious behaviour under certain environmental conditions' (Albinali et al. 2009). In all of these framings, autists appear as figures who are—as a consequence of this dearth of outer awareness and communication—fundamentally *lacking*; lacking control over self, lacking the ability to engage in inference, lacking, in other words, in agency and the ability to choose. An autist is not a person—an autist is a machine, one whose misfiring outputs betray faults in their wiring.

Discussion

In my work here, I have examined the discursive framing of autistic communication and selfhood that is deployed by AIAD research, and the autist-employing startup Daivergent. In doing so, I suggested that in both cases, work follows a normative approach in describing autists as lacking in communication and sociality—and further, lacking in a sense of self. Autists are portrayed and perceived as unpersons: as inhuman, and as lacking in agency and autonomy. Next, I discuss the material and conceptual implications of AI research perpetuating and internalising this logic.

The Consequences of Normative Frames

Discourses around autism both structure society and are an important output of sociotechnical systems. These narratives have material consequences, leading us to ask what effects a normative framing of autism has on autists when played out in AI development practices. At one level we can treat the perpetuation of these discourses as a *reinforcement*: as reaffirming the 'truth' of autists as asocial and as inhuman. Profound and disturbing phenomena can be observed by looking at how autists are already treated in other sites as a result of this discursive framing. Some of it is interactional and quotidien, for example how the treatment of autistic sociality and communication as invalid and less-than creates heightened feelings of stigma and 'negative difference and feeling lesser', leading to the 'exhausting' work of hiding one's otherness, simulating normativity, for the fear of ostracisation should one be detected (Hodge et al. 2019).

Other material consequences are far more tangible and obviously violent. As a result of autistic communication being seen as an oxymoron, approaches to repairing communication failures between autists and non-autists are ones of *normalisation*: 'fixing' autistic people, rather than accepting the need for mutual adaptation. Such 'repair' is frequently violent, featuring—in the case of Applied Behavioural Analysis, the standard 'treatment' for autists—training centred on 'aversives': responding to autists stimming, communicating non-normatively or 'acting out' through withdrawing access to food, social interaction, or touch. Children may be subject to aversives 'in the forms of time-outs (often in closets, cells or segregated rooms), Tabasco sauce on one's tongue, spray bottles filled with vinegar, forced proximity to a cold or hot surface, physical restraint, screams directed at the child, and so on' all for something as simple as refusing to touch one's nose (Yergeau 2018, p.97). The most extreme form of this (or the logical conclusion of it, depending on one's level of cynicism) can be seen in the form of the Judge Rothenberg Center, located in Massachusetts, which uses 'aversives' such as straightjacketing, electrocution to the point of third-degree burns, and the inhalation of ammonia. Despite wide publicity, the centre has never been shut down (Adams and Erevelles 2017).

An immediate reaction to this is one of horror; what monstrosity! What inhumanity! But 'inhumanity' is the point; of *course* these are the therapies, of *course* the centre has not been shut down: those subject to these tortures are not *people*. They cannot consent, in the sense that they cannot say 'no'; what they say (if they can say anything), even about the treatment of their own body, cannot be taken seriously. As Adams & Erevelles point out, it is not that in the ensuing lawsuits autistic people did not testify as to their experience and assaults, it is that their voices were not taken as communicating valid knowledge compared to the normative credibility of doctors (Adams and Erevelles 2017). In a society that treats autism as a problem to be corrected, reinforcing notions of autistic incapability and non-agency—precisely as Daivergent and AIAD are—reinforces the legitimacy of violent and coercive interventions, because one can no more coerce an autist than a rock.

More directly: what does this construction of autism mean for AIAD patients, or Daivergent employees? If autists cannot validly know, whose perspective is foregrounded in the event that an autist disputes the outcome of a diagnostic algorithm? Whose perspective is foregrounded in the event that an autist disputes the morality of this algorithmic work in general? When autistic employment is oriented around assumptions of roboticism and machinic lack of self, what happens when autistic employees *display* autonomy? It is hard to imagine an AI company that sees autists as asocial or non-agentic as taking seriously, for example, attempts to unionise: a union of autists would be a contradiction in terms.

The Consequences of Normative Ethics

In this case study, we have seen how discourses of autists as asocial and non-agentic produce material harms—but it seems to me we should also ask what flaws they highlight in AI ethics frameworks for addressing those self-same harms. AI practitioners are discursively framing certain populations as non-human and non-agentic in parallel with ethical frameworks that depend on humanity and agency for addressing harms and they are doing so under a normative, default set of values that (as discussed before) seem widely agreed upon as a good starting point if not an entire system for achieving justice.

If autistic people are being constructed by AI practitioners as incapable of agency and full humanity in an ethical framework that treats agency and full humanity as mutually dependent, and both as necessary prerequisites for participating in the frameworks that address injustices, then we have an impasse. If our approach to ethics is simultaneously established around notions of communication, credibility, and recognition, and framing autistic people as lacking in those things, there is no viable way for autists to participate in processes that are frequently treated as the panacea to any injustice this domain generates. Autists will be subject to both discursive and material violence, and the former will strip us of the ability to viably dispute either.

One immediate solution to this might appear to be to move the goalposts—to declare that discursively framing autists as less-than-human is wrong. This would certainly help, although the issue is far more widespread than one of discourses within AI ethics. However, maintaining rigid boundaries around who counts as a person and as a knower is nearly ubiquitous in normative philosophy more generally. As the feminist philosopher Lisa Schwartzman highlights in her critique of liberalism, liberal philosophy often treats people as 'fully rational, mutually-independent decision makers' (Schwartzman 2006). Correspondingly, individuals who do not meet these conditions are not people—they are denied access to decision-making processes and modes of political or ethical engagement. Such an approach is frequently criticised, on the same grounds as my concerns with AI ethics. And in both cases, because the resulting ethical frameworks *assume* such status, they are frequently 'strangely silent about the predicaments of outsiders' (O'Neill 2000, p.4); as Lauren Davy notes in reviewing the work of John Rawls, disability is 'relegated. . .to a footnote. . .a problem to be worked out later when all other matters of justice are settled' (Davy 2019, p.105).

Critiques of these approaches, and the uneven distribution of what counts as rationality and interdependence, provide a set of ideas to ameliorate the resulting harms. In particular, the work of José Medina and Miranda Fricker highlights the need to engage in work that includes not only openness and self-criticality in how we interpret people and perspectives on an individual basis, but the construction of forms of 'hermeneutical resistance': ways of knowing and communicating that actively push back against monolithic ideas of personhood and knowledge (Medina 2013; Fricker 2007). In the case of AI ethics, this might look like actively pushing back against proposals for monolithic conceptions of justice, or mechanisms for achieving it, while developing more polyphonic and adaptive approaches.

More broadly, we might consider different ways of conceptualising personhood altogether. A feminist 'care ethics' approach to personhood and knowledge might treat not just disabled people but all people as dependent on communities, infrastructure, and relationships (Davy 2019). This approach could be furthered through an examination of the 'posthumanist ethics' of Karen Barad and others, as Drage and Frabetti explore in this volume, and which (as adroitly explained by Natasha Mauthner) 'seeks to conceptualize ontological, epistemological, and ethical agency without recourse to the human subject' (Mauthner 2019, pp.680–681).

Still, I am cautious and cognizant that these suggestions are ultimately efforts to *ameliorate* dehumanisation in the structure of mechanisms for correcting injustice. As demonstrated by the work of agonistic theorists in political philosophy, there is no singular approach that will 'solve' the question of otherness and silencing (Honig 2016; Mouffe 2000). Regardless of where we draw the line with regards to personhood, knowledge, and access to justice, we are still drawing a line, marking some as legitimate and some as not. While I make the pragmatic demand that AI ethicists consider the discursive impact of our technologies, and the weakness of our frameworks when confronted with disabled perspectives, I do not believe that we can escape the silencing and perpetuation of injustices altogether.

But what we can do is confront dehumanisation in how we theorise justice and the mechanisms that we design in pursuit of it. We can understand harm as an inescapable consequence of efforts to reduce it, and view those efforts as ultimately contingent and open to challenge. Most broadly, then, my demand is not simply for a consideration of discursive harm and disabled voices, but a more wide-ranging insistence that we avoid the fatal mistake of treating any mechanism or set of principles as settled.

One approach would be to make sure that we simply treat autistic people as people while refusing conceptions of personhood that leave some individuals dehumanised and unable to access frameworks for addressing harms, whether or not those individuals are autistic. Instead, I advocate that critical attention be paid not only to the immediate barriers to accessing justice, but the status we give to 'personhood' in the first place. This includes, as discussed, a greater attentiveness to the conditions under which we evaluate knowledge and communication, as well as efforts to acknowledge the relational—rather than hyper-individualised—nature of the ethical agent. I encourage researchers and practitioners concerned about disability justice specifically, or weaknesses in our ethical frameworks more generally, to consider these possibilities.

My intention here is not to demand some particular universal ethic to replace the current one; I am unsure whether *any* universal approach can resolve these issues rather than replicate them in new forms. Instead, my goal is simply to encourage a conversation about how violence too-often depends on our willingness to treat the terms of 'humanity' and 'personhood' uncritically, and accept them as a prerequisite for ethical attention. To this end, I want to underscore how vital it is that we retain and reinforce that critical lens—that we avoid treating any term of art or scheme of justice as unquestionable and 'settled law'. Who can play the game is a vital question within feminist theory, critical disability studies, and AI ethics independently, let alone in intersect. But an equally important question to ask is whether the dice are loaded. Working towards justice requires us to continually ask them both.

References

Adams, D.L. and Nirmala Erevelles. (2017) 'Unexpected Spaces Of Confinement: Aversive Technologies, Intellectual Disability, and "Bare Life"'. *Punishment & Society* 19(3): 348–365.

Albinali, Fahd, Matthew S. Goodwin, and Stephen S. Intille. (2009) 'Recognizing Stereotypical Motor Movements in the Laboratory and classroom: A Case Study with Children on the Autism Spectrum'. In: *Proceedings of the 11th International Conference on Ubiquitous Computing. ACM*, pp. 71–80.

McHoul Alec and Wendy Grace. (2015) *A Foucault Primer: Discourse, Power and the Subject*. Routledge.

Altay, Osman and Mustafa Ulas. (2018) 'Prediction of the Autism Spectrum Disorder Diagnosis with Linear Discriminant Analysis Classifier and k-Nearest Neighbor in Children'. In: *2018 6th International Symposium on Digital Forensic and Security (ISDFS). IEEE*, pp. 1–4.

Belek, Ben (2017) I feel, therefore I matter: Emotional rhetoric and autism self-advocacy. *Anthropology Now* 9(2): pp. 57–69.

Bellamy, Rachel K. E., Kuntal Dey, Michael Hind, Samuel C.Hoffman, Stephanie Houde, Kalapriya Kannan, Pranay Lohia, Jacquelyn Martino, Sameep Mehta, Aleksandra Mojsilovic, Seema Nagar, Karthikeyan Natesan Ramamurthy, John Richards, Diptikalyan Saha, Prasanna Sattigeri, Moninder Singh, Kush R. Varshney, and Yunfeng Zhang. (2018) 'AI Fairness 360: An Extensible Toolkit for Detecting, Understanding, and Mitigating Unwanted Algorithmic Bias'. arXiv preprint at arXiv:1810.01943.

Bennett, Cynthia L. and Os Keyes. (2019) 'What Is the Point of Fairness? Disability, AI and the Complexity of Justice'. arXiv preprint at arXiv:1908.01024.

Billawala, Alshaba, and Gregor Wolbring, (2014) 'Analyzing the Discourse Surrounding Autism in the New York Times Using an Ableism Lens'. *Disability Studies Quarterly* 34(1): DOI: https://doi.org/10.18061/dsq.v34i1.3348.

Bivens, Rena and Anna Shah Hoque. (2018) 'Programming Sex, Gender, and Sexuality: Infrastructural Failures in the "Feminist" Dating App Bumble'. *Canadian Journal of Communication* 43(3): 441–459.

Bovery, Matthieu, Dawson, Geraldine, Hashemi, Jordan, and Sapiro, Guillermo (2019) 'A Scalable Off-The-Shelf Framework for Measuring Patterns of Attention in Young Children and its Application in Autism Spectrum Disorder'. IEEE Transactions on Affective Computing 12(3): 722–731.

Buolamwini, Joy and Timnit Gebru. (2018) 'Gender Shades: Intersectional Accuracy Disparities in Commercial Gender Classification'. In: *Proceedings of the Conference on Fairness, Accountability and Transparency*, 77–91.

Chouldechova, Alexandra (2017) 'Fair Prediction with Disparate Impact: A Study of Bias in Recidivism Prediction Instruments'. *Big Data* 5(2): 153–163.

Congdon, Matthew (2018) '"Knower" as an Ethical Concept: From Epistemic Agency to Mutual Recognition'. *Feminist Philosophy Quarterly* 4(4): DOI: https://doi.org/10.5206/fpq/2018.4.6228.

Daivergent (2019a) 'Interested Candidates'. *Daivergent* URL https://web:archive:org/web/20191114170952/ https://daivergent:com/work-readiness/candidates.

Daivergent (2019b) 'Interested Organizations'. *Daivergent* URL https://web:archive:org/web/20191114171242/ https://daivergent:com/work-readiness/organizations.

Davy, Laura (2019) 'Between an Ethic of Care and an Ethic of Autonomy: Negotiating Relational Autonomy, Disability, and Dependency'. *Angelaki* 24(3): 101–114.

De Jaegher, Hanne (2013) 'Embodiment and Sense-Making in Autism'. *Frontiers in Integrative Neuroscience* 7(15): DOI: https://doi.org/10.3389/fnint.2013.00015.

Demo, Anne Teresa (2017) 'Hacking Agency: Apps, Autism, and Neurodiversity'. *Quarterly Journal of Speech* 103(3): 277–300.

Dinishak, Janette (2016) The Deficit View and its Critics. *Disability Studies Quarterly* 36(4).

Duffy, John and Rebecca Dorner. (2011) The Pathos of 'Mindblindness': Autism, Science, and Sadness in 'Theory Of Mind' Narratives. *Journal of Literary & Cultural Disability Studies* 5(2): 201–215.

Elman, Julie P. (2018) '"Find Your Fit": Wearable Technology and the Cultural Politics of Disability'. *New Media & Society* 20(10): 3760–3777.

Fricker, Miranda. (2007) *Epistemic Injustice: Power and the Ethics of Knowing.* Oxford: Oxford University Press.

Galer, Susan (2019) 'Mother's Wish Spurs Childhood Friends to Pioneer Autism Talent Pool Startup'. *Forbes.* URL https://www.forbes.com/sites/sap/2019/03/28/mothers-wish-spurs-childhood-friends-to-pioneer-autism-talent-pool-startup.

Goode, Luke (2018) 'Life, But Not As We Know It: AI and the Popular Imagination'. *Culture Unbound: Journal of Current Cultural Research* 10(2): 185–207.

Hampton, Lelia Marie (2021). 'Black Feminist Musings on Algorithmic Oppression'. In *Conference on Fairness, Accountability, and Transparency (FAccT '21), March 3–10, 2021, Virtual Event, Canada. ACM, New York, NY, USA*, 12 pages.

Hasan, Che Zawiyah Che, Rozita Jailani, and Md Tahir Nooritawati. (2018) 'ANN and SVM Classifiers in Identifying Autism Spectrum Disorder Gait Based on Three-Dimensional Ground Reaction Forces'. In: *TENCON 2018–2018 IEEE Region 10 Conference. IEEE*, pp. 2436–2440.

Hashemi, Jordan, Dawson, Geraldine, Carpenter, Kimberly L., Campbell, Kathleen, Qiu, Qiang, Espinosa, Steven, Marsan, Samuel, Baker, Jeffrey P., Egger, Helen L., and Sapiro, Guillermo (2018) 'Computer Vision Analysis for Quantification of Autism Risk Behaviors'. *IEEE Transactions on Affective Computing* 12(1): 215–226.

Heasman, Brett and Alex Gillespie. (2019) 'Neurodivergent Intersubjectivity: Distinctive Features of How Autistic People Create Shared Understanding'. *Autism* 23(4): 910–921.

Hodge. Nick, Emma J. Rice, and Lisa Reidy. (2019) '"They're Told All the Time They're Different": How Educators Understand Development of Sense of Self for Autistic Pupils'. *Disability & Society* 34(9–10): 1–26.

Hoffmann, Anna Lauren (2019) 'Where Fairness Fails: Data, Algorithms, and the Limits of Antidiscrimination Discourse'. *Information, Communication & Society* 22(7): 900–915.

Honig, Bonnie. (2016) *Political Theory and the Displacement of Politics.* Cornell University Press.

Hutchinson, Ben and Mitchell, Margaret (2019) '50 Years of Test (Un) Fairness: Lessons for Machine Learning'. In: *Proceedings of the Conference on Fairness, Accountability, and Transparency. ACM*, pp. 49–58.

Irani, Atefeh, Hadi Moradi, and Leila Kashani Vahid. (2018) 'Autism Screening Using a Video Game Based on Emotions'. In: *Proceedings of the 2018 2nd National and 1st International Digital Games Research Conference: Trends, Technologies, and Applications (DGRC). IEEE*, pp. 40–45.

Jiang, Ming and Qi Zhao, (2017) 'Learning Visual Attention to Identify People with Autism Spectrum Disorder'. In: *Proceedings of the IEEE International Conference on Computer Vision*. pp. 3267–3276.

Jobin, Anna, Marcello Ienca, and Effy Vayena, (2019) 'The Global Landscape of AI Ethics Guidelines'. *Nature Machine Intelligence* 1(9): 389–399.

Kadet, Anne (2019) 'Startup Touts Unique Talent Pool: Workers with Autism'. *Wall Street Journal* https://www:wsj:com/articles/startup-touts-unique-talent-pool-workers-with-autism–1536069600.

Kennett, Jeanette (2002) 'Autism, Empathy and Moral Agency'. *The Philosophical Quarterly* 52(208): 340– 357.

Kung, Michelle (2019) 'How Daivergent is Providing Autistic People with a Stepping-Stone to a Career in Tech'. *AWS Startups Blog* https://aws:amazon:com/blogs/startups/how-daivergent-provides-career-help-to-autistic-people/.

Levy, Ari (2019) 'Why this Investor Left the Pursuit of Billion-Dollar Exits to Help Employers Hire People with Autism'. *CNBC* https://www:cnbc:com/2019/08/25/brian-jacobs-started-moai-capital-to-invest-in-autism-employment:html.

Lillywhite, Aspen and Gregor Wolbring. (2019) 'Coverage of Ethics Within the Artificial Intelligence and Machine Learning Academic Literature: The Case of Disabled People'. *Assistive Technology* 33(3): 1–7.

Lloyd, Genevieve (2002) *The Man of Reason: 'Male' and 'Female' in Western Philosophy*. Routledge.

Longino, Helen E. (1990) *Science as Social Knowledge: Values and Objectivity in Scientific Inquiry*. Princeton University Press.

Lund, Henrik Hautop (2009) 'Modular Playware as a Playful Diagnosis Tool for Autistic Children'. In: *2009 IEEE International Conference on Rehabilitation Robotics*. IEEE, pp. 899–904.

Mauthner Natasha S. (2019) 'Toward a Posthumanist Ethics of Qualitative Research in a Big Data Era'. *American Behavioral Scientist* 63(6): 669–698.

McGeer, Victoria (2008) 'Varieties of Moral Agency: Lessons from Autism (and Psychopathy)'. *Moral Psychology* 3: 227–57.

Medina, José. (2013) *The Epistemology of Resistance: Gender and Racial Oppression, Epistemic Injustice, and the Social Imagination*. Oxford University Press.

Meyer, Michael. (2001) Between Theory, Method, and Politics: Positioning of the Approaches to CDA. In Methods of Critical Discourse Analysis, eds Michael Meyer and Ruth Wodack, pp. 14–31. SAGE.

Milton, Damian E. M. (2012) 'On the Ontological Status of Autism: The 'Double Empathy Problem'. *Disability & Society* 27(6): 883–887.

Morris, Meredith Ringel (2019) 'AI and Accessibility: A Discussion of Ethical Considerations'. arXiv preprint at arXiv:1908.08939

Mouffe, Chantal. (2000) *The Democratic Paradox*. Verso.

O'Neill, O (2000) *Bounds of Justice*. Cambridge University Press.

O'Reilly, Michelle, Jessica Nina Lester, and Nikki Kiyimba. (2019) Autism in the Twentieth Century: An Evolution of a Controversial Condition. In: *Healthy Minds in the Twentieth Century*, eds Steven J. Taylor and Alice Brumby, pp. 137–165. Springer.

Omar, Kazi Shahrukh, Prodipta Mondal, Nabila Shahnaz Khan, Md. Rezaul Karim Rizvi, and Md. Nazrul Islam. (2019) 'A Machine Learning Approach to Predict Autism Spectrum Disorder'. In: *2019 International Conference on Electrical, Computer and Communication Engineering (ECCE)*. IEEE, pp. 1–6.

Pahwa, Anjali, Gaurav Aggarwal, and Ashutosh Sharma. (2016) 'A Machine Learning Approach for Identification & Diagnosing Features of Neurodevelopmental Disorders using Speech and Spoken Sentences'. In: *2016 International Conference on Computing, Communication and Automation (ICCCA)*. IEEE, pp. 377–382.

Reddington, Sarah and Deborah Price. (2016) 'Cyborg and Autism: Exploring New Social Articulations via Posthuman Connections'. *International Journal of Qualitative Studies in Education* 29(7): 882– 892.

Reisigl, Martin and Ruth Wodak. (2005) *Discourse and Discrimination: Rhetorics of Racism and Antisemitism*. Routledge.

Roberts, Jacqueline M. A. (1989) 'Echolalia and Comprehension in Autistic Children'. *Journal of Autism and Developmental Disorders* 19(2): pp. 271–281.

Schwartzman, Lisa H. (2006) *Challenging Liberalism: Feminism as Political Critique*. Penn State Press.

Selbst, Andrew D., Danah Boyd, Sorelle A. Friedler, Suresh Venkatasubramanian, and Janet Vertesi. (2019) 'Fairness and Abstraction in Sociotechnical Systems'. In: *Proceedings of the Conference on Fairness, Accountability, and Transparency*. ACM, pp. 59–68.

Speicher, Till, Hoda Heidari, Nina Grgic-Hlaca, P. Gummadi Krishna, Adish Singla, Adrian Weller, and Muhammad Bilal Zafar. (2018) 'A Unified Approach to Quantifying Algorithmic Unfairness: Measuring Individual &Group Unfairness via Inequality Indices'. In: *Proceedings of the 24th ACM SIGKDD International Conference on Knowledge Discovery & Data Mining*. ACM, pp. 2239–2248.

Spiel, Katta, Christopher Frauenberger, and Geraldine Fitzpatrick. (2017) 'Experiences Of Autistic Children with Technologies'. *International Journal of Child-Computer Interaction* 11: 50–61.

Tariq, Qandeel, Jena Daniels, Jessey Nicole Schwartz, Peter Washington, Haik Kalantarian, and Dennis Paul Wall. (2018) 'Mobile Detection of Autism Through Machine Learning on Home Video: A Development and Prospective Validation Study'. *PLoS Medicine* 15(11): e1002705.

Thabtah, Fadi (2019) 'An Accessible and Efficient Autism Screening Method for Behavioural Data and Predictive Analyses'. *Health Informatics Journal* 25(4): 1739–1755.

Thapaliya, Sashi, Sampath Jayarathna, and Mark Jaime. (2018) 'Evaluating the EEG and Eye Movements for Autism Spectrum Disorder'. In: *2018 IEEE International Conference on Big Data (Big Data)*. IEEE, pp. 2328–2336.

Trewin, Shari (2018) 'AI Fairness for People with Disabilities: Point of View'. arXiv preprint at arXiv:1811.10670.

Tuana, Nancy (2017) Feminist Epistemology: The Subject of Knowledge. In: *The Routledge Handbook of Epistemic Injustice*, eds Ian James Kidd, José Medina, and Gaile Pohlhaus, Jr., pp. 125–138. Routledge.

Uluyagmur-Ozturk, Mahiya, Ayse Rodopman Arman, Seval Sultan Yilmaz, Onur Tugce Poyraz Findik, Genc Herdem Aslan, Gresa Carkaxhiu-Bulut, Yazgan M. Yanki, Umut Teker, and Zehra Cataltepe. (2016) 'ADHD and ASD Classification Based on Emotion Recognition Data'. In: *2016 15th IEEE International Conference on Machine Learning and Applications (ICMLA). IEEE*, pp. 810–813.

Van Dijk, Teun A. (1993) 'Principles of Critical Discourse Analysis'. *Discourse & Society* 4(2): 249–283.

Verhoeff, Berend (2012) 'What is This Thing Called Autism? A Critical Analysis of the Tenacious Search for Autism's Essence'. *BioSocieties* 7(4): 410–432.

Wagner, Ben (2018) Ethics as an Escape from Regulation: From Ethics-Washing to Ethics-Shopping. In: *Being Profiled: Cogitas Ergo*, eds Emre Bayamlioglu, Irina Baraliuc, Liisa Albertha, Wilhelmina Janssens, and Mireille Hildebrandt, pp. 84–90. Sum Amsterdam University Press, Amsterdam.

Waidzunas, Tom and Steven Epstein. (2015) 'For Men Arousal is Orientation': Bodily Truthing, Technosexual Scripts, and the Materialization of Sexualities Through the Phallometric Test'. *Social Studies of Science* 45(2): 187–213.

Wedyan, Mohammad and Adel Al-Jumaily. (2016) 'Upper Limb Motor Coordination Based Early Diagnosis in High Risk Subjects for Autism'. In: *2016 IEEE Symposium Series on Computational Intelligence (SSCI). IEEE*, pp. 1–8.

Welz, Erika (2019) 'There are Some Jobs where Candidates on the Autism Spectrum Excel'. *New York Post* URL https://nypost:com/2019/04/07/there-are-some-jobs-where-candidates-on-the-autism-spectrum-excel/.

Whittlestone, Jess, Rune Nyrup, Anna Alexandrova, and Stephen Cave. (2019) 'The Role and Limits of Principles in AI Ethics: Towards a Focus on Tensions'. In: *Proceedings of the 2019 AAAI/ACM Conference on AI, Ethics, and Society. ACM*, pp. 195–200.

Yergeau, Melanie (2013) 'Clinically Significant Disturbance: On Theorists who Theorize Theory of Mind'. *Disability Studies Quarterly* 33(4): DOI: https://doi.org/10.18061/dsq.v33i4.3876.

Yergeau, Melanie (2018) *Authoring Autism: On Rhetoric and Neurological Queerness*. Duke University Press.

18

Digital Ageism, Algorithmic Bias, and Feminist Critical Theory

Rune Nyrup, Charlene H. Chu, and Elena Falco

Introduction

Ageism is a pervasive structure of injustice with severe impacts on those affected. It diminishes older people's access to healthcare and employment, restricts their ability to express their sexuality, exacerbates social isolation, and may increase their risk of abuse and exploitation (World Health Organization 2021). Digital technology often mediates ageism and reinforces its negative impacts (Cutler 2005; McDonough 2016; Mannheim et al. 2019): older people are less likely to use digital technologies; they are often assumed—including by themselves—to be less capable of using and enjoying new technologies; they are frequently excluded and ignored in design processes; and ageism affects how data on older people are collected and compiled (UN Human Rights Council 2020; World Health Organization 2021). Machine learning, and other forms of artificial intelligence (AI), risk entrenching or even exacerbating these patterns. However, little attention in the AI ethics literature has, to date, been dedicated specifically to ageism.[1]

The present chapter seeks to address this gap. We have two aims: first, to argue that the interplay between ageism, AI and their impacts on older people need to be studied as a phenomenon in its own right, rather than merely a generic instance of algorithmic bias; second, to highlight what kinds of future work are needed to mitigate these impacts. To achieve these aims, we utilise a specific feminist position as our perspective on AI, namely what Sally Haslanger calls *materialist feminism*. This view emphasises the tightly coupled roles that material resources and social meanings play in producing social injustice. It highlights the interactions between technical issues—such as age-related biases in algorithms and datasets—and the complex material and social structures that produce and mediate their impacts.

---

[1] Chu et al. (2022a) surveyed the documents in the AI Ethics Guidelines Global Inventory (AlgorithmWatch 2021) and found that only 34 (23.3 percent) out of 146 examined documents even mention age-related bias, while only 12 (8.2 percent) discuss bias against older adults beyond a mere mention.

Rune Nyrup, Charlene H. Chu, and Elena Falco, *Digital Ageism, Algorithmic Bias, and Feminist Critical Theory*.
In: *Feminist AI*. Edited by: Jude Browne, Stephen Cave, Eleanor Drage, and Kerry McInerney, Oxford University Press.
© Oxford University Press (2023). DOI: 10.1093/oso/9780192889898.003.0018

Understanding these interactions is not just important for counteracting ageism itself, but also for broader feminist and intersectional reasons. Ageism scholars have long emphasised and documented the interactions between ageism and sexism, racism, classism, transphobia, and other patterns of discrimination, to modulate and exacerbate the effects of each. To successfully combat any of these injustices, we need to understand each of them.

We situate our argument within two recent trends in the algorithmic bias literature. First, researchers have begun to document and address AI-driven discrimination based on characteristics beyond race and gender, such as disability (Whittaker et al. 2019; Keyes, Chapter 17 in this volume) or genderqueer identities (e.g. Hicks 2019; Albert and Delano 2021). Second, researchers increasingly contextualise algorithmic bias within the social and political structures that influence the design of technologies and shape their impacts (Kalluri 2020; Hampton 2021). Similarly, we argue age-based algorithmic bias raises distinctive challenges, which should be understood in the context of prevalent ageist attitudes within society and associated patterns of digital exclusion.

Building on nascent efforts to study ageism in relation to AI, we develop a model of how ageism interacts with digital technologies and their social contexts. We call this phenomenon *digital ageism* (Chu et al. 2022a). We draw on feminist and anti-racist philosophy—through the lens of Sally Haslanger's recent work—to argue that age-based algorithmic bias needs to be situated within a broader structure of pervasive ageist attitudes, institutional policies, and patterns of digital exclusion. The compound effects of these factors are to exclude older people from the development and design of AI technologies and narrowly focus the application of such technologies towards health and social care. This creates feedback loops—or *cycles of injustice* (Whittlestone et al. 2019)—where negative stereotypes of ageing as primarily associated with decline, poor health, and increased dependency are reinforced, available datasets continue to be skewed towards such applications, and the lack of technologies serving other needs and desires of older people further re-entrenches digital exclusion.[2]

We begin by outlining how ageism has been characterised in previous scholarship, before discussing some examples of how it manifests in machine learning. Then we introduce some ideas from Haslanger's work, which we use to develop an analysis of the technology-mediated cycles of injustice that digital ageism gives rise to. We conclude by discussing the possibilities for counteracting and disrupting those cycles.

---

[2] For further discussion of the connection between design work and the needs of users, see Costanza-Chock (Chapter 21 in this volume).

What Is Ageism?

Ageism is defined as prejudice, discrimination, and oppression based on age. The concept was introduced by the psychiatrist and gerontologist Robert Butler in 1969, and has since been further elaborated by Butler and other scholars. The approach known as critical gerontology has been especially fruitful in articulating the various forms and meanings of ageism, as well as complicating the notion of 'age' itself. We start by outlining how we understand age in this chapter, before looking more closely at ageism and its relation to other forms of oppression.

At first blush, age may seem to refer to a natural fact, namely that of having existed a certain number of years (chronological age). However, for most aspects of ageism, this is not primarily what is at stake.[3] Rather, age should here be understood as a social categorisation, typically expressed in terms of the more or less fine-grained 'life stages' or 'age groups' we routinely divide people into: childhood, adolescence, middle-age, old age and so on. Much like critical studies of race and gender, critical gerontologists have emphasised that such age groups are *social* categories, constructed on the basis of—but not identical to—certain biological factors (Bytheway and Johnson, 1990). As such, they pack a host of additional social meanings: capabilities and desires are attributed to people based on their age, and there are strong expectations for what activities people in different age groups should engage in and what 'milestones' they ought to have achieved. Consequently, most of us articulate our identities partly based on (or in contrast to) these categories. These social meanings sometimes have a relation to biology (e.g. puberty and procreation), sometimes not (e.g. wearing short shorts). But even when they do, age categories and the associated expectations are to at least as large an extent determined by wider social structures (Katz 1996). Their boundaries and implications are by no means fixed or obvious.[4]

Like other social categories, age is not less 'real' or efficacious for being constructed in this way. Age is a concept that structures our perception of social reality and ourselves. Life-stage thinking influences our actions and desires, as well as how many aspects of society are structured and regulated. Ageism can be thought of as the negative side of these perceptions and interactions. It refers to patterns in the way that age categories structure our perceptions and interactions with each other that result in discrimination and oppression.

---

[3] Exceptions include laws or policies explicitly defined in terms of age limits. Even then, how these are applied in practice will also depend on social perceptions of age.

[4] This is reflected in our terminology. Following standard practice, we mostly use the comparative forms 'older' and 'younger'. This should be understood as referring to chronological age, recognising that age in this sense is a continuum without sharp or fixed boundaries. We use the categorical forms ('old', 'young', etc.) to refer specifically to a person's perceived social categorisation, without any presumption as to their chronological age.

Butler originally defined ageism as 'prejudice by one age group toward other age groups' (1969, p.243). Although this formulation allows for ageism to refer to any kind of intergenerational discrimination, the term ageism is mainly associated with injustices affecting older people. Thus, when Butler discusses the affective dimension of ageism, he refers to aversion to perceived characteristics of old age, such as decay, death, and unproductiveness. Authors who address the oppression of younger people specifically therefore often use alternative vocabulary (e.g. 'childism'; Young-Bruehl 2009).

Needless to say, any form of age-based injustice and oppression must be called out and opposed. However, while a unified approach is sometimes productive[5]— for example, for emphasising shared struggles and building solidarity between age groups—there are also important differences in the kinds of age-based injustices each group faces, which can be important to keep in view. The present context is a case in point: older people face distinct challenges in accessing digital technologies that are well-suited to their needs and aspirations, and are subject to distinctive stereotypes and prejudices regarding their ability and willingness to adopt new technologies. How these interact with emerging AI technologies and associated algorithmic biases will be our focus in this chapter.

At the individual level, ageism manifests as beliefs, affective responses, and behaviour (Levy and Banaji 2002). Ageist beliefs take the form of stereotypes regarding older people and old age. Older people are typically framed as helpless, incompetent or unpleasant (Bytheway 1990; Iversen et al. 2012; Levy and Macdonald 2016), while old age is associated with poor health, mental and physical decline, loneliness, and social isolation (Katz 1996; Levy and Macdonald 2002; Levy and Banaji 2002). The affective dimensions range from feelings of disgust related to ideas of decay (Butler 1969), fear and anxiety around one's own death, and decline (Greenberg et al. 2002), to pity, frustration, or impatience with older people's purported helplessness (Nelson 2009). Ageist behaviours include patronising language, paternalism, and outright abuse (Nelson 2009; Levy and Mcdonald 2016). At the social level, ageism manifests as systematic patterns of discrimination against older people, for instance in the workplace (Bytheway 1990; Levy and Banaji 2002; Palmore 2015) or healthcare decisions (Butler 1969; 1980), and in negative portrayals of ageing and older people in mainstream culture (Bytheway 1990; Nelson 2009). Finally, institutional ageism concerns law and institutional policies, such as lack of protection from discrimination (Butler 1980), and compulsory retirement (Iversen et al. 2012).

Ageism is distinct from many other kinds of oppression. First, ageism is something almost all people can be affected by, at least if they live long enough (Bytheway 1990; Palmore 2015). At the same time, and somewhat paradoxically, ageism is often felt to be a less serious injustice than racism or sexism (Bidanure

---

[5] This is for instance the approach adopted by the WHO report on ageism.

2018). Even overt expressions of ageism are often widely accepted, to the point of rendering it mostly invisible (Bytheway 1990; Levy and Banaji 2002). Similarly, at the institutional level, there remain contexts where differential treatment based on age is still legal or at least tolerated (Iversen et al. 2012; Gosseries 2014; Bidanure 2018).

This is not to say that ageism should be studied and addressed in isolation from other forms of oppression. Already, in 1969, Butler adopted a stance that we would today call intersectional. The main topic of his essay is the planning of social housing for the elderly in Chevy Chase, Florida, which elicited resistance from the local community. Butler notes how the race and class of the future inhabitants of the development (mostly Black and poor) played a major role in constructing them as undesirable: marginalised groups tend to earn less than dominant groups and are thus more likely to rely on State support. This played into the trope of older people as undeserving drains on collective resources, who should have been more provident in their youth (Butler 1969). More generally, ageism can exacerbate other forms of oppression by reinforcing other negative norms and stereotypes. It is well-documented that women are viewed more negatively for appearing old, while at the same time being judged more harshly for 'not acting their age' (Chrisler 2019). Here sexist norms that connect women's worth to their attractiveness interact with ageist conceptions of old age as repulsive and ugly. Similarly, ageist stereotypes of older people as warm but incompetent reinforce similar stereotypes of women as kind and nurturing, making older women more likely to be treated in a patronising or infantilising way (*Ibid.*). Ageism can also promote paternalistic attitudes which erase the agency of older people. For instance, older trans people are often discriminated against, misgendered, and excluded from systems of categorisation, while any sign of resistance is read under the ageist assumption of cognitive decline, as opposed to self-affirmation (Ansara 2015; Toze 2019).

However, ageism cannot be reduced to merely a manifestation of these other forms of oppression. The trope of the demanding older person also affects people who are in other respects relatively privileged. For example, the phenomenon of New Ageism, which can be informally summed up as hate towards the 'boomer' generation, relies on portraying older people as greedy (Carney 2017; Walker 2012). 'Grey power' is construed as political influence held by older people, who allegedly vote as a bloc in ways portrayed as antagonistic to, and aimed at taking resources away from, younger people; this happened, for example, during the Brexit vote (Carney 2017). Such (arguably unfounded—see Walker 2012) antagonisms erase diversity within older generations. For instance, the 'grey power' narrative obscures the fact that older women tend to vote progressively and often participated in emancipatory politics in the 1960s and 1970s (Carney 2017). Similarly, while 'over 65s' are often reported as the most pro-Brexit age group, more fine-grained analyses suggest that those who grew up during the Second World

War are significantly more pro-EU membership than the immediately following generations (Devine 2019).

Finally, the 'grey power' narratives also illustrate an important aspect of the interaction between ageism and other forms of oppression, which Audre Lorde highlighted in her classic essay 'Age, Race, Class and Sex':

> As we move toward creating a society within which we can each flourish, ageism is another distortion of relationship which interferes without vision. The 'generation gap' is an important tool for any repressive society. If the young members of a community view the older members as contemptible or suspect or excess, they will never be able to join hands and examine the living memories of the communities, nor ask the all important question, 'Why?' This gives rise to a historical amnesia that keeps us working to invent the wheel every time we have to go to the store for bread.
>
> (Lorde 1984, pp.116–117).

That race and gender antagonisms can be used to drive wedges between otherwise natural allies is a familiar idea from intersectional feminism and the Black Marxist tradition.[6] What Lorde furthermore highlights is ageism's power to erase progress, by impeding the intergenerational transmission of knowledge, forcing each new generation of activists to expend time and energy 'working to invent the wheel'.

Age-Related Algorithmic Bias and Digital Exclusion

Having discussed what ageism looks like in general, we will now explore how it manifests in relation to algorithmic bias. We pick this as our focus, as it is a type of harm from AI that has been discussed extensively in relation to other marginalised demographics, but whose interaction with ageism has yet to be systematically explored.

As is by now well-documented, machine learning models will often encode social biases reflected in the training data. Ageism is no exception. For instance, Díaz et al. (2018) studied fifteen different text-based sentiment analysis tools. They found, among other things, that sentences containing *young* adjectives were 66 percent more likely to be scored as expressing a positive sentiment than sentences containing *old* adjectives, when controlling for other sentential content. Díaz et al. also studied word embedding models (e.g, GloVe, word2vec), finding that these encoded implicit negative associations with old age and ageing. Briefly, word embedding models represent semantic associations in terms of vector addition and similarity, which makes them surprisingly good at solving analogy

---

[6] On the latter, see also Hampton (Chapter 8 in this volume).

puzzles, for example, that 'Japan' is the answer to 'Paris stands to France as Tokyo stands to x'. However, they also tend to reproduce stereotypical associations, so that 'man stands to woman as computer programmer stands to x' yields 'homemaker' (Bolukbasi et al. 2016). Similarly, Díaz et al. (2018) found that in these models, 'young stands to old as stubborn stands to x' yields 'obstinate', while 'old stands to young as stubborn stands to x' yields 'courageous'.

Another type of age-based bias involves algorithms that achieve high accuracy for younger people but whose performance drop for older people. Rosales and Fernández-Ardèvol (2019) discuss the example of age prediction tools based on social media use, which tend to work better for younger people. They identify several potential sources of this bias: limited representation of older people in training data, reflecting the fact that they use digital media less often or for a narrower range of purposes; using overly broad age ranges for older people (e.g. 'over 40s'), thus losing information about differences within this group; and design choices relying on assumptions that are inaccurate for older adults. For example, some age prediction algorithms rely on the average age of users' social network. This works well for younger users, but underestimates the age of older users, who often use online platforms to connect with younger friends and relatives but connect with people of their own generation through other means. Biases in age-estimation has also been documented in computer vision systems (Clapes et al. 2018). In this case, the bias was further exacerbated by gender, such that men's estimated age was always closer to the real age than women's.

Finally, Schrouff et al. (2022) have recently demonstrated that the robustness of machine learning algorithms (i.e. how well they perform in novel contexts) may correlate with age. For instance, they used a dataset collected in the United States to train and validate a model to predict common skin conditions based on images of the affected area plus some demographic metadata. They then compared its performance on two unseen datasets: a sample from the US dataset that had been set aside before training, and a new dataset collected from patients in New Zealand and Australia. While the algorithm's accuracy on the unseen US data was roughly equal across age brackets, this was not the case for the new dataset. While accuracy remained almost the same for patients aged 18–29, it dropped by more than twenty percentage points for those aged 65+.[7] This discrepancy arises because of so-called distribution shifts: that is, differences in the underlying causal processes that generated the two datasets make correlations found in the training data less predictive—and differentially so—of data from new contexts. Schrouff et al. note that currently existing technical bias mitigation techniques cannot guarantee against this type of complex distribution shifts. Instead, they recommend remedies based on design choices at different stages of the machine learning pipeline.

---

[7] These are average numbers across all conditions. For a few conditions, performance for 18–29-year-olds dropped more than for older age groups.

To summarise, in addition to ageist beliefs and attitudes being directly reflected in the data, age-related algorithmic bias also arises from patterns in how older adults use digital technologies, how age-related data are collected and structured (Chu et al. 2022c), and the assumptions and choices designers make. These patterns are in turn the product of broader social structures (cf. previous section, What is Ageism?). The fact that older people tend to use digital technologies to a lesser extent, or in different, often less extensive ways, has been a topic of scholarly attention for several years (Friemel 2016, Neves et al. 2013, Morris et al. 2007). This research is motivated by a concern that older people risk losing access to the benefits of increasingly digitised public and private goods and services. While technology usage is increasing for older people, age continues to correlate negatively with access to computers and the Internet (König et al. 2018).

Researchers have highlighted several factors underlying these differences. First, older people often lack access to relevant devices (computers, mobile phones) or infrastructures (e.g. high-quality Internet connection), typically due to poverty (Neves et al. 2013, Morris and Brading 2007). This is particularly likely to affect older people from racialised or otherwise economically disadvantaged groups. Second, digital technologies are not always well-suited to the physical needs of older adults. limitations of eyesight, hearing, or manual dexterity can reduce older adults' ability to use and enjoy the functionality of digital technologies (Neves et al. 2013, Friemel 2016). Third, epistemic factors, that is, a lack of knowledge or misunderstandings about the functioning of the technology at hand, may lead to a reduced usage of digital devices. For instance, Kong and Woods (2018), in a study of a panic button among Singaporean older people, highlight how participants would become worried due to not knowing when caregivers were available to be called or how long it would take to receive help. More generally, Neves et al. (2013) indicate education as one of the two main predictors of mobile phone and computer usage, along with age. Conversely, many authors argue that improvements in education and learning resources may help improve technology uptake among older people (Gallistl and Nimrod 2020; Vaz de Carvalho et al. 2019).

Fourth, technology use depends on the attitudes and (self-)perceptions of older adults, such as the perception that digital technologies are not 'made for them' or the feeling of being too old to understand technologies. Friemel (2016) highlights how the idea of the Internet being complicated and not easy to navigate discouraged older people themselves from approaching technology (Friemel 2016; Kong and Woods, 2018; Neves et al. 2013, 2018; Morris et al. 2007). Sometimes, the expectation of failure is gendered: older women believe men to be better equipped to understand technology (Kong and Woods 2018), although some studies suggest that gender differences in technology use disappear once one adjusts for education and class (Friemel 2016). Relatedly, older people who are more familiar with technologies (e.g. due to having worked with computers during their careers), or who are supported by their social networks, tend to fare better (Friemel 2016;

Morris 2007; Neves et al. 2013). These categories overlap in revealing ways. Kong and Woods (2018) describe how some of their—mostly low-income—participants were resistant to the panic button because it was perceived as expensive, and worried about handling it improperly and breaking it. Their fear was a result of self-perception, a relationship with money informed by their social background, and anxiety around technology.

These factors are easily framed as concerning the choices and preferences of individuals, given the limitations or constraints they face. However, they can equally—and preferably—be seen as the product of choices made during the research, design, and validation of technologies (Gallistl and Nimrod 2020). Consider the point about technologies being poorly suited for people with sensory impairments. Rather than seeing it as a regrettable consequence of people's physiology, it can equally (and often more appropriately) be seen as a failure of design. Why was this technology seemingly only designed to work well for a young, healthy, non-disabled person? This point, already familiar from disability studies,[8] can also be extended to cover the knowledge and skills that the technology presumes a user to have. For instance, why did the users of the panic button in Kong and Woods' study not know key details about how the device would work for them and their caregivers?

To make sense of the interplay between structural inadequacies, algorithmic bias and ageism, we will now turn to Sally Haslanger's work on critical theory and social practices. Finally, we will apply Haslanger's framework to digital ageism and carve a way forward.

Haslanger on Critical Theory and Social Practices

Haslanger defines critical theory as the attempt to describe and understand social realities, with the express purpose of providing resources for social movements seeking to critique and remedy social injustice (Haslanger 2018, p.231). Haslanger develops a general theory of how social injustice is produced and reinforced. She has applied it to analyse racism and misogyny (e.g. Haslanger 2017b), to identify how these forms of discrimination and oppression may be resisted and counteracted.[9] Our aim is to use Haslanger's theory to develop a similar analysis for digital ageism.

As a materialist feminist, Haslanger emphasises that 'the social world is a material world' (2017a, p.11). By this, she means two things.

---

[8] For more on lessons from the disability justice movement for design, see Costanza-Chock (Chapter 21 in this volume).

[9] Haslanger in turn draws extensively on earlier work in feminist, Marxist and anti-racist critical theory, including Young (1990), Fraser (1997), Hall (1986), and Shelby (2014).

First, she sees social justice as fundamentally a matter of preventing material harms and deprivation. Following Iris Marion Young (1990), she stresses that material harms go beyond economic injustice and violence, but also include marginalisation, powerlessness, and cultural imperialism. While these forms of oppression have social meaning, they are not merely symbolic (Haslanger 2017a, p.11). Rather, they impose social and material constraints that systematically curtail the opportunity of certain people to meet their needs, express their experience, and develop and exercise their capacities to live a fulfilling life (Young 1990, p.37).[10]

Second, Haslanger emphasises that social injustice is not produced by cultural or social factors in isolation. Rather, material conditions and culture both play an irreducible and closely intertwined role in producing and sustaining social injustice. It is this claim she seeks to elucidate.

The central concept in Haslanger's theory is that of a *social practice*. She defines practices as 'patterns of behaviour that enable us to coordinate due to learned skills and locally transmitted information, in response to resources, and whose performances are "mutually accountable" by reference to some shared schemas/social meanings' (2019, p.7; see also 2018, p.245). Practices consist of—and in turn constitute—structures of social relations, both between people ('being a parent of, being an employee of, being a student of'), as well as to things ('cooking, owning, occupying, driving, eating, herding', Haslanger 2019, pp.6–7). Each of these social relations (and many more besides) are made possible and meaningful by the overall structure of relations that make up a given social practice, such as the practice of parenting, of wage labour, or of education.

Practices make possible and condition agency, both at the level of individuals and of groups. At the individual level, practices are 'a site of socially organized agency, a nexus where individual agency is enabled and constrained by social factors' (2018, p.232). Meaningful action, as opposed to mere physical behaviour, is only possible within a social practice. Putting on a hearing aid involves more than just fastening an electronic device on (or in) your ear. It carries meanings, which influence how other people perceive you, how they react to you, what they expect from you, what kinds of spaces you can gain access to, and so on. This is only possible because you are situated within a social practice that attaches differential significance to different types of electronic devices. In a more material sense, it is only possible for you to put on the hearing aid because of your relation to systems of production and distribution, resource extraction and waste disposal, with attendant relations of ownership and employment. As well as making action possible in the first place, these material and social factors constrain what kinds of action you can take. What type of hearing aid (if any) can you afford? Do you worry about looking old, and choose a discreet model as a result? Can it be delivered to a place near you? Do you occupy a position in the social practice such that wearing this

---

[10] For more on Young's work and its relation to AI, see Browne (Chapter 19 in this volume).

type of device makes people think you look smart and tech-savvy? Or will it make people more likely to perceive you as frail and dependent? (Astell et al. 2020).

At the group level, social practices play a vital role in facilitating communication and coordination. Typically, their central function is to coordinate the production and distribution of what Haslanger calls *resources*. Resources are anything that is taken to have some kind of (positive or negative) value. They include material as well as more abstract goods: 'time, knowledge, status, authority/power, health/well-being, security, ... toxic waste, menial work, vulnerability' (2018, p.232). Haslanger construes social injustices as arising from this type of coordination, namely when it functions to systematically cause material harms and deprivation to certain groups.

Social practices facilitate coordination through providing us with a suite of readily available *social meanings* that allow us to fluently and efficiently navigate highly complex social relations (2019, pp.6–9). Haslanger construes social meanings broadly to include all sorts of representations and symbols, but also concepts, categories, narratives, and default patterns of inference. These structure the assumptions and expectations we have for each other: who needs or deserves what resources, what kinds of things are valuable (i.e. what counts as a resource in the first place) and to whom, who is supposed to do what, what is the 'right' (or 'normal', 'natural') way to do things, and so on. Since we have been socialised to fluently read and communicate these assumptions to each other, social meanings allow us to generally know what to expect of others and what they expect of us. This not only influences individual decisions directly but also tends to result in material and institutional structures which facilitate decisions that conform to these expectations.

Finally, Haslanger argues that social practices—and their resulting injustices—tend to be stable and self-reinforcing, as disposing of them would cause unwelcome disruption (2017a, pp.17–18; 2018, pp.246–247). Social practices enable smooth coordination and allow us to find value and meaning. Disrupting practices causes friction and deprives us of tools to make sense of the world and ourselves. Thus, for most of us most of the time, we slide into compliance. Furthermore, social practices tend to create patterns of behaviour and distributions of resources that seemingly confirm and reinforce the social meanings that themselves structure the social practice: the majority of care work is in fact done by women; older people often struggle with technology. This in turn makes it easy to conclude—mistakenly—that such patterns are natural, inevitable, or even justified (Haslanger 2012, ch.17). As Haslanger puts it, we often learn such assumptions by 'looking around' and noticing how the world seems to work (2017a, p.15).

In summary, Haslanger's theory of social practices provides a general framework for understanding how social injustice arises from the interaction of material and cultural factors and prevails even in the face of efforts to root them out. We now explore how these ideas apply to the context of ageism and digital technologies.

Digital Ageism and Technology-Mediated Cycles of Injustice

Previous analyses of algorithmic bias have emphasised that its causes are many and complex (Friedman and Nissenbaum 1996; Barocas and Selbst 2016; Danks and London 2017). They include technical factors, such as biased data or design constraints, the fact that seemingly innocuous proxies may correlate with patterns of discrimination, and mismatches between designers' and users' assumptions, to name a few. All of these are in turn shaped by wider structural factors, from who is involved in the design of technologies, who has the resources and capabilities to access and benefit from new technologies, to how, where, about whom and for what purpose data is collected. This can lead to the aforementioned cycles of injustice (Whittlestone et al. 2019).

When algorithmic bias, together with these complex underlying causes, interact with patterns of ageism within society, we refer to the resulting causal nexus as *digital ageism* (Chu et al. 2022a). Using Haslanger's theory we can now start to analyse some of the technology-mediated cycles of injustice that are produced by and sustain digital ageism. We propose the following diagram to illustrate the principal routes through which these arise (Figure 18.1). We start from Haslanger's 'social practices' (dashed arrows to the left), that is, the self-reinforcing structures of resources and social meanings that shape our individual and collective agency. Under resources we include the material resources available to older people, especially their ability to buy or otherwise access digital technologies and infrastructure. It also refers to the broader distribution of resources within society such as what kinds of data are readily available to different organisations and existing incentives within the economy. The relevant social meanings include stereotypical beliefs about older people and their relation to technology, as well as the affective dimensions of ageism. Relevant stereotypes include that older people struggle with digital technology, are resistant to change and slow to adopt innovations, and are simply less interested in new technology. The affective dimensions include feelings of frustration or impatience or a sense that older people are 'difficult to work with'. Finally, these stereotypes and affective responses interact with assumptions about what digital technologies for older people look like: healthcare devices, companionship robots, cognitive support tools, and so on; in short, technologies designed to help or support older people to manage vulnerabilities and address problems, rather than sources of joy, fulfilment, or fun.

To this structure we add a third factor, 'institutions', which refers to law and regulation but also more or less formal guidelines and policies of relevant companies and organisations. Although Haslanger tends to focus on the first two factors, she sometimes distinguishes them explicitly from law and policies and seems to regard all three as distinct routes through which changes to social practices can

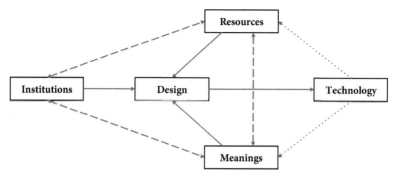

Figure 18.1 Cycles of injustice in technology-mediated discrimination.

be pursued (Haslanger 2018, p.247; Haslanger 2014, pp.17, 20). In any case, it is particularly relevant to include institutions when discussing ageism, as there are often fewer restrictions on differential treatment of older people (as mentioned in the section Age-Related Algorithmic Bias and Digital Exclusion).

Social practices shape the decision-making that goes into the development, design, and deployment of new technologies (solid arrows). This can happen at the level of individual decision-making (e.g. what assumptions do designers make about the needs and capabilities of older users?), structural conditions (who is involved in the design process? What kinds of data are readily available?), economic incentives (what technologies are easiest to market? What types of research and development receive the most funding?), or most likely some combination thereof. As discussed in the section, Haslanger on Critical Theory and Social Practices, these are all factors that contribute to creating technologies that are biased against older adults.

Biased technologies feed back into the social practice through two principal routes (dotted arrows): impacts on resource allocation and impacts on social meanings.[11]

With regards to resource allocation, biased algorithms can directly and unfairly deprive older users of valuable resources, such as healthcare or job opportunities. For instance, if the accuracy of diagnostic algorithms, such as those studied by Schrouff et al. (2022), drops for older adults, they could lose out on timely medical interventions. Direct impacts on resource allocation can also take the form of technologies that function less smoothly or are less satisfying for older users, say,

---

[11] Figure 18.1 does not include a direct arrow from technology to institutions, as we assume most impacts on institutions will be mediated by human decision-makers and impacts on resources and social meanings. We are some way away from widespread fully automated policymaking. Even if a few counter-examples exist, we regard this simplification as benign. Our diagram is only intended to capture the principal lines of influence.

because they predict their age wrongly or misinterpret the sentiment of the text describing them. Added to general factors (material, physical, epistemic, etc.) that constrain older people's ability to benefit from digital technologies, these biases can contribute to depriving older people of the social goods brought by digital technologies.

With regards to social meanings, obvious impacts include the kinds of representational biases that Díaz et al. (2018) document in the case of natural language processing. Extant datasets reflect society's ageist bias, which these algorithms risk reproducing and thereby confirming. For instance, if sentiment analysis tools systematically classify text describing older people as having negative connotations, this could easily reinforce those connotations in the minds of users. Similarly, if word embedding models are used to generate autocomplete suggestions and systematically generates less favourable adjectives for older people, this would again reinforce negative associations.

However, these direct impacts are not the end of the story. The elements that make up social practices are deeply intertwined: social meanings affect resource allocation and vice versa. Negative associations with older adults that biased technologies reinforce in the minds of individuals will in turn impact materially significant decisions that those individuals make. For instance, technologies that portray older people as 'obstinate' can contribute to perceptions that they are inflexible, difficult, or undesirable to work with. If these stereotypes are reinforced in the minds of hiring committee members, older people lose out on job opportunities.

Conversely, if older people have fewer resources available for enjoying and benefitting from digital technology—such as smoothly functioning technologies that are suited to their needs and interests—it will naturally make them less interested in such technologies or more concerned about their ability to use them. This can in turn reinforce the previously mentioned stereotypes, both to designers and older users themselves. This is an example of Haslanger's point that stereotypes are often reinforced by people 'looking around' and noticing certain patterns in society—though, of course, this erroneously assumes that such patterns reflect intrinsic features of what older people 'are like', rather than external structures and circumstances.

Finally, these now-reinforced social meanings and patterns of resource allocation may once again feed back into future design processes. For instance, if researchers and designers are more liable to become impatient with older people or assume that they are uninterested in digital technology, they are less likely to include them in the design process. If older people have less-satisfying technologies available to them and therefore use them less, this can skew data collection. And so on, through further cycles of injustice.

Disrupting Digital Ageism

What can be done to disrupt cycles of injustice and counteract digital ageism? We suggest that more focused attention and action on multiple fronts are necessary (Chu et al. 2022b). Interventions directly targeting the design node in Figure 18.1, such as better methods for detecting and mitigating algorithmic bias against older people, are certainly important but unlikely to be effective on their own. At the level of resources, an important intervention is to proactively create and curate better datasets. This would involve expanding existing dataset to include more older people, and relabelling data so older people or not represented by single overly broad labels (e.g. '55+'), especially in contexts and use cases beyond health and social care. At the level of social meanings, making researchers and designers aware of common ageist assumptions affecting technology design will be important, as well as increasing the participation of older people in the design process itself.

Increasing participation will, we suspect, present specific challenges.[12] What needs to be overcome is not just designers' assumptions and affective responses in relation to working with older adults, but also older people's internalised ageism and self-perceptions relating to technology. As we have emphasised, these will often be informed by past and ongoing experiences of interacting with technology. Furthermore, unlike many other minoritised groups, most older people are retired or close to retirement, so will lack a natural incentive to pursue careers as researchers or designers. In addition, for technologies in the development phase, older people will have less of a chance of benefitting from the research they contribute to. Thus, we need to carefully consider what incentives and motivations older people have for being more actively involved in the design of technology.

To conclude, we have argued in this chapter that digital ageism should receive more attention within AI ethics, as a problem in its own right, rather than a generic instance of algorithmic bias. Ageism involves characteristic patterns of resource allocation and social meanings, whose interactions with biased AI systems give rise to distinctive technology-mediated cycles of injustice. We have outlined some proposals for counteracting these, though we emphasise that these are necessarily at the level of speculative proposals. Further systematic analyses of how age-related bias is encoded or amplified in AI systems, as well as their corresponding societal and legal implications, are needed (Chu et al. 2022a, 2022b). More detailed knowledge of how best to counteract digital ageism will likely only come from this type of in-depth analysis, as well as the experience of trying to implement these or other

---

[12] In addition to general challenges to meaningful inclusion and empathy in design (Bennett and Rosner 2019).

proposals in practice. As digital technologies increasingly impact our lives and society, this will be crucial not just to avoid reproducing existing ageist stereotypes and practices, but also for combatting other forms of injustice and oppression that ageism tends to reinforce and exacerbate, such as sexism, racism, classicism, and transphobia.

References

Albert, Kendra and Maggie Delano. (2021) 'This Whole Thing Smacks of Gender: Algorithmic Exclusion in Bioimpedance-based Body Composition Analysis'. *FAccT '21: Proceedings of the 2021 ACM Conference on Fairness, Accountability, and Transparency.*

AlgorithmWatch. (2021). *AI Ethics Guidelines Global Inventory.* https://algorithmwatch.org/en/ai-ethics-guidelines-global-inventory/

Ansara, Y. Gavriel. (2015) 'Challenging Cisgenderism in the Ageing and Aged Care Sector: Meeting the Needs of Older People of Trans and/or Non-Binary Experience'. *Australasian Journal of Ageing* 34: 14–18.

Astell, Arlene J., Colleen McGrath, and Erica Dove. (2020). 'That's for Old So and So's!': Does Identity Influence Older Adults' Technology Adoption Decisions?' *Ageing and Society*, 40(7): 1550–1576. doi:10.1017/S0144686X19000230

Barocas, Solon and Andrew Selbst. (2016) 'Big Data's Disparate Impact'. *California Law Review* 104: 671–732.

Bennett, Cynthia L., and Daniela K. Rosner. (2019). 'The Promise of Empathy: Design, Disability, and Knowing the "Other"'. Proceedings of the 2019 CHI Conference on Human Factors in Computing Systems (CHI '19). Association for Computing Machinery, New York, NY, USA, Paper 298, 1–13. https://doi.org/10.1145/3290605.3300528.

Bidanure, Julia. (2018) Discrimination and Age. In *The Routledge Handbook of the Ethics of Discrimination*, ed. Kasper Lippert-Rasmussen. London: Routledge.

Bolukbasi, Tolga, Kai-Wei Chang, James Zou, Venkatesh Saligrama, and Adam Kalai. (2016) 'Man is to Computer Programmer as Woman is to Homemaker? Debiasing Word Embeddings'. In *Advances in Neural Information Processing Systems 29 (NIPS 2016)*, eds D. Lee and M. Sugiyama and U. Luxburg and I. Guyon and R. Garnett, pp. 1–25.

Butler, Robert N. (1969) 'Age-ism: Another Form of Bigotry'. *The Gerontologist* 9: 243–246.

Butler, Robert. N. (1980) 'Ageism: A Foreword'. *Journal of Social Issues* 36: 8–11.

Bytheway, Bill and Julia Johnson. (1990) 'On Defining Ageism'. *Critical Social Policy* 10: 27–39.

Carney, Gemma M. (2017) 'Toward a Gender Politics of Aging'. *Journal of Women and Aging* 30: 1–17.

Carvalho, Carlos V., Pedro de Cano, José. M. Roa, Anna Wanka, and Franz Kolland. (2019) 'Overcoming the Silver Generation Digital Gap'. *Journal of Universal Computer Science* 25: 1625–1643.

Chrisler, Joan C. (2019) Sexism and Ageism. In *Encyclopedia of Gerontology and Population Aging* eds D. Gu and M. Dupre. Springer. https://doi.org/10.1007/978-3-319-69892-2_603-1

Chu, Charlene H., Simon Donato-Woodger, Kathleen Leslie, Corinne Bernett, Shehroz Khan, Rune Nyrup, and Amanda Grenier. (2022c) 'Examining the Technology-mediated Cycles of Injustice that Contribute to Digital Ageism'. *The 15th International PErvasive Technologies Related to Assistive Environments (PETRA) Conference. 1–2 July 2022, Corfu, Greece.*

Chu, Charlene H., Kathleen Leslie, Jiamin Shi, Rune Nyrup, Andria Bianchi, Sherhoz S. Khan, Samira A. Rahimi, Alexandra Lyn, and Amanda Grenier. (2022b) 'Ageism and Artificial Intelligence: Protocol for a Scoping Review'. *JMIR Research Protocols* 11(6): e33211, doi: 10.2196/33211

Chu, Charlene H., Rune Nyrup, Kathleen Leslie, Jiamin Shi, Andria Bianchi, Alexandra Lyn, Molly McNichol, Sherhoz Khan, Samira Rahimi, and Amanda Grenier. (2022a) 'Digital Ageism: Challenges and Opportunities in Artificial Intelligence for Older Adults'. *The Gerontologist* 62(7): 947–955.

Clapes, Albert, Ozan Bilici, Dariia Temirova, Egils Avots, Gholamreza Anbarjafari, and Escalera Sergio. (2018). 'From Apparent to Real Age: Gender, Age, Ethnic, Makeup, and Expression Bias Analysis in Real Age Estimation'. *Proceedings of the IEEE Conference on Computer Vision and Pattern Recognition (CVPR) Workshops*, pp. 2373–2382.

Cutler, Stephen J. (2005) 'Ageism and Technology: A Reciprocal Influence'. *Generations* 29: 67–72.

Danks, David, and Alex J. London. (2017). 'Algorithmic Bias in Autonomous Systems'. *IJCAI International Joint Conference on Artificial Intelligence*, pp. 4691–4697.

Devine, Kieran. (2019) 'Not All the "Over 65s" Are in Favour of Brexit – Britain's Wartime Generation are Almost as pro-EU as Millennials'. LSE Comment 21 March, https://blogs.lse.ac.uk/europpblog/2019/03/21/not-all-the-over-65s-are-in-fav our-of-brexit-britains-wartime-generation-are-almost-as-pro-eu-as-millennials/

Díaz, Mark, Isaac Johnson, Amanda Lazar, Anne M. Piper, and A D. Gergle (2018) 'Addressing Age-Related Bias in Sentiment Analysis'. *CHI '18: Proceedings of the 2018 CHI Conference on Human Factors in Computing Systems.*

Fraser, Nancy. (1997) 'Heterosexism, Misrecognition, and Capitalism: A Response to Judith Butler'. *Social Text* 15: 279–289.

Friedman, Batya, and Helen Nissenbaum. (1996) 'Bias in Computer Systems'. *ACM Transactions on Information Systems* 14: 330–347.

Friemel, Thomas N. (2016) 'The Digital Divide Has Grown Old: Determinants of a Digital Divide Among Seniors'. *New Media and Society* 18: 313–331.

Gallistl, Vera and Galit Nimrod. (2020) 'Media-Based Leisure and Wellbeing: A Study of Older Internet Users'. *Leisure Studies* 39: 251–265.

Gosseries, Axel. (2014) 'What Makes Age Discrimination Special? A Philosophical Look at the ECJ Case Law'. *Netherlands Journal of Legal Philosophy* 43: 59–80.

Greenberg, Jeff, Jeff Schimel, and Andy Martens. (2002). Ageism: Denying the face of the future. In *Ageism: Stereotyping and Prejudice against Older Persons*, ed. T.D. Nelson, pp. 24–48. Cambridge, MA: MIT Press.

Hall, Stuart (1986). 'The Problem of Ideology—Marxism Without Guarantees'. *Journal of Communication Inquiry* 10: 28–44.

Hampton, Lelia. (2021) 'Black Feminist Musings on Algorithmic Oppression'. *FAccT '21: Proceedings of the 2021 ACM Conference on Fairness, Accountability, and Trans- parency.*

Haslanger, Sally. (2012) *Resisting Reality: Social Construction and Social Critique.* Oxford: Oxford University Press.

Haslanger, Sally. (2014) 'Social Meaning and Philosophical Method'. *Eastern APA Presidential Address, 2013. Published in APA Proceedings and Addresses,* Vol. 88, pp. 16–37.

Haslanger, Sally. (2017a) *Critical Theory and Practice (The 2015 Spinoza Lectures),* Amsterdam: Koninklijke Van Gorcum.

Haslanger, Sally. (2017b) 'Racism, Ideology and Social Movements'. *Res Philosophica* 94: 1–22.

Haslanger, Sally. (2018) 'What Is a Social Practice?' *Royal Institute of Philosophy Supplement* 82: 231–247.

Haslanger, Sally. (2019) 'Cognition as a Social Skill'. *Australasian Philosophy Review* 3: 5–25.

Hicks, Mar. (2019) 'Hacking the Cis-Tem'. *IEEE Annals of the History of Computing* 41: 20–33.

Iversen, Thomas. N., Lars Larsen, and Per E. Solem. (2012) 'A Conceptual Analysis of Ageism'. *Nordic Psychology* 61: 4–22.

Kalluri, Pratyusha (2020) 'Don't Ask if AI is Good or Fair, Ask How it Shifts Power'. *Nature* 583: 169.

Katz, Stephen. (1996) *Disciplining Age: The Formation of Gerontological Knowledge.* Charlottesville, Virginia: The University Press of Virginia.

Kong, Lily and Orlando Woods. (2018) 'Smart Eldercare in Singapore: Negotiating Agency and Apathy at the Margins'. *Journal of Aging Studies* 47: 1–9.

König, Ronny, Alexander Seifert, and Michael Doh. (2018) 'Internet Use Among Older Europeans: An Analysis based on SHARE Data'. *Universal Access in the Information Society* 17: 621–633.

Levy, Becca R. and Mahrazin R. Banaji. (2002) Implicit Ageism. In *Ageism: Stereotyping and Prejudice against Older Persons,* ed. T.D. Nelson, pp. 49–76. Cambridge, MA: MIT Press.

Levy, Sheri R. and Jamie L. Macdonald. (2016) 'Progress on Understanding Ageism'. *Journal of Social Issues* 72: 5–25.

Lorde, Audre. (1984) *Sister Outsider: Essays and speeches.* Berkeley: Crossing Press.

Mannheim, Ittay, Ella Schwartz, Wanyu Xi, Sandra C. Buttigieg, Mary McDonnell-Naughton, Eveline J. M. Wouters, and Yvonne van Zaalen. (2019) 'Inclusion of Older Adults in the Research and Design of Digital Technology'. *International Journal of Environmental Research and Public Health* 16: 3718.

McDonough, Carol C. (2016) 'The Effect of Ageism on the Digital Divide Among Older Adults'. *HSOA Journal of Gerontology & Geriatric Medicine* 2: 008.

Morris, Anne and Helena Brading. (2007) 'E-Literacy and the Grey Digital Divide: A Review with Recommendations'. Unpublished manuscript, available via: https://pdfs.semanticscholar.org/6aac/c7e39dcd2f81eec1a250878f9a533c47978f.pdf (accessed 15 March 2022).

Nelson, Todd D. (2009) Ageism. In *Handboook of Prejudice, Stereotyping and Discrimination,* ed. T.D. Nelson. New York: Psychology Press.

Neves, Barbara. B., Fausto Amaro, and Jaime R. S. Fonseca. (2013) 'Coming of (Old) Age in the Digital Age: ICT Usage and Non-Usage Among Older Adults'. *Sociological Research Online* 18: 22–35.

Neves, Barbara. B., Jenny Waycott, and Sue Malta. (2018) 'Old and Afraid of New Communication Technologies? Reconceptualising and Contesting the "Age-Based Digital Divide"'. *Journal of Sociology* 2: 236–248.

Palmore, Erdman (2015) 'Ageism Comes of Age'. *The Journals of Gerontology: Series B* 70: 873–875

Rosales, Andrea, and Mireia Fernández-Ardèvol. (2019) 'Structural Ageism in Big Data Approaches'. *Nordicom Review* 40: 51–64.

Schrouff, Jessica, Natalie Harris, Oluwasanmi Koyejo, Ibrahim Alabdulmohsin, Eva Schnider, Krista Opsahl-Ong, Alex Brown, Subhrajit Roy, Diana Mincu, Christina Chen, Awa Dieng, Yuan Liu, Vivek Natarajan, Alan Karthikesalingam, Katherine Heller, Silvia Chiappa, and Alexander D'Amour. (2022) 'Maintaining Fairness Across Distribution Shift: Do We have Viable Solutions for Real-World Applications?' Preprint at *arXiv*, 2202.01034

Shelby, Tommie. (2014) 'Racism, Moralism, and Social Criticism'. *Du Bois Review* 11: 57–74.

Toze, Michael. (2019) 'Developing a Critical Trans Gerontology'. *British Journal of Sociology* 70: 1490–1509.

UN Human Rights Council. (2020) 'Report of the Independent Expert on the Enjoyment of All Human Rights by Older Persons'. 9 July (A/ HRC/45/14). New York, NY, USA: United Nations Human Rights Council.

Walker, Alan. (2012) 'The New Ageism'. *Political Quarterly* 83: 812–819.

Whittaker, Meredith, Meryl Alper, Cynthia L. Bennett, Sara Hendren, Liz Kaziunas, Mara Mills, Meredith R. Morris, Joy Rankin, Emily Rogers, Marcel Salas, and Sarah M. West. (2019) *Disability, Bias, and AI*. New York: AI Now Institute. https://ainowinstitute.org/disabilitybiasai-2019.pdf

Whittlestone, Jessica, Rune Nyrup, Anna Alexandrova, Kanta Dihal, and Stephen Cave. (2019) *Ethical and Societal Implications of Algorithms, Data, and Artificial Intelligence: A Roadmap for Research*. London: Nuffield Foundation.

World Health Organisation. (2021) *Global Report on Ageism*. Geneva, Switzerland: World Health Organization.

Young, Iris Marion. (1990) *Justice and the Politics of Difference*. Princeton: Princeton University Press.

Young-Bruehl, Elisabeth. (2009) 'Childism—Prejudice against Children'. *Contemporary Psychoanalysis* 45: 251–265.

19

AI and Structural Injustice

A Feminist Perspective

Jude Browne

'The common story: you start analyzing how a certain ML [Machine Learning] model works and at some point you see the familiar phrase "and then some magic happens". This is called the black box of AI, where algorithms make predictions, but the underlying explanation remains unknown and untraceable'.

(Sciforce 2020).

'The problem with structural injustice is that we cannot trace... how the actions of one particular individual, or even one particular collective agent, such as a firm, has directly produced harm to other specific individuals'.

(Young 2011, p.96).

In this chapter, I draw on the seminal work of feminist theorist Iris Marion Young to consider how political responses to structural injustice, exacerbated by AI,[1] might be developed. My interpretation of Young's work is centred on what I see as an overlooked feature of her account of structural injustice - 'untraceability'. Drawing on Arendt's concept of 'thoughtlessness' and the work of other scholars such as Benjamin, I explore the relationship between structural injustice and algorithmic decision making,[2] which, itself, is becoming increasingly untraceable. I argue that

---

[1] Artificial Intelligence (AI) refers to a range of technologies in computer science which perform tasks that typically require human intelligence. Machine Learning (ML) is a subcategory of AI and is defined by the capacity to learn from data without explicit instruction. In this chapter, I join the other authors in this volume as well as those I cite and many others in the fields of gender studies, critical race studies and critical disability studies in calling attention to the wide range of injustices (structural and otherwise) brought about by AI. This chapter is based on a much larger piece of work in Browne (forthcoming).

[2] Watson and Floridi (2020, p.9216) are right to point out that there is no real sense in which machines take 'decisions'. This, they say, is an 'anthropomorphic trope granting statistical models a degree of autonomy that dangerously downplays the true role of human agency in sociotechnical systems'. Because machines cannot 'think' in the way that humans do (Floridi et al. 2009, Floridi 2017),

Jude Browne, *AI and Structural Injustice*. In: *Feminist AI*. Edited by: Jude Browne, Stephen Cave, Eleanor Drage, and Kerry McInerney, Oxford University Press. © Oxford University Press (2023). DOI: 10.1093/oso/9780192889898.003.0019

while fault-based regulation of AI is vital for combating injustice exacerbated by AI, too little thought goes into addressing the structural dynamics of AI's impact on society which requires a different sort of approach. I conclude by offering some suggestions on how we might think about mitigating the exacerbation of structural injustice posed by AI. In particular, I advocate adopting several elements of the mini-public approach within the regulatory public-body landscape of democratic governance systems to form a new 'AI Public Body' (using the UK as my example).

What Is Structural Injustice?

Structural injustice is a complex topic with many varying interpretations.[3] At its centre, however, is the work of the feminist political theorist Iris Marion Young. One of the core feminist insights of her work was that marginalised individuals and groups, such as those who are gendered, racialised and dehumanised, invariably find themselves at the sharp end of a type of injustice that society tends to take no responsibility for—namely, structural injustice. What follows is my interpretation of how her scholarship helps us to think through how best to address some of the troubling structural dynamics of AI (Young 1990, 2003, 2006, 2011).

Structural injustice is distinct from the traditional ways in which we tend to think about injustice as grounded in blame, culpability, guilt, and fault. For the sake of clarity, I'll call this sort of injustice 'fault-based injustice'. Fault-based injustices are grounded in liability and can be direct, indirect, intentional, or unintentional and understood as either a legal wrong or the sort of moral wrong that transgresses the moral expectations of a given context or society. Key to fault-based injustice is traceability. Whether inadvertently or wilfully, there is always an identifiable agent or agents (individuals, groups, institutions, or states) whose actions can be causally traced to a particular injustice. As Young (2011, p.98) argues,

> A concept of responsibility as guilt, blame or liability is indispensable for a legal system and for a sense of moral right that respects agents as individuals and expects them to behave in respectful ways towards others. When applying this model of responsibility, there should be clear rules of evidence, not only for demonstrating the causal connection between this agent and a harm, but also for evaluating the intentions, motives, and consequences of the actions.

we are ultimately responsible for the outcomes generated by AI even when we cannot understand the processes by which AI provides the data we ask of it. That said, for reasons of clarity I will use the phrase 'algorithmic decisions' in this chapter with Watson's point in mind nevertheless.

[3] See for example, Browne and McKeown (forthcoming 2023).

It is, for example, vital to determine who is at fault when a self-driving car kills a pedestrian.[4] However, Young argues that fault-based injustice is not a productive way of understanding structural injustice, which she sees as distinct and more complex than fault-based injustice. Rather than by traceable acts of wrong-doing, structural injustice is generated by a vast array of structural processes that emanate from the complex, multitudinous, and accumulated 'everyday' actions of individuals, groups, and institutions operating 'within given institutional rules and accepted norms' (Young, 2011, p.53).[5]

These structural processes that act as the background conditions to structural injustice are inherently intersectional and generated by relations of social position that shape the opportunities and life prospects including both material and social resources, of everyone in those positions.[6] The various strands of structural processes that affect particular gendered or racialised groups for example are relational, co-constituted and mutually reinforce each other.

> [S]tructures refer to the relation of social positions that condition the opportunities and life prospects of the persons located in those positions. This positioning occurs because of the way that actions and interactions reinforce the rules and resources available for other actions and interactions involving people in other structural positions. The unintended consequences of the confluence of many actions often produce and reinforce opportunities and constraints, and these often make their mark on the physical conditions of future actions, as well as on the habits and expectations of actors. This mutually reinforcing process means that the positional relations and the way they condition individual lives are difficult to change.
>
> (Young 2003, p.6).

On Young's view, everyone who is connected in virtue of their *contributions* to structural injustice is responsible, not by having traceably caused or intended injustice, but because they have acted in ways that have contributed to the

---

[4] To use a recent AI example that caught a great deal of media attention, Elaine Herzberg tragically died when hit by a self-driving Uber car that failed to stop as she wheeled her bike across the road in Tempe, Arizona in 2018. Despite being a self-driving car, the Uber car required a human 'safety driver' at all times. The safety driver Rafael Vasquez on that occasion, had been streaming an episode of the television show 'The Voice' at the time and was accused of not watching the road with full attention. Ms Vasquez was found guilty of negligent homicide and Uber ceased its programme of self-driving cars. See BBC 2020 for more details https://www.bbc.co.uk/news/technology-54175359#:~:text='Visually%20distracted,the%20vehicle%20in%20an%20emergency

[5] What is meant here as 'untraceable' is that whilst we are able to speculate about our collective behaviour contributing to the background conditions of structural injustice, it is not possible to meaningfully draw a line between an individual's actions and the plight of another (if it were, then such a case would fall into the fault-based injustice category).

[6] Here Young acknowledges the work of many others such as Bourdieu (1984) Kutz (2000), Sartre (1976), and Sewell (2005).

background structural processes that enable structural injustice through their participation in the seemingly neutral activities of everyday life. The untraceability inherent in Young's account of structural injustice is, I suggest, overlooked in the existing literature engaging with her work and the discourse on structural injustice more broadly. In my interpretation of Young's work, I highlight untraceability as a defining feature of structural injustice.[7] I have suggested elsewhere[8] that it is helpful to understand structural injustice as the consequence of 'structural actions' that are to be found amongst legitimate (i.e. legally and morally accepted) pursuits of private interest.[9] This clarification helps us to draw the distinction between, on the one hand, actions that render individuals and institutions liable for direct or indirect wrong-doing (either legal or moral wrong-doing) and on the other, those legitimate pursuits of private interest which, in an amorphous, untraceable way, contribute to the conditions that serve as the background to structurally harmful outcomes. The purpose of Young's distinction, as I see it, is to capture a wider set of responsibilities for injustice than those included in the remit of liability. Because compounding structural processes result from individual, group, and institutional actions at the macro-level, they are impossible for the individual alone to change.[10] What is needed then, in addition to fault-based approaches, is a macro-level collective focus on negative structural outcomes and their background conditions.

Before exploring this idea further, I first turn to the relationship between AI and injustice in general terms.

AI, Thoughtlessness, and Injustice

'[W]e can safely dispense with the question of whether AI will have an impact; the pertinent questions now are by whom, how, where, and when....'.

(Cowls and Floridi 2018, p.1).

---

[7] See Browne (forthcoming) for fuller discussion.

[8] See footnote 7.

[9] This account inevitably relies on some degree of relativity across different societies and individuals who will judge liability and structural injustices differently according to their legal and moral norms. Structural injustice is of course a moral issue but what is meant by Young as 'moral' is a 'moral norm' within a given context that if transgressed brings about a charge of liability. I emphasise here that both fault-based and structural approaches are vital to combatting injustice.

[10] Unlike Young whose theory of structural injustice does not give us a way of epistemologically moving from the structural to the legally or morally liable, I argue elsewhere (Browne, forthcoming) that these categories can in fact be transitory. As new epistemologies come along, not least through AI, we will be able to understand that which was previously unknowable and untraceable, (for example our contributions to climate change have moved from untraceable structural actions to traceable breaches of moral norms and, in some cases, legal prohibitions over the past 50 years).

A great deal has been made of the fact that today many decisions are allocated to algorithms without direct human involvement: performing medical diagnoses, determining whether or not a person is credit-worthy, who is the right candidate for particular jobs or for how long someone ought to be sentenced to prison.[11] Certainly, in a world oriented towards the accumulation of resources, targeted economic growth and attendant strategies of efficiency, the human is a poor substitute for the vast computational power of ML and the speed, precision and regularity of automation in so many spheres of human activity. However, largely driven by public fears about the unfeeling rationality of AI and the problems it may cause, mechanisms for tracing AI-generated harms to culpable individuals (including 'humans in the loop'[12]) or companies, is often required to bolster confidence in AI competence.[13] I shall argue this is an all too limited approach in the context of structural injustice.

Although Young did not write about the relationship between injustice and technology, one of her greatest influences was the work of Hannah Arendt[14] who harboured a deep suspicion of automation and AI. Writing soon after Turing posed the question 'Can machines think?' (1950, p.433), Arendt (1958/1998) recognised that automation and AI were seen as crucial in accelerating progress but was unconvinced that these new technologies ought to be thought of as the dominant characteristic of prosperity. For Arendt, the most important feature of humanity is its plurality emanating from the distinctiveness of each of us and the unpredictable political and creative possibilities such a distinctiveness brings to social transformation. Technologies that threaten our plurality are, therefore, a threat to the potential for social transformation and the capacity to correct injustices.

Set against a backdrop of a great many technological advancements that are to be celebrated,[15] the increasing dependency on AI outputs brings with it a decreasing ability to know any different to that which is produced by a given algorithm. As algorithms tend to reduce the complexities of human society to small numbers of data points on a scale incomparably larger than ever seen before they consolidate stereotypes and homogenised discourses about what is, in reality, a hugely

---

[11] As the likes of Haraway's (1985), Hayles' (1999), and Bradiotti's (2013) seminal work led us to imagine, humans have become co-constituted by technology. Depending on your view, we have become more than, or less than, human.

[12] See also Wilcox's critique of 'human in the loop' in Chapter 6 in this volume.

[13] This has, for example, been a central approach to the new European Union AI Act (Enarsson et al 2021) as well as the new UNESCO Recommendations on the ethics of artificial intelligence https://unesdoc.unesco.org/ark:/48223/pf0000379920

[14] Young certainly did not agree with Arendt on many topics and indeed Arendt's work is in places deeply problematic on questions of identity (notably gender and race). Somewhat ironically, however, Arendt's work on what it means to take up political responsibility serves to help us understand the threat of technologies such as AI to the plurality of humanity.

[15] See Taddeo and Floridi (2018) for example.

diverse and heterogeneous social world.[16] On this point, I find Arendt's concept of 'thoughtlessness' instructive.

> [T]houghtlessness—the heedless recklessness or hopeless confusion or compla-
> cent repetition of 'truths' which have become trivial and empty—seems to me to
> be among the outstanding characteristics of our time.
>
> (Arendt 1958/1998, p.5).

One way of reading this is not as absent-mindedness or negligence but rather a sort of habitual disengagement from speculating on the plight of those negatively affected by a community's institutions and collective behaviour—in our present context, we can think of the thoughtless adoption and acceptance of algorithmic decision making by way of example. For Arendt thoughtlessness is an affront to the flourishing of a pluralistic humanity and as Schiff (2014), explains: 'Plurality is not just about multiplicity, about the coexistence of many rather than the existence of just one. It is more fundamentally about connections between us, connections that make it intelligible to say that we share the world in common' (p.54). Young did not make explicit use of Arendt's idea of 'thoughtlessness' for her account of struc-tural injustice.[17] However, I find it to be acutely relevant. Under the pressures of everyday life, 'thoughtlessness' as an unquestioning habitual abidance to norms and social rhythms, the consequences of which accumulate at the macro-level and collectively contribute in large part to the background processes that cause struc-tural injustice.[18] These sorts of 'thoughtless' structural actions are enhanced by our increasing tendency to rely on AI which, as I shall discuss, is itself often untraceable despite the AI Tech Sector's best efforts.

Before reflecting further on the specific question of structural injustice and AI, I shall begin by considering the general relationship between AI and fault-based injustice. Consolidation of human bias is now a common story of AI (see Keyes, Chapter 17, and Nyrup, Chu, and Falco, Chapter 18 in this volume). Take AI recruitment technology, for example, which has already transformed the human resources sector in many countries. High-profile companies are often faced with very large numbers of applications for jobs and are increasingly attracted to what is known as 'intelligent screening software'. This sort of AI uses companies' exist-ing employee data as its training ground and calculates which existing employees have become 'successful' within the company and which have not. Such ranking is

---

[16] Chun (2021) makes an important point that ML's 'truth' is no more than selected, repeated data.

[17] Although the concept is mentioned in passing in a description of Arendt's work (Young 2011, p.84 and p.87).

[18] I am not suggesting here that Arendt's concept of 'thoughtlessness' is constrained to what I have described as structural actions. Indeed Schiff (2012) explains that there are several different ways in which Arendt uses the concept. For my purposes here, I think of structural actions as contained *within* the concept of thoughtlessness alongside other of its forms such as (fault-based) wilful blindness to injustice that Arendt sought to highlight.

built upon metrics of education, skills, experience, and, most importantly, key performance indicators set by the company. From those findings, the algorithm can
map certain characteristics to the CV data submitted by new applicants as well as
extend the applicant's data with a trawl of their social media profiles and any other
publicly available personal data. From there the algorithm can weed out those who
are unsuitably qualified or unlikely to become as successful as the 'ideal employee'
already working for the company (Ruby-Merlin and Jayam 2018; Saundarya et al.
2018).

Additional to intelligent screening software is 'digitised interviewing', whereby
candidates respond to automated questioning and their answers are documented
by their choice of words, speech patterns, facial expressions, and the like (Drage
and Mackereth 2022). These responses are then assessed in terms of 'fit' for specific
duties of the advertised role as well as the employing organisation's culture including their 'willingness to learn' and 'personal stability'.[19] On the up-side, it is clear
that a great deal of time is saved through using these technologies, especially in the
case of large numbers of applicants. The argument might be made that any of the
technology's biases are no worse than those applicants chosen by other humans—
indeed algorithms can be set to 'ignore', for example, ethnicity, gender, and age.
However, it is not difficult to see how the plurality of professional profiles becomes
reduced to a very small number of candidate profiles after several rounds of intelligent screening software that might be better known as 'consolidation of human
bias software'.[20] Leading technology firm Amazon, for example, abandoned its
home-grown AI recruitment technology when it could find no way to stop its algorithm from systematically downgrading women's profiles for technical jobs.[21] In
line with most AI recruitment technologies, Amazon's algorithm assessed all the
CVs received over the past decade alongside those who had been employed. From
that data it deduced the most 'successful' candidate profile. Given the low numbers
of women working in technical roles in Amazon it was not altogether surprising
that the algorithm's ideal candidate was male, thereby compounding existing sex
segregation patterns within the company (Lavanchy 2018).

In a similar vein, O'Neil's (2016) work shows how algorithmic technology
used in fields such as insurance, education, health, and policing systematically
disadvantages the poorer cohorts of society:

> Employers... are increasingly using credit scores to evaluate potential hires. Those
> who pay their bills promptly, the thinking goes, are more likely to show up to work

[19] See Hirevue promotional materials, for example https://www.hirevue.com/blog/hiring/hirevue-hiring-intelligence
[20] See also the discussion of AI and pseudoscientific practices by Agüera y Arcas, Mitchell, and Todorov, Chapter 13 in this volume.
[21] It is worth noting here that simply abandoning the algorithm to return to human bias on which the algorithm was trained in the first place is unlikely to yield better results in this example.

on time and follow the rules. In fact, there are plenty of responsible people and good workers who suffer misfortune and see their credit scores fall. But the belief that bad credit correlates with bad job performance leaves those with low scores less likely to find work. Joblessness pushes them toward poverty, which further worsens their scores, making it even harder for them to land a job. It's a downward spiral.

(p.17)[22]

This sort of 'evaluation' is what both Katz (2021) and Fraser (2021) mean when they discuss the provocative idea of the management of 'human waste'—those who are subjected to the increasing precarity of neo-liberal societies and systematically 'expelled from any possibility of steady employment' (Fraser 2021, p.162).

Alongside McInerney, Hampton, and Amrute in this volume (Chapters 7, 8, and 11, respectively) scholars such as Noble (2018), Buolamwini and Gebru (2018), Benjamin (2019) and Katz (2020) have also shown the far reaching racialised dimensions of AI and the reduction of racialised peoples to crude stereotypes and mis-identifications. Like other technologies and scientific practices, AI is starting to play a central role in the production and codification of racialised understandings of the human body (Felt et al. 2017). Benjamin, for example, discusses the Google search algorithm, used by hundreds of millions of people daily to understand and explore, virtually, the world in which they live. Benjamin describes how Google users searching for 'three black teenagers', were presented with 'criminal mugshots' (2019, p.93). In this way, Google captures and gives form to what many people trust and assume are fair and relevant descriptions of what they have searched for, by learning and replicating from big data snapshots of human behaviour and opinions that are biased and racist. While Tech companies might argue that their algorithms are only reflecting back amassed human views and actions, some algorithmic products clearly serve to promote discriminatory stereotypes that can be traced to a given company and called to account.

As we can see from these examples, the picture that emerges of how AI exacerbates injustice is a complex one. While AI-generated fault-based injustices can be traced back to particular designers and Tech companies, the fact that we are perhaps becoming more 'thoughtless' in our dependence on AI is more of a structural concern. Benjamin's work on the automation of racial discrimination, for example, helps us to see how AI-generated injustice does not only manifest in forms that are traceable to liable agents of fault but also operates on the macro structural scale;

By pulling back the curtain and drawing attention to forms of coded inequity, not only do we become more aware of the social dimensions of technology but we can work together against the emergence of a digital caste system that relies

---

[22] Also see for example Eubanks (2018).

on our naivety when it comes to the neutrality of technology. This problem extends beyond obvious forms of criminalization and surveillance. It includes an elaborate social and technical apparatus that governs all areas of life.

(Benjamin 2019, p.11)

Here I see a link between Benjamin's reach beyond traceable mechanisms of 'criminalization and surveillance' to 'apparatus that governs all areas of life', and Young's reach beyond 'fault-based injustice' to the structural, whereby societal outcomes are determined by macro relational forces. While individual and or group agent-centric actions are of course vital to understanding and acting on in the pursuit of justice, they do not explain the full extent of how either AI compounds racial injustice or how other forms of structural injustice are generated by the structural actions of ordinary people going about their everyday business. The argument here is that the level of causal complexity between individual action and structural injustice becomes so intricate and convoluted on a mass scale that it is no longer meaningfully traceable with any existing epistemological tools. To give an example: 'in what meaningful ways could we trace all the individuals who are implicated in the global demand for general technologies which result in increased international R&D that, in turn, produce new products, some of which may create unintended negative consequences for particular groups?' I think Young's view would have been that although there may be some traceable elements of fault in this example (which ought to be addressed), it certainly would not make sense to attempt to trace the causal chain back to the initial individuals. Rather, it is more productive to think broadly about what can be done to reset the background conditions that facilitate structural injustice. I now turn to the increasing untraceability of AI as an exacerbating factor.

AI and Untraceability

'No one really knows how the most advanced algorithms do what they do.'

(Knight 2017)

Opacity is key to the AI sector. Not only does it give AI technology an air of rational neutrality, but also no firm wants to share the inner workings of its algorithms in such a competitive market. However, beyond the protection of trade secrets, there is a different sort of increasing opacity in AI—often referred to as a lack of AI 'explainability' or 'interpretability'. AI systems are not programmed to recognise certain salient features of the world, rather they learn through adjusting millions of parameters until they attain the requisite level of accuracy in a given task. More often than not, these 'deep-learning' processes are not comprehensible in ways that are meaningful to us—that is to say, they are untraceable. We can only understand the data inputs and outputs rather than the algorithmic decisions themselves, even with the help of other algorithms. This is what is meant by the 'black box' of neural networks (Nielly 2020; von Eschenbach 2021). As the Uber AI designer Jason Yosinski (2017) explains '[w]e build amazing models, but we don't quite understand them. And every year, this gap is going to get a bit larger'.

Ironically, as we give or leak more and more of our data to what Lawrence (2015) has called a system of 'digital servitude' to the institutions, firms and governments that hold and utilise them, we become more transparent and controllable as the algorithms become more opaque and independent. By increasingly deferring to algorithmic decisions, not only are we relinquishing or avoiding accountability but perhaps most importantly, we are diminishing our capacity to reflect on and judge our norms and assumptions per se. Indeed, we might think of this as the technological 'impoverishment of the sphere of reflexivity' D'Agostino and Durante (2018) or in Arendtian terms, an increasingly important case of 'thoughtlessness'.

The imperative to rely on algorithms is strong. Indeed, Floridi (2014, 2020) has argued algorithms have become core to what most people think of as elements of human well-being (medical predictions, environmental planning, transport logistics, complex financial data management, etc.). In this sense, Arendt was sceptical about humanity's capacity to think: '[I]t could be that we...will forever be unable to understand, that is, to think and speak about the things which nevertheless we are able to do'. (Arendt 1958/1998, p.3). Similarly, Delacroix (2021) argues that much like muscle atrophy that comes from lack of physical activity, '[r]eliance upon non-ambiguous systems—whose opaque, multi-objective optimisation processes makes any effort of critical engagement redundant—will affect the extent to which we are made to flex our "normative muscles" in the longer term' (pp.12–13).

As we shall see in the next section even as algorithmic decision making becomes more untraceable, much of the current focus on countering the downsides of AI manifests in discourses of 'Responsible AI'[23] that requires 'algorithmic transparency' or the presence of a 'human in the loop' to ensure liability if and when things go wrong. I argue that this is far from a sufficient approach.

Political Responsibility for AI

Perhaps unsurprisingly, tech companies tend to favour, where possible, self-regulation based on a set of similar principles or ethical codes. As Field et al. (2020, p.4) report, 'seemingly every organization with a connection to technology policy

---

[23] See for example Google's 'Responsible AI' pledge: https://ai.google/responsibilities/responsible-ai-practices/; Microsoft's 'Responsible AI' pledge: https://www.microsoft.com/en-us/ai/responsible-ai?activetab=pivot1%3aprimaryr6; And PWC's 'Responsible AI toolkit' https://www.pwc.com/gx/en/issues/data-and-analytics/artificial-intelligence/what-is-responsible-ai.html

has authored or endorsed a set of principles for AI'.[24] The ethical approach, oriented towards human well-being, is of course important and more recently has been supplemented by legislation. The EU is at the forefront of legislating AI, with, for example, the General Data Protection Regulation and, should it be ratified, the new EU AI Act.[25] This new Act introduces a liability-based regulatory approach that is specifically designed not to impede AI markets by restricting regulation to 'those concrete situations where there is a justified cause for concern' (EC 2021 Sec1.1.1). The Act requires, for example, that 'high-risk AI'[26] adheres to a strict set of restrictions in the interest of the public, which the Act defines as 'health, safety, consumer protection and the protection of other fundamental rights' (EC 2021 Sec 3.5). Here the focus is on the protection of individuals' privacy, safety and security, transparency, and fairness as well as non-discriminatory programmes and professional responsibility by AI designers.[27] Providers of high-risk AI must devise systems of risk management that can 'identify known and foreseeable risks associated with the AI system.... Residual risks must be "acceptable" and communicated to users' (EC 2021, Art.9.4). There is no real guidance however on how all this should be finely interpreted in the increasing cases where the process of algorithmic decision making is untraceable. The only protection that comes from this legislation is that which is traceably liable 'known and foreseeable' (EC 2021Art 9.2a). Those who breach the rules or fail to comply are liable. However, what use is this approach if the opacity of algorithmic decision making becomes an exacerbating feature of the background conditions to untraceable structural injustice—that is, where there is no identifiable culprit?[28] This seems an intractable problem to analyse and solve. Rather than restricting ourselves to the legal tools of seeking out agents of blame, what would it take to make us less thoughtless about what we are actually doing when technology is developed and used in such a way that it contributes to the background conditions of structural injustice? Part of the solution, even in the presence of such an all-encompassing technological environment, is to rail against normative atrophy? So how might we foster political responsibility for AI?

---

[24] The latest of these is the new UNESCO Recommendations on the ethics of artificial intelligence https://unesdoc.unesco.org/ark:/48223/pf0000379920

[25] 'A legal framework for trustworthy AI. The proposal is based on EU values and fundamental rights and aims to give people and other users the confidence to embrace AI-based solutions, whilst encouraging businesses to develop them'. (EC 2021 Sec1.1.1) https://eur-lex.europa.eu/legal-content/EN/TXT/?uri=CELEX%3A52021PC0206

[26] The weight of the Act focuses on 'high risk AI' such as biometric identification and categorisation; For a full description see https://www.jdsupra.com/legalnews/european-commission-s-proposed-3823933/

[27] For a critique of professional responsibility by AI designers see Costanza-Chock, Chapter 21 in this volume.

[28] This is a different consideration to that which Floridi (2016) raises about 'faultless responsibility' ending in criminal behaviour (this might be akin, for example, to indirect discrimination) which will be grounded in the traceable actions of a liable agent.

Thinking back to the earlier discussion of structural injustice, Young's work has shown us that structural injustice requires a different sort of responsibility to liability. Were we to apply this additional form of responsibility to AI, it would be a primarily forward-looking exploration of how AI might generate and perpetuate structural injustice and speculative consideration of what can be done about it collectively. This is a very different focus from the narrow remit of 'those concrete situations where there is a justified cause for concern' set out in current laws and regulation designed to offset AI-generated harm. Once we move beyond the liability approach of protecting the public from discriminatory algorithmic outputs and the like, we come into the structural realm of dynamics that are not readily traceable but about which we ought to speculate politically nevertheless. This is not something that could be achieved by any fault-based legislation,[29] as discussed earlier: 'it is not possible to trace which specific actions of which specific agents cause which specific parts of the structural processes of their outcomes' (Young 2003, p.7).

In the United Kingdom, the principal AI advisory body to Government is the Centre for Data Ethics and Innovation (CDEI).[30] This is a body of technology experts whose role it is to make recommendations on the governance of AI and to identify ways in which the UK Government can support the development of digital technologies such as automated decisions systems, robotics and forms of artificial intelligence.[31] To its credit the CDEI has attempted to engage the general public in a far more comprehensive way than many of the other bodies of Government through its 'tracker survey' which monitors how public attitudes to the use of data and data-driven technologies change over time.[32] Its methods include working with a nationally representative sample of 4000 individuals (as well as a further 200 without access to the internet) to collect citizens' views. While a wholly valuable exercise in generating news topics of broader discussion around AI, the problem with this sort of polling as a way of engaging the public, is that it only tells you what the public already think about a very narrow set of questions

---

[29] See Amrute's critique of current approaches to techno-ethics in this volume.

[30] The UK government has been explicit that it will not set up a regulatory public body for AI governance: AI and Public Standards (Weardale 2020, p.47) and House of Lords Select Committee on AI (2018, p.386). Instead, the Government introduced a range of advisory bodies related to the Department of Digital, Culture, Media, and Sport and or the Department of Business Energy and Industrial Strategy bodies. In addition to the principal body, the Centre for Data Ethics and Innovation is the Office for AI (charged with overseeing the implementation of the National AI strategy—see here for details: www.gov.uk/government/organisations/office-for-artificial-intelligence) and the AI Council—a high level industry led advisory board—https://www.gov.uk/government/groups/ai-council.

[31] https://www.gov.uk/government/publications/advisory-board-of-the-centre-for-data-ethics-and-innovation/advisory-board-of-the-centre-for-data-ethics-and-innovation

[32] See for example: https://www.gov.uk/government/publications/public-attitudes-to-data-and-ai-tracker-survey. Whilst it is common for public bodies to welcome public consultation this often amounts to no more than industry submissions and self-selecting groups and individuals who have a particular vested interest (see Browne forthcoming).

posed. There is little room for engagement with alternative experiences, debate or what Young called 'self-transcendence' (Young 1997, p.66)—the development of a willingness to be open to a different way of conceiving of and solving collective problems altogether.

So, what might an alternative approach be that gets us closer to a 'structural perspective'? We might draw some inspiration from an experimental form of political deliberation, the mini-public (Fishkin 2018), which creates the opportunity for the public to think and gives policymakers a much richer sense of what the plurality of experiences connected to the structural consequences of AI might be. Common features of the general mini-public model are as follows: the selection of members is made randomly through, for example, the electoral register by an independent polling company according to gender, age, location, and social class; none of the members are politicians; administration and running costs are met by the state; no member is paid (although expenses such as travel and childcare are met); members are exposed to competing expert evidence and submissions by other interested parties such as individual members of the public, social movements, industry experts, NGOs, charities, and the state and so on; there is a facilitated discussion and debate that is streamed live to the general public, and an anonymous private vote on various elements of the proceedings at the end. The mini-public approach is not concerned with collecting pre-existing public attitudes but rather is a speculative exercise in thinking about how everyday occurrences contribute to the background conditions of everyone's varying prospects. The plurality of those involved brings an infinitely wider set of experiences and concerns than those of industry experts or politicians and opens the scope of enquiry to much bigger structural questions. This approach aligns with Young's idea of taking up political responsibility to address structural injustice from a range of different perspectives and experiences. While Young was suspicious of deliberative models of decision making that presupposed sameness and consensus, her aim was not to reject dialogical politics altogether but rather to be productively critical (Young 2001, 2011). In particular, Young wanted to stress that that various forms of meaningful communication usually rejected from governance mechanisms as too personal, irrational and irrelevant, could in fact be extremely helpful in understanding the complex experiences and perspectives of those on the periphery—those who tend to suffer most from structural injustices. That is to say, those experiences and stories that give us a richer picture of the structural dynamics at play in the everyday.

Fishkin's work on mini-publics shows how they tend to operate outside of governance mechanisms (Fishkin 2018) at large expense and over long time periods but here I suggest including a lay element within our everyday policy making structures directly.

'A New Lay-Centric AI Public Body'

A very different sort of approach to devising policy would be a new specialised lay-centric public body—I shall call it the 'AI Public Body'.

Once imagined, the shape of such a body is not difficult to sketch. The AI Public Body would retain many of the features of the current model of a public body—a Government-funded recommendatory body, operating at arm's length so as to be independent of political steering with subsequent regulatory powers on settled specified remits. Central to the function of a public body would be a range of expertise emanating from specialists with differing perspectives and the ability to explain and interpret the underlying technicalities of any particular question or concern (much like the existing UK CDEI). However, if we look to what was once widely considered as the 'gold standard' designs for public bodies, the 1984 Warnock Report[33] which led to the Human Fertilisation and Embryology Authority in 1990, we can see an argument for lay-inclusion chosen for their 'ordinariness as a member of the public' rather than their expertise or special insight to the technology in question. In the Warnock report, however, the recommendation of lay-inclusion was focused solely on the Chair. While this recommendation led to a range of impressive incumbents, all those chosen by the Government were well known public intellectuals or industry leaders rather than lay members characterised by 'ordinariness'. In the new model of the public body—I have a more fundamental connection to the public in mind. While the Chair must rightly have the professional skills to choreograph a complex decision-making body, an additional set of lay members would be selected, much like jury service, based on a range of demographic census data (age, ethnicity, gender, geography, socio-economic status, etc.) and there would be a permanent rotation of lay members into the decision-making design of the public body. Such an approach is not capable of achieving perfect representation of the plurality of humanity, nor the citizenry nor the many forms of situated knowledge that emanate from structural injustice, but it is the most practical link to a more diverse set of concerns and interests that if built into the deliberations of AI regulation and policy, will ensure, at the very least, that a focus on the background conditions of structural injustice is more likely to override the private interests of industry as is the current state of affairs. The potential of lay input is to attempt to bring a focus on the structural dynamics of AI and look beyond the usual remit of establishing liability as the best form of AI governance.

The attraction of such a model is that the possibility of speculating on the structural dynamics of AI is more likely than within the standard liability-based

---

[33] See Franklin (2013) and Browne (2018).

approach which narrows attention and the use of political resources to searching for traceable agents of fault. On substantive questions such as how personal data ought to be collected and how its use be governed, or how much analysis should be done on the biased outcomes of algorithms before their assessments and predictions become the bases of policy, or how ought the Government to plan to counter the socio-economic effects of automation of certain labour market tasks, it is highly likely that a group of citizens would draw substantially different conclusions to those of industry experts or politicians. I argue that this is the key to creating a very different sort of public-body approach to AI-generated structural injustice than the models we currently have in play. It is a simple idea but to bring a range of lay voices firmly into the mix of discussions around the fundamental questions of AI's structural impact on society would be a profound change to the way public bodies currently operate and approaches to AI governance are approached.

Conclusion

In this chapter, I have attempted to demonstrate how feminist scholarship helps us to think productively about why a liability approach to regulating AI is not sufficient to mitigate against its capacity to exacerbate structural injustice. I have argued that we need to construct new forms of lay-centric fora into our democratic governance systems to bring a broader set of interests into the process of rebalancing the background conditions of structural injustice. Including the public in deliberations on AI is not, however, a short-cut to good decision making. The field of ML is so complex and technical that we certainly need specialists and industry experts to help design the best policy going forward. However, I argue that without a lay-centric orientation in policy deliberations, the outcomes will tend towards a sole focus on liability mitigation which is necessary to enhance faith in market operations and address traceable fault-based injustices but little else. Rather what is also needed is a broader speculative approach to structural accounts of AI and its influences on society. My sense is that Warnock had something of the same instinct about reproductive technologies back in the 1980s.[34]

Much of the current focus within AI governance debates is centred on creating 'Responsible AI'[35] as if the ultimate political goal is to ground AI in traceable liability. This is of course a vital task but it does not suffice as an approach to AI's relationship to structural injustice which takes on a much more complex and amorphous shape, populated by diverse experiences and perspectives. Indeed,

[34] See Browne (2018, 2020) for a fuller discussion.
[35] See for example Google's 'Responsible AI' pledge: https://ai.google/responsibilities/responsible-ai-practices/; Microsoft's 'Responsible AI' pledge: https://www.microsoft.com/en-us/ai/responsible-ai?activetab=pivot1%3aprimaryr6; And PWC's 'Responsible AI toolkit' https://www.pwc.com/gx/en/issues/data-and-analytics/artificial-intelligence/what-is-responsible-ai.html

to default to the liability-based approach to AI governance is as 'thoughtless' as assuming a default to AI is always best. What is additionally required is a collective political endeavour that is grounded in the situated knowledge of a pluralistic public and an active resistance to normative atrophy.

References

Allen, Danielle S. (2016) Toward a Connected Society. In *Our Compelling Interests: The Value of Diversity for Democracy and a Prosperous Society,* eds E. Lewis and N. Cantor, pp. 71–105. Princeton: Princeton University Press.

Arendt, Hannah (1958/1998) *The Human Condition.* Chicago: University of Chicago Press.

Benjamin, Ruha (2019) *Race After Technology: Abolitionist Tools for the New Jim Code.* Cambridge: Polity Press.

Bourdieu, Pierre (1984) *Distinction: A Social Critique of the Judgement of Taste,* trans Richard Nice. Cambridge MA: Harvard University Press.

Braidotti, Rosi (2013) *The Posthuman.* Cambridge: Polity Press.

Browne, Jude. (2018) 'Technology, Fertility and Public Policy: A Structural Perspective on Human Egg Freezing and Gender Equality'. *Social Politics: International Studies in Gender, State & Society* 25(2): 149–168.

Browne, Jude. (2020) 'The Regulatory Gift: Politics, Regulation and Governance'. *Regulation & Governance* 14: 203–218.

Browne, Jude (forthcoming) The Limits of Liability Politics (working title).

Browne, Jude and Maeve McKeown. (forthcoming 2023) *What is Structural Injustice?* Oxford University Press.

Buolamwini, Joy and Timnit Gebru. (2018) 'Gender Shades: Intersectional Accuracy Disparities in Commercial Gender Classification'. *Proceedings of Machine Learning Research.* Vol. 81: pp. 1–15 http://proceedings.mlr.press/v81/buolamwini18a/buolamwini18a.pdf

Cabinet Office (2010) *Public Body Review Published.* The Right Honourable Lord Maude of Horsham. UK Cabinet Office Press Release. London. HMG https://www.gov.uk/government/news/public-body-review-published

Chun, Wendy Hui Kyong (2021) *Discriminating Data: Correlation, Neighborhoods, and the New Politics of Recognition.* Cambridge MA: MIT Press.

Cowls, Josh and Luciano Floridi. (2018) 'Prolegomena to a White Paper on an Ethical Framework for a Good AI Society'. *SSRN Electronic Journal* https://www.researchgate.net/publication/326474044_Prolegomena_to_a_White_Paper_on_an_Ethical_Framework_for_a_Good_AI_Society

D'Agostino, Marcello and Massimo Durante. (2018) 'Introduction: The Governance of Algorithms'. *Philosophy & Technology* 31(4): 31.

Delacroix, Sylvie (2021) 'Diachronic Interpretability & Machine Learning Systems'. *Journal of Cross-disciplinary Research in Computational Law* First View: https://papers.ssrn.com/sol3/papers.cfm?abstract_id=3728606

Drage, Eleanor and Kerry Mackereth. (2022) 'Does AI de-bias recruitment? Race, Gender and AI's 'Eradication of Difference Between Groups''. Philosophy and Technology, 35 (89).

Enarsson, Therese, Lena Enqvist, and Markus Naarttijärvi. (2021) 'Approaching the Human in the Loop – Legal Perspectives on Hybrid Human/Algorithmic

Decision-Making in Three Contexts'. *Information & Communications Technology* Early View: https://www.tandfonline.com/doi/citedby/10.1080/13600834. 2021.1958860?scroll=top&needAccess=true

Eubanks, Virginia (2018) Automating Inequality: How High-Tech Tools Profile, Police and Punish the Poor. New York: St Martin's Publishing Group.

Felt, Ulrike, Rayvon Fouché, Clark Miller, and Laurel Smith-Doerr. (eds) (2017) *Handbook of Science and Technology Studies*. Cambridge, Massachusetts: MIT Press

Fishkin, James (2018) *Democracy When the People Are Thinking*. Oxford: Oxford University Press.

Fjeld, Jessica, Nele Achten, Hannah Hilligoss, Adam Nagy, and Madhulika Srikumar. (2020) 'Principled Artificial Intelligence: Mapping Consensus in Ethical and Rights-based Approaches to Principles for AI'. Research Publication No. 2020–1 *The Berkman Klein Center for Internet & Society Research Publication Series*: Harvard University https://cyber.harvard.edu/publication/2020/principled-ai

Floridi, Luciano (2014) *The Fourth Revolution: How the Infosphere is Reshaping Human Reality*. Oxford: Oxford University Press.

Floridi, Luciano (2016) 'Faultless Responsibility: On the Nature and Allocation of Moral Responsibility for Distributed Moral Actions'. *Royal Society's Philosophical Transactions A: Mathematical, Physical and Engineering Sciences* 374(2083): 1–22. https://doi.org/10.1098/rsta.2016.0112.

Floridi, Luciano (2017) 'Digital's Cleaving Power and its Consequences'. *Philosophy and Technology* 30(2): 123–129.

Floridi, Luciano (2020) 'Artificial Intelligence as a Public Service: Learning from Amsterdam and Helsinki'. *Philosophy & Technology* 33: 541–546.

Floridi, Luciano, Mariarosaria Taddeo, and Matteo Turilli. (2009) 'Turing's Imitation Game: Still an Impossible Challenge for All Machines and Some Judges— An Evaluation of the 2008 Loebner Contest'. *Minds and Machines* 19(1): 145–150.

Franklin, Sarah (2013) 'The HFEA in Context'. *Reproductive Biomedicine Online* 26(4): 310–312.

Fraser, Nancy (2021) 'Gender, Capital and Care'. In *Why Gender?*, ed. Jude Browne, pp. 144–169. Cambridge: Cambridge University Press.

Haraway, D. (1985/2000) A Cyborg Manifesto: Science, Technology and Socialist-Feminism in the Late Twentieth Century. In The Gendered Cyborg: A Reader, eds G. Kirkup et al., pp. 50–57. London: Routledge.

Hayles, Katherine (1999) How We Became Posthuman: Virtual Bodies in Cybernetics, Literature, and Informatics. Chicago: The University of Chicago Press.

Hui Kyong, Chun (2021) Discriminating Data: Correlation, Neighborhoods, and the New Politics of Recognition. Massachusetts: MIT Press.

Katz, Cindi (2021) 'Aspiration Management Gender, Race, Class and the Child as Waste'. In *Why Gender?*, ed. Jude Browne, pp. 170–193. Cambridge: Cambridge University Press.

Katz, Yarden (2020) Artificial Whiteness: Politics and Ideology in Artificial Intelligence. New York: Columbia University Press.

Knight, Will (2017) 'The Dark Secret at the Heart of AI (April 11th)'. *MIT Technology Review* https://www.technologyreview.com/2017/04/11/5113/the-dark-secret-at-the-heart-of-ai/

Kutz, Christopher (2000) *Complicity: Ethics and Law for a Collective Age*. Cambridge: Cambridge University Press.

Lavanchy, Maude (2018) Amazon's sexist hiring algorithm could still be better than a human. *Quartz*. 7 November, https://qz.com/work/1454396/amazons-sexist-hiring-algorithm-could-still-be-better-than-a-human/

Lawrence, Neil (2015) The information barons threaten our autonomy and our privacy. *The Guardian* Monday 16th November, https://www.theguardian.com/media-network/2015/nov/16/information-barons-threaten-autonomy-privacy-online

McKeown, Maeve (2021) 'Structural Injustice'. *Philosophy Compass* 16(7): Early View https://onlinelibrary.wiley.com/doi/10.1111/phc3.12757

Nielly, Cyprien (2020) 'Can We Let Algorithm Take Decisions We Cannot Explain? Towards Data'. *Science* https://towardsdatascience.com/can-we-let-algorithm-take-decisions-we-cannot-explain-a4e8e51e2060

Noble, Safiya Umoja (2018) *Algorithms of Oppression: How Search Engines Reinforce Racism*. New York: NYU Press.

O'Neil, Cathy (2016) *Weapons of Math Destruction*. *New York*: Crown Publishing Group.

ONS (2019) *Which Occupations are at Highest Risk of Being Automated?* 25 March. HMG. London: Office of National Statistics. https://www.ons.gov.uk/employmentandlabourmarket/peopleinwork/employmentandemployeetypes/articles/whichoccupationsareathighestriskofbeingautomated/2019-03-25

Ruby-Merlin. P. and R. Jayam. (2018) 'Artificial Intelligence in Human Resource Management'. *International Journal of Pure and Applied Mathematics* 119(17): 1891–1895

Sartre, Jean-Paul (1976) *Critique of Dialectical Reason*, trans Alan Sheridan-Smith. London: New Left Books.

Saundarya, Rajesh, Umasanker Kandaswamy, and Anju Rakesh (2018) 'The Impact of Artificial Intelligence in Talent Acquisition Lifecycle of Organizations. A Global Perspective'. *International Journal of Engineering Development and Research* 6(2): 2321–9939.

Schiff, Jacob (2012) 'The Varieties of Thoughtlessness and the Limits of Thinking'. *European Journal of Political Theory* 12(2): 99–115.

Schiff, Jade (2014) Burdens of Political Responsibility: Narrative and The Cultivation of Responsiveness. Cambridge: Cambridge University Press.

Sciforce (2020) 'Introduction to the White-Box AI: the Concept of Interpretability'. *Sciforce* https://medium.com/sciforce/introduction-to-the-white-box-ai-the-concept-of-interpretability-5a31e1058611

Sewell, William H. (2005) *Logics of History: Social Theory and Social Transformations*. Chicago: University of Chicago Press.

Taddeo, Mariarosaria and Luciano Floridi. (2018) 'How AI Can Be a Force for Good'. *Science* 361(6404): 751–752

Turing, Alan (1950) 'Computing Machinery and Intelligence'. *Mind* 49: 433–460.

von Eschenbach, Wareen (2021) 'Transparency and the Black Box Problem: Why We Do Not Trust AI'. Philosophy and Technology 34(4): 1607–1622.

Watson, David and Luciano Floridi. (2020) 'The Explanation Game: A Formal Framework for Interpretable Machine Learning'. Synthese 198: 9211–9242.

Weardale, Lord. (2020) 'Artificial Intelligence and Public Standards A Review by the Committee on Standards in Public Life'. The Committee of Standards in Public Life. UK Government. https://assets.publishing.service.gov.uk/government/uploads/

system/uploads/attachment_data/file/868284/Web_Version_AI_and_Public_
Standards.PDF.

Yosinski, Jason (2017) quoted by Voosen, Paul (2017) 'How AI Detectives are Cracking
Open the Black Box of Deep Learning'. *Science* 29 August, science.sciencemag.org/

Young, Iris Marion (1990) *Justice and the Politics of Difference*. Princeton, NJ, USA:
Princeton University Press.

Young, Iris Marion (1997) *Intersecting Voices: Dilemmas of Gender, Political Philosophy
and Policy*. Princeton, NJ, USA: Princeton University Press.

Young, Iris Marion (2001) 'Activist Challenges to Deliberative Democracy'. *Political
Theory* 29(5): 670–690

Young, Iris Marion (2003) 'The Lindley Lecture'. 5 May. University of Kansas:
https://kuscholarworks.ku.edu/bitstream/handle/1808/12416/politicalresponsibi-
lityandstructuralinjustice-2003.pdf?sequence=1

Young, Iris Marion (2006) 'Responsibility and Global Justice: A Social Connection
Model'. *Social Philosophy and Policy* 23(1): 102–130.

Young, Iris Marion (2011) *Responsibility for Justice*. Oxford: Oxford University Press.

20

Afrofeminist Data Futures

Neema Iyer, Chenai Chair, and Garnett Achieng

In the quest for gender equality and societal change, the transformative role of data, when applied accordingly, can be used to challenge dominant power imbalances and create social impact in communities.[1] This research focuses specifically on African feminist movements working towards social justice. It explores the collection, sharing and use of digital data for social transformation. In this research project, we defined data as distinct pieces of information, stored as values of quantitative and qualitative variables, which can be machine-readable, human-readable or both. Through a mixed-methods approach that centres these movements, we determine the extent of data use, the opportunities and the challenges of working with data, as well as present recommendations for social media companies to better contribute to the data ecosystems in the African context.

Introduction

In May 2013, the United Nations (UN) coined the term 'data revolution' to usher in a new era where international agencies, governments, civil society organisations (CSOs), and the private sector would commit to the improvement of the quality and availability of data. Better data and statistics would enhance progress tracking and accountability, and promote evidence-based decision making (UN Data Revolution, 2013). A true data revolution includes transformative steps such as 'improvements in how data is produced and used; closing data gaps to prevent discrimination; building capacity and data literacy in "small data" and Big Data analytics; modernising systems of data collection; liberating data to promote transparency and accountability; and developing new targets and indicators' (UN Data Revolution, 2016). Big Data is defined as extremely large and complex data sets—structured and unstructured—that grow at ever-increasing rates. Big Data may be analysed computationally to reveal patterns, trends, and associations, especially relating to human behaviour and interactions. The Data Conversation in the African Region Data is seen as a powerful tool to address global challenges as it can

---

[1] The full paper that this chapter is based on was published in March 2022 in English, French, and Portuguese by Pollicy, accessible here: http://archive.pollicy.org/feministdata/.

Neema Iyer, Chenai Chair, and Garnett Achieng, *Afrofeminist Data Futures*. In: *Feminist AI*. Edited by: Jude Browne, Stephen Cave, Eleanor Drage, and Kerry McInerney, Oxford University Press. © Oxford University Press (2023).
DOI: 10.1093/oso/9780192889898.003.0020

offer new insights into areas as diverse as health research, education, and climate change (boyd and Crawford 2012, p.674).

In addition, what is counted often becomes the basis for policymaking and resource allocation (D'Ignazio and Klein 2020). While analysed data is not a panacea to solve all problems, it is a way to know the depth (qualitative) as well as breadth (quantitative) of a phenomenon. For example, basic population estimates often do not exist for LGBTQIA+ (Lesbian, Gay, Bisexual, Transgender, Queer (or Questioning), Intersex, and Asexual) persons and some forced migrants, allowing some governments to deny their very existence—that is, rendering them invisible. Data collected on marginalised groups makes them visible and puts forth a case to provide for their needs. On the other hand, through the manifestation of surveillance, data enables invasions of privacy, decreases civil freedoms, and increases state and corporate control (boyd and Crawford 2012, p.674). This theory can be seen on the continent as some African governments that have managed to digitise their data collection have focused more on 'siphoning citizen data while keeping the state opaque and making civilians toe the line than in improving the services that states provide to citizens' (Nyabola 2018, p.70). In this case, these governments may have improved in the collection of citizen data but not necessarily at managing it or harnessing it for positive potential. Our desk review highlights the data infrastructures on the continent, the context of data practices, and the availability of data to understand the extent of the gaps and opportunities across Africa. We nuance this exploration with a gender perspective grounded in data feminism, data justice and African feminist movements.

Data Infrastructures on the Continent

In response to the UN's call for a data revolution, governments, CSOs, and the private sector have turned to digitisation as a way of collecting and storing data. The premise is that digitisation translates to better services and products. Despite this shift, many African countries are still lagging. They lack well-functioning civil registration and vital statistics systems that often act as the foundations that digital data infrastructures are built on (Data 2X 2019b). Statisticians have found it difficult to track how well African countries are moving towards their 2030 UN sustainable development goals because of this absence of data. On average, African governments collect statistics covering only about a third of the relevant data needed to track this progress (Pilling 2019). Gender data has also remained under-collected, as large gender data gaps exist in both national and international databases. A 2019 Data 2X study of national databases in fifteen African countries, including leading economic and digital hubs such as Kenya, Nigeria and South Africa, found that sex-disaggregated data were available for only 52% of the gender-relevant indicators. Large gender data gaps existed in all fifteen countries, with these gaps unevenly distributed across the indicators. For instance, no indicator in the environmental domain had gender-disaggregated data at the international database level (Data 2x 2019b). The importance of gender data and sex-disaggregated data has largely been ignored because of the lesser value that some societies place on women and girls (Temin and Roca 2016, p.268). Moreover, where gender data may be available, its interpretation and analysis may be biased because the production of gender data is not a simple exercise in counting women and men (Ladysmith et al. 2020). Gender data demands that researchers firmly comprehend how bias and power dynamics are embedded in the study design, sampling methodologies, data collection, and raw data itself. All researchers are interpreters of data and ideally must account for the biases in their understanding of the data (boyd and Crawford 2012, p.663). As a result of these gaps and biases, many issues unique or predominantly related to African women remain poorly understood (Temin and Roca, 2016, p.268).

Data Protection across the Continent

Data protection in Africa can still be described as in its nascent stage, as many African states still do not have a data protection law or have not fully implemented such laws yet (Ilori 2020). Out of the fifty-five states on the continent, twenty-eight countries have a data protection law, of which fifteen have set up data protection authorities (DPAs) to enforce the law. DPAs are independent public authorities that monitor and supervise, through investigative and corrective powers, the application of the data protection law. They provide expert advice on data protection issues and handle complaints. Yet, even African countries that have enacted a data protection law still fall short of protecting citizens' data for a number of reasons. For instance, Kenya, Uganda, Botswana, Equatorial Guinea, Seychelles, and Madagascar are examples of countries that have passed data protection laws and are yet to set up their DPAs (Ilori 2020). The absence of the regulator to enforce the law creates a unidirectional data system, where citizens cannot hold governments and private institutions accountable for the mismanagement of citizen data (Nyabola 2018, p.71). Another issue is the lack of standard structures that ensure the independence of DPAs in Africa (Ilori 2020). Senegal, however, set the stage for other countries by initiating legal reform to address the gaps identified in their data protection laws (Robertson 2020). These reforms will address the need for more independence for the Commission on Personal Data, among other issues. In addition, gendered issues of data protection are often ignored, but as Chair (2020) argues, there is a need to nuance gendered data protection issues, especially for vulnerable groups.

Data Practices Rooted in Colonialism

The practices of data extraction and use must be explored within the context of power dynamics and historical events that are rooted in colonialism (see Hampton, this volume). While African governments have lagged in setting up data infrastructures and passing data protection laws, the private sector has found welcoming soil in this void. Africa has been touted as a treasure trove of untapped data, and large technology companies are rushing to set up digital infrastructures for their profit-making. This move has been described as imperialist, with scholars likening it to the Scramble and Partition of Africa and referring to it as 'digital colonialism'. Digital colonialism is the decentralised extraction and control of data from citizens with or without their explicit consent through communication networks that are predominantly developed and owned by Western tech companies (Coleman 2019). While Western companies are not the only ones using extractive means to obtain data, a significant proportion of Africa's digital infrastructure is controlled by Western technology powers such as Amazon, Google, Facebook, and Uber (Abebe et al. 2021). Furthermore, companies extract, mine and profit from data from Africans without their explicit consent and knowledge of what the data is used for. One such case was when Guinness Transporters, which operates in Uganda as SafeBoda, sold unsuspecting clients' data to Clever Tap, a third-party US company (Kasemiire 2021). An investigation by the National Information Technology Authority—Uganda (NITA-U) found that SafeBoda disclosed users' email addresses, telephone numbers, first and last names, mobile device operating system, application version and type, as well as user login status (Kasemiire 2021). Ultimately, while the data mined by corporations could be repurposed to benefit other entities, the links between science, state, and corporations on data sharing are relatively weak on the continent. In cases where the private sector has collaborated with governments, the partnerships have not been entirely beneficial to citizens. For instance, while Huawei Technologies is responsible for up to 70 percent of Africa's telecommunications network, it has also laid the ground for the surveillance of citizens by authoritarian governments under their Safe City projects (Kidera, 2020). In Uganda and Zambia, allegations against Huawei Technologies claim that they aided government surveillance of political opponents by intercepting their encrypted communications and social media, and using cell data to track their whereabouts (Parkinson et al. 2019). In both cases, this surveillance led to the arrest of politicians and bloggers (Parkinson et al. 2019).

Nuancing Gender in Data: Women, Datafication, and Dataveillance

Nonetheless, the use of digital data such as Big Data can add nuance to our understanding of women and girls' lives by providing information that is highly granular in both space and time and offering insights on aspects of life that are often difficult to quantify and capture in standard types of data collection (Data 2X 2019a). For instance, national socio-economic surveys typically offer information about the status of the family as an entirety, ignoring inequalities within the household and different family structures. Information gathered from mobile phone use, meanwhile, can help us learn more about the wellbeing of millions of individual women and girls. However, datafication (the transformation of social action into online quantified data, thus allowing for real-time tracking and predictive analysis) simultaneously poses a particular risk to women and girls' privacy (Data 2X 2019a; Van Dijck 2014, p.200). When existing social relationships are already patriarchal, then surveillance (and other) technologies tend to amplify those tensions and inequalities. In addition, the power of data to sort, categorise, and intervene has not been deliberately connected to social justice and feminist agendas (Taylor 2017). Data use has largely remained technical, with research focusing more on promoting corporations and states' ability to use data for profit and surveillance (Taylor 2017). Additionally, societal norms restrict women and girls' ability to voice their opinions over their rights, such as privacy standards, a concern arising from data collection (Data 2X 2019a; World Wide Web Foundation 2020). They may also have poor access to legal services to protect their consumer rights to privacy and may be excluded from participating in the public debate around issues like ethical private sector use of individual data. In a recent study by Pollicy involving 3306 women from five African countries, 95 percent of the Ugandan respondents and 86 percent of the Senegalese respondents reported not knowing of any laws and policies existing to protect them online (Iyer et al. 2020).

Data Feminism // Data and the Myths Surrounding Data Neutrality

Data created, processed, and interpreted under unequal power relations by humans and/or human-made algorithms potentially reproduce the same exclusions, discriminations, and normative expectations present in societies (Shephard 2019). Since data practices have been rooted in patriarchy and colonialism, power and gender relations manifest in data practices, especially regarding how data is generated, analysed, and interpreted (Tamale 2020). In light of this, it is important to identify gaps, bias, and how factors such as racism, sexism, classism, homophobia, and transphobia intersect to discriminate and further marginalise those underrepresented and otherwise othered in data (Shephard 2019). This volume includes important work on data feminism by Catherine D'Ignazio and Lauren Klein that responds to this need by creating a way of thinking about data, both their uses and their limits, that is informed by direct experience, by a commitment to action, and by intersectional feminist thought (D'Ignazio and Klein 2020). Framing data within feminist movements requires thinking about data justice.

This study's approach to data justice is drawn from the work of Linnet Taylor, who frames it as fairness in the way people are made visible, represented, and treated as a result of their production of digital data (Taylor 2017). Data justice is posited as a way of determining ethical paths in a datafied world. It is anchored in three pillars as follows: (in)visibility, (dis)engagement with technology, and anti-discrimination.

Last, Afrofeminism is an important lens through which to examine digital colonisation and unjust data practices, specifically in the African continent. Afrofeminism is a branch of feminism that distinctly seeks to create its own theories and discourses linked to the diversity of African experiences (Tamale 2020). It works to reclaim the rich histories of Black women in challenging all forms of domination (Tamale 2020). African feminists' understanding of feminism places systems embedded in exploitative and oppressive structures such as patriarchy, colonialism, and imperialism at the centre of their analysis (AWDF 2007; Tamale 2020).

African Women and Afrofeminist Data Futures

Globally, women in sub-Saharan Africa are the least likely to be online. Only 28 percent of them are connected and as a result, they have a minimal digital footprint (Iglesias 2020). This leads to their exclusion from the positive aspects of our ongoing digital revolution. African women are marginalised from the technology industry and lack funds and technical expertise to utilise data for feminist causes. Yet, when it comes to conversations about datafication and the digital revolution, they are often instrumentalised as a vulnerable target group rather than a stakeholder group with a crucial perspective on the kind of internet access that guarantees rights rather than restricts them (Feministinternet.org n.d.). Additionally, the shift to digital data has ushered in an era where the emphasis on computational analysis and machine learning as core (and qualitatively superior) ways of understanding the social world moulds the way people relate to information and knowledge (Milan and Velden 2016, p.58). As a result, alternative sources of data that African feminists use in their work, such as personal accounts and Indigenous knowledge systems, are seen as inferior. Fortunately, data feminism offers a framework for African women to imagine and build Afrofeminist data futures. Data feminism insists that the most complete knowledge comes from synthesising multiple perspectives, with priority given to local, Indigenous, and experiential ways of knowing (D'Ignazio and Klein 2020). African women are experts on their own contexts and experiences with data and datafication and need to have the opportunity to envision alternatives to the current algorithmic order (Tamale 2020). Africa sits on the outer edge of the geopolitical margins, and marginalised groups within Africa, such as women, have an even more peripheral worldview (Tamale 2020). Therefore, in building alliances with anguished netizens worldwide, African women should provide useful insights into the affected landscape. An Afrofeminist data future would be one where African women have the right to privacy and full control over personal data and information online at all levels—a form of data justice. African women, just like grassroots data activists, understand the need for engagement with data but resist the massive data collection done by individuals, non-state actors, corporations, and states (Milan and Velden 2016, p.58). They also acknowledge the 'paradox of exposure'— that datafication carries with it its own risks and potential for harm because of the increased visibility that collecting data on these populations might bring them (D'Ignazio and Klein 2020,). Historically, surveillance has been employed as a patriarchal tool used to control and restrict women's bodies, speech, and activism (FeministInternet.org n.d.). African women and LGBTQIA+ communities are especially vulnerable to violations of privacy, including cultural and family practices of surveillance. They must be empowered with effective legal and technical tools and a clear language with which to talk about data rights (Ada Lovelace Institute 2020).

A Feminist Journey

Mapping Feminist Movements in Sub-Saharan Africa

Using a combination of the researchers' networks, the advisory board and online searches, over 120 feminist organisations were identified across sub-Saharan Africa. This list is not comprehensive and does not include all feminist organisations. The researchers aim to make this list publicly available and regularly updated.

Types of Organiser

To reach a broad array of feminist organisers, four main groups of organisers were selected:

1. Individual activists
2. Grassroots movements or collectives
3. Small-to-medium sized CSOs (more than five and fewer than twenty staff)
4. Regional/large-scale organisations (more than twenty staff).

Key Sectors for Feminist Organising

Furthermore, seven key sectors were identified:

1. Sexuality and sexual/reproductive health
2. Socio-economic rights focus particularly employment, subsistence, unpaid care, and work

3. Education, with an emphasis on STEM pipelines
4. Civic and political rights—political participation and representation of women
5. Cultural rights—championing against traditional norms and culture, for example land rights and female genital mutilation
6. Environmental issues

The Role of Social Media Platforms for Organising

Social media has become an integral part of organizing feminist movements across the continent. Every movement we spoke to as part of this research makes use of social media in their work. Social media has become a space for organising, and is particularly crucial in oppressive regimes where the right to assemble is hindered.

> 'In a country like Angola, where meeting to discuss certain social and political issue is often discouraged by the authorities, social networks allow us not only to meet, but also to have access to our target audience'.
>
> *Interview Respondent*

Social media is now a place of learning and exchange. Women who previously had no access or exposure to feminist thinking are now able to learn from one another and question patriarchal norms:

> Currently, it has been advantageous because, for me at least, I have started to have more insight into feminism and how feminist perspectives are, through the internet.
>
> *Interview Respondent*

Most importantly, social media serves as a space for sisterhood, despite geographical, class, patriarchal, or other barriers:

> One of the advantages is also the possibility to know feminist women. When outside the social networks, we can think that we are few or we are the only ones, but with the social networks, the possibility to know others and make interactions and crossing of information is possible.
>
> *Interview Respondent*

How are feminist movements currently utilising data?

Types of data collected

1. Incidence and prevalence rates
2. Provision of services and clinic in-take
3. Social media metrics and reach
4. Knowledge, perceptions, and behaviours

Types of tools used

1. Facebook products (Facebook, WhatsApp, Instagram)
2. Microsoft Suite products (Excel, Word)
3. Google products (Forms, Sheets, Hangouts)
4. Data analysis platforms (SPSS, STATA, NVIVO)
5. Productivity platforms (Slack, Trello, Notion, Airtable)
6. Data collection platforms (SurveyMonkey, ODK, KoboToolbox, SurveyCTO)
7. Other communication platforms (Twitter, Zoom, Telegram, YouTube)

Purpose for data collection

1. Advocacy and awareness raising
2. Program and impact measurement
3. Policy influence
4. Fundraising and needs assessments

Challenges in Data Use

Through our interviews, focus group discussions and personal experiences, ten major challenges to the full utilisation of data by feminist movements in Africa were identified. These challenges represent the immediate impediments in the conceptualisation, collection, analysis, and dissemination of data (See Figure 20.1).

Data Literacy

A major challenge amongst the movements we interviewed in effectively utilising data for advancing feminist causes is a lack of technical skills in developing survey tools and data collection and analysis. While some movements are able to

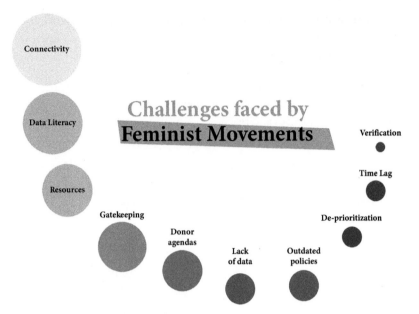

Figure 20.1 Challenges Faced by Feminist Movements

collect data, either using paper-based methods or through mobile devices, they may be unable to move on to the next steps of cleaning, organising, analysis, and visualisation:

> Yes, we did collect our own data but its raw data and we have not managed to process it properly because we don't have any research expertise within the organisation.

Interview Respondent

Additionally, digital tools for data collection and analysis are often only available in English and other Western languages. Furthermore, secondary datasets, when available openly, whether from international institutions or from national bodies, are in dominant Western languages and are not translated into local languages:

> When you come across data from the World Health Organization or UN Women in Tanzania, it is never in Swahili. They all post data in English, but it is never translated to Swahili. This information could be useful to many Tanzanians, but then, they don't speak English.

Interview Respondent

Our research also showed that amongst regions within Africa, interviewees from Anglophone African countries were better prepared in terms of data literacy and knowledge of the challenges and implications that digital data poses to their work and movements compared to their counterparts in Francophone and Lusophone countries.

Connectivity and Access

According to the latest GSMA report, in 2019, the digital gender gap in mobile internet use for sub-Saharan Africa was 37 percent (GSMA, 2020). These figures range from a digital gender divide as low as 13 percent in South Africa to as high as 48 percent in Uganda. Many women remain disconnected from the internet due to high costs of internet bundles, high costs of devices, lack of service in their communities, lack of digital literacy, fear of online violence, and patriarchal norms that prohibit women from owning mobile phones:

> There's difficulty in accessing the internet. Very few people still have access to the internet. It seems not, but internet access is a bit bourgeois. So we are, in a way, privileged to be able to be here on an online platform.

Interview Respondent

The feminist movements that we spoke to understand that in order to reach women, non-digital approaches must be embedded in their work. Without taking into account these differences in mobile ownership and internet access and solely relying on digital engagement, feminist movements in Africa would be exclusionary. Furthermore, any data sourced from digital platforms will be biased and will not account for the true reality of the situation:

> I think the divide is with rural communities, which are untouched, unresearched, who don't have access to telecommunications like WhatsApp, Facebook.

Interview Respondent

Lack of enabling policies

Many policies, especially gender policies, on the continent have failed to keep up with the changing times and a rapidly advancing digital ecosystem. This is especially evident when it comes to cases of online violence, gendered disinformation and hate speech, and how these issues disproportionately affect women.

We have a gender policy that was developed ten years ago. It is expiring this year and there has not been a review of that policy yet to show how much we have achieved, how much we need to do so that we can start up another target for another policy and these are the goals that we are trying to achieve in the next five or ten years but that has not been done.

Interview Respondent

Furthermore, archaic morality laws, and newer computer misuse laws, have disproportionately punished women for incidents in online spaces. Furthermore, there is a need for policy that supports development of evidence to ensure access to locally relevant data. This is particularly an issue for feminists in non-Anglophone African countries, who struggle to source appropriate data and content for their movements:

We go to Google and search for any subject and there are always Brazilian sites. We find very few resources. It is rare that you find anything about Angola. This is not only not on the gender issue, but it is a generalized thing here in Angola. There is no interest. There are no policies that encourage research. This is the issue.

Interview Respondent

Lack of gender-disaggregated data

Data, when available, are not disaggregated by gender. According to a report by Open Data Watch (Data 2X 2019b), the health sector tends to have the highest proportion of gender-disaggregated data, while environmental indicators have the least. Yet, women often bear the brunt of the negative effects of climate change.

In our advocacy work, we regularly encounter opposition to our information because we have no evidence based on data. We want to have access to disaggregated data on the impact of social policies on women and men.

Similar to a lack of enabling policies and practices by governments, private sector companies do not prioritise the use of gender-disaggregated data. This is evidenced by the mobile network companies, which amass large volumes of data across Africa.

We know for instance many telecom companies require people to register to get a SIM card. On the registration form, you have to indicate whether you are a man or a woman. When you ask them for gender-disaggregated data, who has access to telecom infrastructure, they tell you they don't have that data. They do have it.

It's just that either it's not in an excel sheet, or it's not coded in the right way or it's just that they are not interested.

Focus Group Participant

Time lag between large-scale national surveys

The availability of timely data is a challenge that was brought up by multiple inter-viewees. Large-scale representative surveys are conducted every four to five years, often funded by international agencies and donors.

> Data that is relevant to our work is not available. We are relying on the Multiple Indicator Cluster Survey. It is done periodically by UNICEF together with the statistics agency to measure progress towards the SDGs. They are done periodi-cally, I think every 4 years, so you just have to wait for 4 years. Then, we have the Demographic Health Surveys which happen every 5 years. We don't have quar-terly surveys. We wait for those donor-funded surveys. Our government is in no position to have any national periodic surveys, so they wait, as well.

Interview Respondent

Gatekeeping

Access to data from gatekeepers is an ongoing challenge for feminist movements. Gatekeepers can include CSOs, government agencies, or international institutions. Researchers must deal with extensive bureaucracy to access data from government bodies. Due to competing ecosystems created by donors, CSOs often withhold data from one another. Similarly, private sector companies hold on to their data based on their initial investments made to procure it, or sell the data at exorbitant costs, which can only be afforded by other large private sector organisations.

> This data might be available but it is not accessible to us. It is possible that the government collects some of this information. It is not accessible and when you attempt to reach out to specific government organizations to receive data, it is this long journey of bureaucratic processes. So, we have realized that collecting data on our own is much quicker than requesting from the government.

Interview Respondent

It is also worth noting that there is a gatekeeping of access to information and engagement in spaces of knowledge. There are high costs to accessing publications such as academic journals; some Universities across Africa and research insti-tutions cannot afford the high fees associated with these journals, even though oftentimes the data has been sourced from Africa, with the support of African researchers. This is a form of epistemic violence. Similarly, prior to COVID19, many academic spaces were also off-limits for African researchers who could not procure visas to Western countries or afford the expensive airfare. Now, with almost all activities moving to online spaces, these researchers can finally enter these previously inaccessible spaces.

> In the academic world, we have to publish. It's a competition to publish. How can we publish without access to sources of information, including books or if we don't have money to buy it. I have occasionally used illegal sites to download data that I needed. We use Russian sites a lot. They are our Robin Hood. They allow distribution of material that is otherwise kept inaccessible somewhere else. But some of these sites are now being banned or locked as part of the fight against piracy. I think academic data should be made public and accessible to everybody.

Interview Respondent

Resources

Collecting, analysing, and disseminating data is a labour and resource-intensive process, especially for grassroots organisations. It is often difficult to receive funding for the sole purpose of conducting research. For many feminist movements, the resources that would go towards research might be better spent on providing vital services to the women they work with, such as healthcare, shelter or counselling. Furthermore, as previously mentioned, it can be expensive to hire data scientists to support feminist movements with data analysis. There is also a shortage of data scientists across the continent, particularly feminist data scientists.

> When we think of the impact on our feminist organizations, we see that it drains our resources that could be spent on other key issues. Because, these are issues that government should be responsible for, not the feminist organizations, when it comes to the data.

Interview Respondent

De-prioritisation of feminist causes

Socially, culturally, and economically, the collection of data related to women's issues is often not considered a priority. Government support and interest in collecting representative data on feminist causes remains low across Africa.

For example, on the impact of COVID-19 on women, there is no information available. There may even be some information in a government member's office, but it is not digitized. It is not available online.

Interview Respondent

According to one respondent, governments might allocate funds for research in sectors that are associated with masculine pursuits or directly related to economic growth such as agriculture, business or infrastructure. One respondent said that 'Socially, this might not be looked at as a priority, especially when you are framing it as feminism as opposed to women's empowerment'. Additionally, many funders, non-governmental bodies and similar stakeholders tend to focus on short-term impact and projects that provide measurable outcomes. However, these initiatives have an adverse effect on feminist movements which seek to bring about long-term systemic change (Girard 2019).

Verification and Replicability of Available Data Sources

Owing to challenges related to data availability and accessibility, it is challenging for both governments and international bodies to verify data. In cases where governments do produce statistics concerning feminist causes, for example, low levels of gender-based violence or workplace harassment, they neither make public their methodologies nor the datasets from which these statistics are drawn. This reduces the trustworthiness and credibility of the data produced. We highlighted this issue after having conducted literature reviews of existing or open data, which also found a lack of rigour in research methodologies. One interview respondent shared that,

> We found a lot of these studies were not robust. These studies did not have significant outcomes, did not apply methods that were appropriate for the target population or for the specific aims of the study. So, it was difficult to find any empirical evidence that supported the notion that sexual assault was a problem in the workplace and is a problem that needs to be addressed'. Additionally, feminist movements using data from social media sources also had difficulties in verifying the veracity of the information sourced.

Donor Agendas

One of the core issues discussed during the interviews was the NGO-ization of feminist movements. To receive funding, many grassroots movements must formally register and acquire office spaces and assets. This bogs them down with administrative formalities and takes away from the energy that would fuel their movements. They must engage in a cycle of sourcing funding and catering to the needs of donor agendas, which may often not align with the initial mission of the movement. This is often exacerbated by fundamental differences between Western ideals of feminism and that of African feminisms, which often makes it difficult to stay focused on our context-specific problems such as those related to colonialism, classism and harmful traditional or cultural practices.

Interviewees also raised the issue of donors demanding large amounts of evidence. Given the difficulties that feminist movements face in collecting primary data, accessing or verifying secondary data, garnering resources to analyse data and backlash faced against qualitative data, producing this evidence can be difficult. Respondents emphasised that data was often unnecessary because if one woman has to deal with an issue, it becomes an issue for all women:

> If one woman dies from unsafe abortion, for a feminist that is already an issue. We are not waiting to say how many women, has this been verified, is there evidence for your advocacy. Feminism comes from a place of passion. Facts and figures are not very relatable in that kind of context. But, as the movement has gone out to seek resources from donors and as we engage policymakers to say, 'Let's make a change around these areas', we are constantly being asked but what evidence do you have.

Interview Respondent

Furthermore, as previously mentioned, philanthropy bodies fund women's groups in ways that undermine the entire movement by focusing on short-term outcomes, burdening them with administrative work, and promoting divisiveness and competition within the movement through the grant/funding structures, rather than focusing on building coalitions (Girard, 2019). Despite good intentions, research has shown that time-bound, project-based funding has fractured grassroots movements and stalled the progress of feminist groups in developing countries. In the long run, these women's groups become unattractive for future funding because of restrictive budgets and projects that eventually may fail to show long-term progress, thus de-legitimising the entire movement.

> I just feel like we as a feminist movement, we know what to do. But, the problem is we are restricted by the donor funds. In most cases, you write a proposal to a donor and it's on data collection methods for the betterment of the trans community, and the donor tells you that it's not their priority.

A Roadmap for Strengthening the Feminist Data Ecofeminism

A number of recommendations were suggested by these feminist activists and movements. They have been grouped into potential short-term and long-term actions to be taken by key stakeholders in the data ecosystem.

SHORT TERM

Independent and Intersectional Data Centres

Challenges and concerns for feminist movements surrounding the effective use of data include political manipulation, the lack of an ethics-based approach in research methodologies, and issues of trust and consent. Furthermore, feminist causes are often not prioritised by governmental bodies. A decentralised approach to data collection could address these concerns. Governments should consider setting up a number of non-partisan and independent data centres that are accountable to the citizens, whereby data is (i) open, (ii) shared in accessible formats, (iii) verifiable, and (iv) replicable. Additionally, it is important to consider an intersectional approach to these data centres. We must ensure that these data centres question power dynamics arising from patriarchy, classism, sexism, racism, ableism etc. This may be through the inclusive participation of women when developing data centres and taking into account societal power dimensions that may close off marginalised groups from accessing these centres. For example, one respondent shared how there is an extreme dearth of information, not just within women's issues but even more so on people with disabilities.

'When it comes to a woman who has a disability, there is a different specificity for women. This was when we started to look for some material that I could really understand what the reality of women is, here in Angola, for women with disabilities. I found no content. I found nothing. Not even for the institutions that deal with issues related to people with disabilities. This information is not disseminated very much. Almost no research is done. No content is produced.'

Interview Respondent

Effective data collaboratives

Many feminist movements focus on similar thematic areas and provide similar services within the same region or within other regions in a country, or even across the continent. These movements could learn from each other but also contribute their data to larger feminist datasets. Such knowledge-sharing benefits other organisations—especially grassroots movements without the resources to conduct their own research. Feminist movements can self-govern to ensure that the data is collected ethically, is standardised across indicators and is based on feminist principles that prioritise the needs of women.

> We see that a lot of feminist organisations do the same thing. We offer similar services. We use similar tools. We have similar webinars. But, we are not even sharing information with one another. It's a disadvantage to all of us because now we are not able to pool resources together and make sure that we are clogging the gaps. What we are doing instead is duplicating our efforts, which is not valuable for anyone.
>
> *Interview Respondent*

Building trust with feminist movements

Trust is vital between social networks and feminist movements to ensure that data may be used and feminist organisations may work with social networks on improving their publicly available data. The issue of trust is significant, given the practices of data collection and process contextualised in the data colonialism context and women's experiences of dataveillance. Recommendations for building trust between feminist movements and social networks include listening to feminists, hiring feminists, looking at diverse business models that allow for co-ownership of knowledge, and being accessible to assist movements to understand practices on the platform that may stifle the engagement of movements. Trust is vital for engagement with movements so that the engagement is mutually beneficial.

> I think beyond listening and thinking through why they need to support this feminist movement, we should ask why they don't listen to feminists when we tell them what we want.
>
> *Interview Respondent*

Funding for data training initiatives and feminist technologists

Funders, partners and technology companies should focus their efforts and funds on supporting and developing data training initiatives on the continent. By focusing training only on individual organisations, the learnings often end there or are lost when those staff members move on from that organisation. There is a shortage of such initiatives across the continent, and especially those that take into account feminist research methodologies. However, programs such as Code for Africa's WanaData and Data Science Nigeria have made significant progress in training data scientists and journalists. Similarly, more efforts should be focused on improving the pipeline of women in STEM fields. It is rare to come across female developers, and feminist female developers are even rarer. We need to continue to examine the intersection of technology, gender and ethics.

Funding for feminist research

Feminist research seeks to explore ontological and epistemological concerns about traditional research methods by examining underlying assumptions regarding the power dynamics of who is considered the 'knower' versus the 'known'. Epistemological violence happens when social science research subjects are 'othered' and data highlights their problems and inferiority; consider how Western NGOs promote visuals of poverty porn in their fundraising advertisements. This also de-legitimises knowledge that does not fit the Western normative ideals (Tandon, 2018). Similar concerns have been raised around how data is collected, processed and interpreted that may subsequently be used to advocate for change. Feminist movements must be funded to conduct research from a decolonial, feminist lens.

LONG TERM

Build appreciation for different forms of data

Decolonial research values, reclaims and foregrounds Indigenous voices and ways of knowing and utilises Indigenous methods of transferring knowledge like storytelling, participatory, hands-on learning, community-based learning, and collaborative enquiry (Tamale 2020). African feminists and feminist organisations are already participating in this form of research by creating platforms where they share stories and profiles of women and gender diverse persons to make visible their impact and the complexities of their experiences. African feminists urge the deployment of innovative and subversive critical tools in African research and training. South African scholar Pumla Dineo, for instance, recommends the use of visual arts in articulating topics like sexuality (Tamale 2020). Qualitative data both poses a significant challenge to feminist movements and is collected proficiently by them, notably in the form of storytelling and case studies. There is a need to combine different forms of data—qualitative, quantitative, and Big Data—to influence policy change and practices.

Strengthen women's safety online

Technology platforms, government institutions, and civil organisations have been steadily providing more educational resources on digital hygiene and security tools over the past few years, however, use and access to these resources remain limited. Many women do not know where to access information related to digital security. Digital security resources must be adapted to local contexts and languages, as well as mainstreamed into educational curricula. Social media platforms must place more emphasis on protecting women on their platforms. They must engage with Indigenous content moderators who understand the nuance and context of local cultures and linguistics, and improve the effectiveness of reporting mechanisms. Policy advocacy, legal approaches, and law enforcement could strengthen laws against online harassment and are a viable pathway to preventing perpetrators from committing online gender-based violence. However, precautions must be taken to ensure that regulation does not lead to the stifling of freedom of expression. Law enforcement personnel must be trained using a gender-sensitive digital safety curriculum to address complaints of online gender-based violence and to provide timely technical assistance, counselling and support to women. Along with the engagement of safety personnel, countries must adopt data protection and privacy laws and put committees and mechanisms in place to implement these laws. Last, many digital hygiene solutions put the onus of security upon the shoulders of victims. Research shows that few interventions are aimed at preventing primary and secondary perpetrators from acting violently in the first place. It would be worthwhile to teach new (and established) users of the internet how to conduct themselves in digital environments in a way that is grounded in empathy and kindness.

Future of feminist data governance

Feminist data governance must take into account the power imbalances that exist between who provides and who collects data. It should direct attention towards the invisible labour involved in producing data, whether that means an individual's content and knowledge, or the burden placed upon citizens and civil society to collect data for the betterment of society. Even the landscape of social services is rapidly becoming dependent on data systems where technological determinism overrides a nuanced analysis of contexts and power dynamics at play (Fotopoulou 2019). Technologists must move beyond providing illegible Terms and Conditions or tweaking consent mechanisms on digital platforms to think more holistically about how data flows could distribute costs and benefits fairly across society and uphold the values of social domains, such as health, democracy, and balance lifestyles (Berinato 2018).

Conclusion

Data, in its various forms, can play a vital role in feminist movements fighting for a more just society. In imagining Afrofeminist data futures, the context of colonial practices, power imbalances and lack of feminist data indicate a need to address challenges within the data ecosystem. Afrofeminist data futures may only be possible when those who hold the data ensure that data is shared in a transparent and accountable manner and aligned with feminist principles. It also requires supporting solutions to the challenges and concerns highlighted in this research. This research is a first step in understanding the role of data in feminist movements. More research and significant investment are needed to explore best practices to support data use and developing data practices that work to ensure data justice across Africa.

Acknowledgements

We would like to acknowledge the support of Chinonye 'Chi Chi' Egbulem and Zenaida Machado in conducting our interviews and focus group discussions in French and Portuguese, respectively. We would like to thank the many African feminists that we spoke to and who generously gave us their time, knowledge and perspectives to shape this research paper. Finally, we would like to thank Aïda Ndiaye, Mazuba Haanyama, and Brownen Raff for their support and feedback.

References

Abebe, Rediet, Kehinde Aruleba, Abeba Birhane, Sara Kingsley, George Obaido, Sekou L.Remy, and Swathi Sadagopan. (2021) 'Narratives and Counternarratives on Data Sharing in Africa. Conference on Fairness, Accountability, and Transparency (FAccT '21)'. 3–10 March 2021, Virtual Event, Canada, pp. 329–341. Retrieved from https://dl.acm.org/doi/10.1145/3442188.3445897.

Ada Lovelace Institute. (2020) 'Rethinking Data. Changing Narratives, Practices and Regulations'. Retrieved from https://www.adalovelaceinstitute.org/wp-content/uploads/2020/01/Rethinking-Data-Prospectus-Print-Ada-Lovelace-Institute-2019.pdf.

AWDF. (2007) 'Charter of Feminist Principles for African Feminists'. African Women's Development Fund. Retrieved from https://awdf.org/wp-content/uploads/AFFFeminist-Charter-Digital-â-English.pdf 4.

Berinato, Scott. (2018) 'Why Data Privacy Based on Consent is Impossible'. Retrieved from https://hbr.org/2018/09/stop-thinking-about-consent-it-isnt-possible-and-it-isnt-right.

boyd, danah and Kate Crawford. (2012) 'Critical Questions for Big Data'. *Information, Communication & Society* 15(5): 662–679. https://doi.org/10.1080/1369118x.2012.6788786.

Chair, Chenai. (2020) 'My Data Rights: Feminist Reading of the Right to Privacy and Data Protection in the Age of AI'. Retrieved from https://mydatarights.africa/wp-content/uploads/2020/12/mydatarights_policy-paper-2020.pdf.

Coleman, Danielle. (2019) 'Digital Colonialism: The 21st Century Scramble for Africa through the Extraction and Control of User Data and the Limitations of Data Protection Laws'. *Michigan Journal of Race and Law* 24: 1–24. Retrieved from https://repository.law.umich.edu/mjrl/vol24/iss2/6/

D'Ignazio, Catherine and Lauren F. Klein. (2020) *Data Feminism (Strong Ideas)*. Massachusetts, USA: The MIT Press.

D'Ignazio, Catherine and Lauren F. Klein. (2020) 'Seven Intersectional Feminist Principles for Equitable and Actionable COVID-19 Data'. *Big Data & Society* 7(2): 205395172094254. https://doi.org/10.1177/20539517209425448816.

Data 2X. (2019a, November) 'Big Data, Big Impact? Towards Gender-Sensitive Data Systems'. Retrieved from https://data2x.org/wp-content/uploads/2019/11/BigDataBigImpactReportWR.pdf.

Data 2X. (2019b) 'Bridging the Gap: Mapping Gender Data Availability in Africa'. March. Retrieved from https://13data2x.org/resource-center/bridging-the-gap-mappinggender-data-availability-in-africa/.

Degli Esposti, Sara (2014) 'When Big Data Meets Dataveillance: The Hidden Side of Analytics'. *Surveillance & Society* 12(2): 209–225. https://doi.org/10.24908/ss.v12i2.5113.

FeministInternet.org. (n.d.) 'Privacy & Data | Feminist Principles of the Internet'. Retrieved from https://feministinternet.org/en/principle/privacy-data.

Fotopoulou, Aristea (2019) 'Understanding Citizen Data Practices from a Feminist Perspective: Embodiment and the Ethics of Care'. In Citizen Media and Practice: Currents, Connections, Challenges, eds H. Stephansen and E. Trere, p. 1. Retrieved from https://core.ac.uk/download/pdf/188259135.pdf

Girard, Francoise. (2019) 'Philanthropy for the Women's Movement, Not Just "Empowerment"'. 4 November Retrieved from https://ssir.org/articles/entry/philanthropy_for_the_womens_movement_not_just_empowerment.

GSMA (2020) 'The Mobile Gender Gap: Africa'. Retrieved from https://www.gsma.com/mobilefordevelopment/wp-content/uploads/2020/05/GSMA-The-Mobile-Gender-GapReport-2020.pdf on 4 February 2021.

Iglesias, Carlos. (2020) 'The Gender Gap in Internet Access: Using a Women-Centred Method'. 29 October. Retrieved from https://webfoundation.org/2020/03/the-gender-gap-in-internetaccess-using-a-women-centred-method/.

Ilori, Tomiwa. (2020) 'Data Protection in Africa and the COVID-19 Pandemic: Old Problems, New Challenges and Multistakeholder Solutions. Association for Progressive Communications (APC)'. June. Retrieved from https://www.apc.org/en/pubs/data-protectionafrica-and-covid-19-pandemic-old-problems-new-challenges-and-multistakeholder.

Iyer, Neema, Bonnita Nyamwire, and Sandra Nabulega. (2020) 'Alternate Realities, Alternate Internets: African Feminist Research for a Feminist Internet'. August. *Policy*. https://www.apc.org/en/pubs/alternate-realities-alternate-internets-african-feminist-research-feminist-internet.

Kasemiire, Christine. (2021) 'SafeBoda Illegally Shared Users' Data with US Company - NITA-U'. 8 February. Retrieved from https://www.monitor.co.ug/uganda/business/finance/safebodaillegally-shared-users-data-with-us-company-nita-u—3283228.

Kidera, Momoko. (2020) 'Huawei's Deep Roots put Africa Beyond Reach of US Crackdown'. 14 August. Retrieved from https://asia.nikkei.com/Spotlight/Huawei-crackdown/Huaweis-deep-roots-put-Africa-beyond-reach-of-US-crackdown90.

Ladysmith, Cookson, Tara Patricia, Julia Zulver, Lorena Fuentes, and Melissa Langworthy. (2020) 'Building Alliances for Gender Equality: How the Tech Community can Strengthen the Gender Data Ecosystem'. Ladysmith. Retrieved from https://drive.google.com/file/d/1bsy_Gx7DHh8K54sRp74rJoWTLY47TdCU/view.

Milan, Stefania and Lonneke van der Velden. (2016) 'The Alternative Epistemologies of Data Activism'. *Digital Culture & Society* 2(2): 57–74. https://doi.org/10.14361/dcs-2016-0205.

Nyabola, Nanjala. (2018) *Digital Democracy, Analogue Politics*. Amsterdam, Netherlands: Adfo Books.

Parkinson, Joe, Nicholas Bariyo, and Josh Chin. (2019) 'Huawei Technicians Helped African Governments Spy on Political Opponents'. 15 August. Retrieved from https://www.wsj.com/articles/huaweitechnicians-helped-african-governments-spy-onpoliticalopponents11565793017.

Pilling, David. (2019) 'African Countries are Missing the Data Needed to Drive Development'. 23 October. Retrieved from https://www.ft.com/content/4686e022f58b11e9b0183ef8794b17c6.

Robertson, Thomas. (2020) 'Senegal to Review Data Protection Law'. 14 January/ Retrieved from https://cipesa.org/2020/01/senegal-to-review-data-protection-law/ #:~:text=Twelve%20years%20after%20being%20among,2008%20Personal%20Data %20Protection%20Law.

Shephard, N. (2019) 'Towards a Feminist Data Future | Gunda-WernerInstitut. 25 November. Retrieved from https://www.gwi-boell.de/en/2019/11/25/towards-feminist-datafuture.

Tamale, Sylvia. (2020) *Decolonization and Afro-Feminism*, 1st edn. Quebec City, Canada: Daraja Press.

Tandon, Ambika. (2018) 'Feminist Methodology in Technology Research'. The Centre for Internet and Society, India. December. Retrieved from https://cisindia.org/ internetgovernance/feminist-methodoloty-in-technology-research.pdf.

Taylor, Linnet. (2017) 'What is Data Justice? The Case for Connecting Digital Rights and Freedoms Globally'. *Big Data & Society* 4(2): 205395171773633. https://doi.org/ 10.1177/2053951717736335.

Temin, Miriam and Eva Roca. (2016) 'Filling the Gender Data Gap'. *Studies in Family Planning* 47(3): 264–269. https://doi.org/10.1111/sifp.70.

UN Data Revolution. (2013) 'What is the Data Revolution? The High Level Panel on the Post-2015 Development Agenda'. August. Retrieved from https://www. post2020hlp.org/wpcontent/uploads/docs/What-is-the-Data-Revolution.pdf.

UN Data Revolution. (2016) What is the 'Data Revolution'? 9 May. Retrieved 20 January 2021, from https://www.undatarevolution.org/data-revolution/#comment– 46064.

Van Dijck, Jose. (2014) 'Datafication, Dataism and Dataveillance: Big Data between Scientific Paradigm and Ideology'. *Surveillance & Society* 12(2): 197–208. https:// doi.org/10.24908/ss.v12i2.4776.

World Wide Web Foundation. (2020) Women's Rights Online: Closing The Digital Gender Gap For A More Equal World'. Retrieved from http://webfoundation.org/ docs/2020/10/Womens-RightsOnline-Report-1.pdf

21

Design Practices

'Nothing About Us Without Us'

Sasha Costanza-Chock

Design Justice: A Feminist Approach to Technology Design

The goal of design justice, a queer intersectional feminist approach to the design of objects, interfaces, the built environment, and a wide range of systems including so-called 'Artificial Intelligence' (AI), is to spur our imaginations about how to move beyond a system of technology design that remains embedded in and constantly reproduces the matrix of domination. A feminist approach to the design of AI systems is crucial, because we need to imagine whether and how such systems might be reorganised around human capabilities, collective liberation, and ecological sustainability. In this chapter, which originally appeared as chapter 2 of *Design Justice: Community-Led-Practices to Build the Worlds We Need,* I explore a series of key questions about how design practices might be reorganised towards such ends.

Designers: Who Gets (Paid) to Do Design?

To begin with, design justice as a framework recognises the universality of design as a human activity. *Design* means to make a mark, make a plan, or problem-solve; all human beings thus participate in design (Papanek 1974). However, though all humans design, not everyone gets paid to do so, since intersectional inequality systematically structures paid professional design work (see Wajcman and Young, Chapter 4 in this volume). This is not optimal, even through the limited lens of business best practices. Indeed, there is a growing managerial literature on the competitive business advantages of employee diversity. Diverse firms and product teams have repeatedly been shown to make better decisions, come up with more competitive products, and better understand potential customers. Racial and gender diversity are linked to increased sales revenue, more customers, and greater relative profits (Herring 2009), although some research complicates this narrative. However, despite steadily increasing interest in establishing a diverse pool of designers, developers, product managers, and other tech workers, the industry

Sasha Costanza-Chock, *Design Practices.* In: *Feminist AI.* Edited by: Jude Browne, Stephen Cave, Eleanor Drage, and Kerry McInerney, Oxford University Press. © Oxford University Press (2023). DOI: 10.1093/oso/9780192889898.003.0021

Figure 21.1 Cover illustration for 'Nothing About Us Without Us: Developing Innovative Technologies For, By and With Disabled Persons' by David Werner (1998) http://www.dinf.ne.jp/doc/english/global/david/dwe001/dwe00101. html.

persistently fails to meaningfully diversify. Sector-wide employment trends are not steadily advancing towards increasing diversity; instead, women and/or B/I/PoC sometimes gain ground, sometimes lose ground (Harkinson 2014; Swift 2010).

Of course, in the business literature structural inequality is rarely mentioned, let alone challenged. Because design justice as a framework includes a call to dismantle the matrix of domination and challenge intersectional, structural inequality, it requires more than a recognition that employment diversity increases capitalist profitability. Although employee diversity is certainly a laudable goal, it remains comfortably within the discourse of (neo)liberal multiculturalism and entrepreneurial citizenship (Irani 2015). In other words, employment in paid design fields is important, but is not the whole picture. Design justice also involves rethinking other aspects of design practice, including the intended design beneficiaries: the 'users'.

User-Centred Design, the 'Unmarked' User, and the Spiral of Exclusion

For whom do we, as a society, design technology?

User-centred design (UCD) refers to a design process that is 'based upon an explicit understanding of users, tasks, and environments; is driven and refined by user-centred evaluation; and addresses the whole user experience. The process involves users throughout the design and development process and it is iterative'.[1] Over time, UCD has become the recommended design approach within many firms, government bodies, and other institutions. However, UCD faces a paradox: it prioritises 'real-world users'. Yet if, for broader reasons of structural inequality, the universe of real-world users falls within a limited range compared to the full breadth of *potential* users, then UCD reproduces exclusion by centring their needs. Put another way, design always involves centring the desires and needs of some users over others. The choice of *which users* are at the centre of any given UCD process is political, and it produces outcomes (designed interfaces, products, processes) that are better for some people than others (sometimes very much better, sometimes only marginally so). This is not in and of itself a problem. The problem is that, too often, this choice is not made explicit.

In addition, designers tend to unconsciously default to imagined users whose experiences are similar to their own. This means that users are most often assumed to be members of the dominant, and hence 'unmarked' group: in the United States, this means (cis) male, white, heterosexual, 'able-bodied', literate, college educated, not a young child, and not elderly, with broadband internet access, with a smartphone, and so on. Most technology product design ends up focused on this relatively small, but potentially highly profitable, subset of humanity. Unfortunately, this produces a spiral of exclusion as design industries centre the most socially and economically powerful users, while other users are systematically excluded on multiple levels: their user stories, preferred platforms, aesthetics, language, and so on are not taken into consideration. This in turn makes them less likely to use the designed product or service. Because they are not among the users, or are only marginally present, their needs, desires, and potential contributions will continue to be ignored, sidelined, or deprioritised.

It is tempting to hope that employment diversity initiatives in the tech sector, if successful over time, will solve this problem. Diversifying the technology workforce, as noted above, is a good move, but unfortunately, it will not automatically produce a more diverse default imagined user. Research shows that unless the gender identity, sexual orientation, race/ethnicity, age, nationality, language, immigration status, and other aspects of user identity are explicitly specified, even diverse design teams tend to default to imagined users who belong to the dominant social group (Hamraie 2013). There is growing awareness of this problem, and several initiatives attempt to address it through intentional focus on designing together with communities that are usually invisibilised. For example, the Trans∗H4CK series of hackathons focuses on trans∗ and gender-non-conforming communities. Contratados.org[2] is a site built by

---

[1] See https://www.usability.gov/what-and-why/user-centered-design.html.
[2] See http://contratados.org/.

the Center for Migrant Rights that operates like Yelp, but for migrant work-ers, to let them review potential employers and recruitment agents, educate them about their rights, and protect them from transnational recruitment scams (Melendez 2014). Such efforts to design together with users from communi-ties that are mostly overlooked by design industries are important. However, they remain small-scale. What is more, individual inclusive design projects can-not, on their own, transform the deeply entrenched systemic factors that mili-tate toward design that constantly centres an extremely limited set of imagined users.

'Stand-in Strategies' to Represent Communities That Are Not Really Included in the Design Process

Well-meaning designers and technologists often agree that including 'diverse' end users in the design process is the ideal. However, many feel that this is usually, sometimes, or mostly impossible to realise in practice. To mitigate the poten-tial problems that come from having no one with lived experience of the design problem actually participate in the design team, researchers and designers have suggested several strategies. Unfortunately, most of these strategies involve creat-ing abstractions about communities that are not really at the table in the design process. Such strategies include design ethnography, focus groups, and a great deal of what passes for participatory design. Here I explore the most widely used 'stand-in' strategy: user personas.

User personas are short, fictional characterisations of product users, often with a name, an image, and a brief description. They are widely used to guide a range of design processes, including UX and UI, graphic design, product devel-opment, architecture, service design, and more (Nielsen 2012). User personas are so widespread that there is even a small sector of firms in the business of provid-ing tools for design teams to generate, manage, and share them. For example, the Userforge website (Figure 21.2) allows rapid random generation of user personas and promises to help design teams 'build empathy and develop focus quickly. Cre-ate realistic representations of your user groups in far less clicks than it would take using design software or word processors, which means you can start prioritizing design decisions and get to the wins sooner'.[3] User personas can be useful tools for communicating project goals, both within teams and firms and to other actors, including funders, investors, the press, and potential users. There is some evidence that user personas help designers stay focused on the intended use case (Guo et al. 2011). In addition, some case-control studies have sought to demonstrate the util-ity of user personas for better design outcomes (Long 2009). If they are developed

[3] See Userforge.com.

Build empathy and develop focus quickly.

Create realistic representations of your user groups in far less clicks than it would take
using design software or word processors, which means you can start prioritizing
design decisions and get to the wins sooner.

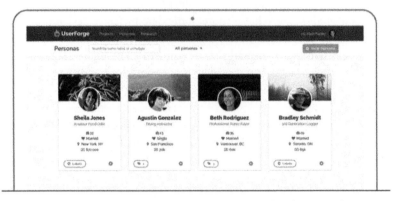

Figure 21.2   User forge user persona generator. Screenshot from Userforge.com.

in ways that are truly grounded in the lived experience of the community of end
users, through careful research or by community members themselves they may
be especially worthwhile. However, there is no systematic study that I was able
to locate that examines whether the use of diverse user personas produces less
discriminatory design outcomes.

Too often, design teams only include 'diverse' user personas at the beginning
of their process, to inform ideation. Occasionally, diverse user stories or personas
are incorporated into other stages of the design process, including user acceptance
testing. However, even if the design team imagines diverse users, creates user personas based on real-world people, and incorporates them throughout the design
process, the team's mental model of the system they are building will inevitably
be quite different from the user's model. Don Norman, one of the most important figures in UCD, notes that in UCD 'the designer expects the user's model to
be identical to the design model. But the designer does not talk directly with the
user—all communication takes place through the system image' (Norman 2006,
p.16).

To make matters worse, far too often user personas are created out of thin air by
members of the design team (if not autogenerated by a service such as Userforge),
based on their own assumptions or stereotypes about groups of people who might
occupy a very different location in the matrix of domination. When this happens,
user personas are literally objectified assumptions about end users. In the worst
case, these objectified assumptions then guide product development to fit stereo-
typed but unvalidated user needs. Sometimes, they may also help designers believe

they are engaged in an inclusive design process, when in reality the personas are representations of designers' unvalidated beliefs about marginalised or oppressed communities. Unsurprisingly, there are no studies that compare this approach to actually including diverse users on the design team.

Disability Simulation Is Discredited; Lived Experience Is Nontransferable

Ultimately, pretending to be another kind of person is not a good solution for design teams that want to minimise discriminatory design outcomes. As Os Keyes argues in Chapter 17 of this volume, the possibility of Disabled technologists bringing their own experience to the table is often foreclosed by the industry's inability to recognise their status as knowers and creators rather than mere users of technology. For example, the supposedly beneficial design practice of 'disability simulation' has been discredited by a recent meta-analysis (Flower et al. 2007). In disability simulation

> a nondisabled person is asked to navigate an environment in a wheelchair in order, supposedly, to gain a better understanding of the experiences of disabled persons. These 'simulations' produce an unrealistic understanding of the life experience of disability for a number of reasons: the nondisabled person does not have the alternate skill sets developed by [Disabled people], and thus overestimates the loss of function which disability presents, and is furthermore likely to think of able-normative solutions rather than solutions more attuned to a [Disabled person's] life experience
>
> (Wittkower 2016, p.7).

For example, abled designers typically focus on an ableist approach to technologically modifying or augmenting the individual bodies of Disabled people to approximate normative mobility style, compared to Disabled people, who may be more interested in architectural and infrastructural changes that fit their own mobility needs. As Wittkower says, ultimately, attempting to imagine other people's experience is 'no substitute for robust engagement with marginalised users and user communities. ... [systematic variation techniques], although worth pursuing, are strongly limited by the difficulty of anticipating and understanding the lived experiences of other' (Wittkower 2016, p.7). A design justice approach goes further still: beyond 'robust engagement', design teams should be led by and/or in other ways be formally accountable to marginalised users.

If You're Not at the Table, You're on the Menu

Design justice does not focus on developing systems to abstract the knowledge, wisdom, and lived experience of community members who are supposed to be the end users of a product. Instead, design justice practitioners focus on trying to ensure that community members are actually included in meaningful ways throughout the design process. Another way to put this is 'If you're not at the table, you're on the menu'.[4] Design justice practitioners flip the 'problem' of how to ensure community participation in a design process on its head to ask instead how design can best be used as a tool to amplify, support, and extend existing community-based processes. This means a willingness to bring design skills to community-defined projects, rather than seeking community participation or buy-in to externally defined projects. Ideally, design justice practitioners do not focus on how to provide incentives that we can dangle to entice community members to participate in a design process that we have already determined and that we control. Instead, design justice compels us to begin by listening to community organisers, learning what they are working on, and asking what the most useful focus of design efforts would be. In this way, design processes can be community-led, rather than designer-or funder-led. Another way to put this might be: do not start by building a new table; start by coming to the table. What is more, in addition to equity (we need more diverse designers, and more diverse imagined users), design justice also emphasises accountability (those most affected by the outcomes should lead design processes) and ownership (communities should materially own design processes and their outputs).

Participatory Design

The proposal to include end users in the design process has a long history. The 'participatory turn' in technology design, or at least the idea that design teams cannot operate in isolation from end users, has become increasingly popular over time in many subfields of design theory and practice. These include participatory design (PD), user-led innovation, UCD, human-centred design (HCD), inclusive design, and codesign, among a growing list of terms and acronyms (Von Hippel 2005; Schuler and Namioka 1993; and Bardzell 2010). Some of these approaches have been adopted by multinational technology companies. For example, in 2017, in a

---

[4] Chris Schweidler from Research Action Design, cofounder of the Research Justice track at AMC, remixed this saying and turned it into a hilarious operating table meme that illustrates it best.

story for *Fast Company* about Airbnb's new inclusive design toolkit,[5] technology journalist Meg Miller writes:

> Microsoft has an inclusive design kit and a general design strategy centred around the philosophy that designing for the most vulnerable among us will result in better products and experiences for all. Google focuses on accessibility practices for their developers for the same reasons. Industry leaders like John Maeda and Kat Holmes have built their careers on speaking on the importance of diversity in the field, and how human centred design should encompass potential users of all different races, genders, and abilities.
>
> (Miller 2017)

Only some of these approaches and practitioners, however, ask key questions about how to do design work in ways that truly respond to, are led by, and ultimately benefit the communities most targeted by intersectional structural inequality.

The question of community accountability and control in supposedly inclusive design processes has recently come to the fore in public conversations about civic tech. Daniel X. O'Neil, one of the key early actors in the field, has written a blistering critique of civic tech's lack of community accountability or connection to existing social movements (O'Neil 2016). Artist, educator, and community technologist Laurenellen McCann calls for technologists to 'build with, not for' (McCann 2015).[6] Both find fault with civic tech's frequent solutionism, disconnection from real-world community needs, and tech-centric ideas about how to address difficult social problems, as well as for ongoing reproduction of white cis-male 'tech bro' culture that alienates women, trans∗ folks, B/I/PoC, Disabled people, and other marginalised communities.[7] This debate is the latest incarnation of a long-standing conversation about the relationship between communities and technology development that has animated shifts in theory, practice, and pedagogy across fields including design, software development, science and technology studies, international development, and many others over the years.

For example, as early as the 1960s, in parallel with the rise of the Non-Aligned Movement (formerly colonised countries across the Global South that hoped to chart a path away from dependency on either the United States or the USSR; Prashad 2013), the *appropriate technology movement* argued that technology should be cheap, simple to maintain and repair, small-scale, compatible with

---

[5] See https://airbnb.design/anotherlens.
[6] McCann (2015) and see http://www.buildwith.org.
[7] See the web magazine *Model View Culture* at modelviewculture.org for excellent summaries of these critiques.

human creativity, and environmentally sustainable (Pursell 1993). Writings by economist E. F. Schumacher (1999) and popular manuals such as Stewart Brand's *Whole Earth Catalog* (Turner 2010) focused attention on small, local economies powered by appropriate technology. Countercultural movements throughout the 1960s spawned thousands of organisations dedicated to locally governed, environmentally sustainable technologies that could be adapted to the contexts within which they were embedded, in opposition to one-size-fits-all megaprojects championed by both Cold War powers as keys to 'international development' (Willoughby 1990).

In Scandinavia, the field of PD was created by trade unionists working with software developers such as Kristen Nygaard. They hoped to redesign industrial processes, software interfaces, and workplace decision-making structures (Gregory 2003). In PD, end users are included throughout. Philosopher of science, technology, and media, Peter Asaro, describes PD as 'an approach to engineering technological systems that seeks to improve them by including future users in the design process. It is motivated primarily by an interest in empowering users, but also by a concern to build systems better suited to user needs' (Asaro 2000, p.345). Like many scholars, Asaro traces the roots of PD to the Norwegian Industrial Democracy Project (NIDP). In the 1960s, Scandinavian designers and researchers were concerned with the ways that the introduction of new technology in a workplace is often used to eliminate jobs, deskill workers, and otherwise benefit the interests of owners and managers over the interests of workers. The collective resources programme of NIDP centred on bringing choices about technology into the collective bargaining process. According to Asaro, British researchers at the Tavistock Institute focused on a parallel strand of research about individual worker empowerment through technology design, known as *sociotechnical systems design*. Asaro also points to the UTOPIA project as the canonical first successful instance of PD. UTOPIA was a collaboration among the Nordic Graphic Workers Union, researchers, and technologists, who worked with newspaper typographers to develop a new layout application. UTOPIA was developed after earlier PD experiments had failed, in part because of the creative limitations of existing technologies. For decades, software developers employing PD have met at the biannual Participatory Design Conference (Bannon, Bardzell, and Bødker 2019). PD has been widely influential and has spread to fields such as architecture and urban planning (Sanoff 2008, computer software (Muller 2003), public services, communications infrastructure, and geographic information systems (Dunn 2007), among others. The Nordic approach to PD is also characterised by an emphasis on the normative value of democratic decision making in the larger technological transformation of work, not only the microlevel pragmatic benefits of improved user interface design. However, in the US context, this broader concern is often lost in translation. Here, PD has sometimes (at worst) been reduced to an extractive process to gather new product ideas (Byrne and Alexander 2006).

From the 1980s through the early 2000s, a parallel set of concepts was developed by scholars such as Eric Von Hippel, whose studies of lead user innovation demonstrated that the vast majority of innovation in any given technological field is performed not by governments or formal research and development branches of corporations, but by technology end users themselves (2005). This insight led to changes in product design approaches across a wide range of fields. Technology appropriation researchers such as Ron Eglash (2004) and Bar, Weber, and Pisani (2016) have shown that user practices of hacking, modifying, remixing, and otherwise making technologies work for their own ends are enacted quite commonly across diverse contexts. Whereas lead user innovation focuses on the hacks that people implement to make technologies serve their needs, and technology appropriation theory centres activities outside of formal product or service design processes, HCD emphasises better understanding of everyday user needs and experiences in professional technology design and development (Steen 2011). By the 1990s, design consultancies such as IDEO emerged to champion (and capitalise on) this approach by selling HCD and design thinking as a service to multinational firms, governments, educators, and NGOs.[8] An extensive community of practitioners and scholars also clusters around the term *codesign*, often used as an umbrella that includes various approaches to PD and HCD. This approach is reflected in the journal *CoDesign*, in annual codesign conferences, and in the appearance of the concept across multiple fields (Sanders and Stappers 2008).

In the tech sector, *lean product development*, an approach that emphasises early and frequent tests of product assumptions with real-world users, has largely replaced top-down 'waterfall' design approaches as established best practice (Ries 2011). This shift has been increasingly influential in civic tech and government tech circles as well. Lean and HCD approaches to civic tech led to innovations such as 18F, a unit within the federal government's General Services Administration that is focused on bringing software development best practices to government, as well as the Chicago User Testing group (CUTgroup), based on the experience of the Smart Chicago Collaborative and meant to promote the inclusion of end users in product design (O'Neil 2013). These approaches certainly increase end user input into key design decisions, but most of them have little to say about community accountability, ownership, profit sharing, or credit for innovation.

Power Dynamics and the Ladder of Participation

Power shapes participation in all design processes, including in PD, and the politics of participation are always intersectionally classed, gendered, and raced. Asaro outlines several challenges in PD projects: for one, it is not enough to have end

---

[8] See IDEO's design toolkit at https://www.ideo.com/post/design-kit.

users simply join design meetings. In a workplace context (or in any context), some users will feel they have more power than others. For example, workers participating in a PD meeting with managers at the table may not feel comfortable saying what they mean or sharing their full experience. The same may be the case in any PD process in which socially dominant group members are in the same room as marginalised folks, but without skilled facilitation. In addition, engineers and professional designers may control the PD process relatively easily, based on their 'expert' knowledge. What is more, according to Asaro, gender inequality shapes participation in design processes: 'In many work contexts, the positions traditionally occupied by women are often viewed as being of lower value by management and unions. This undervaluing of women's work easily overflows into inequalities of participation in design activities, especially when combined with social prejudices that view technological design as a masculine pursuit. Unless gender issues in the design process are recognised and dealt with, there exists a strong possibility of gender inequalities being built into the technology itself' (Asaro 2014, p.346). In the worst case, PD processes may actually normalise cultural violence through seemingly participatory processes. As design scholar and practitioner Ramesh Srinivasan says, 'Foucault points out that cultural violence is perpetuated through seemingly inclusive systems, what one today might describe as liberal or neoliberal. These systems appear democratic, yet in practice they subordinate beliefs and practices not in line with those who manufacture discourse and manipulate media and technology systems to maintain their power and privilege' (Srinivasan 2017, p.117).

Participatory Design, Community Knowledge Extraction, and Non-Extractive Design

Many design approaches that are supposedly more inclusive, participatory, and democratic actually serve an extractive function. Sometimes this is intentional, as in design workshops run by multinational corporations with potential end users, in which the goal is explicitly to generate ideas that will then be turned into products and sold back to consumers.[9] More frequently, the intentions of the designers are good. Well-meaning designers employ PD techniques for a wide range of reasons. For one thing, the process of working with community members is enjoyable. It feels good to elicit design ideas and possibilities from 'nondesigners', it can be quite fun and engaging for everyone involved, and it can feel empowering for both design professionals and community members. Unfortunately, this does not change the fact that in most design processes, the bulk of the benefits end up going to the professional designers and their institutions. Products, patents, processes, credit, visibility, fame: the lion's share goes to the professional design firms and

---

[9] For example, see the work of Jan Chipchase at Nokia: http://janchipchase.com/ content/ essays/nokia-open-studios.

designers. Community members who participate in design processes too often end up providing the raw materials that are processed for value further up the chain. Design justice practitioners are working to rethink extractive design processes and to replace them with approaches that produce community ownership, profit, credit, and visibility.

Legal scholar Barbara L. Bezdek, theorising what she terms *development justice*, notes:

> Sherry Arnstein, writing in 1969 about citizen involvement in planning processes in the United States, at the height of American racial and economic tensions, described a typology of citizen participation arranged as a ladder with increasing degrees of decision-making clout ranging from low to high. The Arnstein rungs ascend from forms of 'window-dressing participation', through cursory information exchange, to the highest levels of partnership in or control of decision-making.
>
> (Bezdek 2013, p.3)

Bezdek revisits the Arnstein rungs and rethinks the rules that govern public participation in urban economic redevelopment projects. She proposes a revised set of principles for civic engagement, and a series of actions toward development justice. Arnstein's ladder might also be useful to further articulate community participation in any design process.

Consider Figure 21.3, in which the *x*-axis represents the design phase (in this case, based on the widely used five-phase model from the Stanford d.school), and the *y*-axis represents the degree of participation by people from the communities most affected by the design project (following Arnstein's ladder). Each horizontal

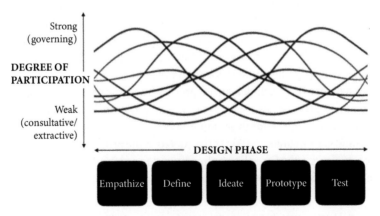

Figure 21.3 Analysis of community participation throughout the design process.

Source: Author.

wavy line represents a (hypothetical) visual shorthand for how community participation unfolds across the life cycle of an individual design project. Put aside for
the moment the fact that design does not really proceed along a linear path from
phase to phase and that there are many, many different design process models.[10]
In reality, phases have porous boundaries and are revisited multiple times during
the project life cycle. The point is to encourage a more complex understanding of
participation and to emphasise that very few design processes are characterised
by community control throughout. A version of this diagram may be a useful
heuristic for thinking through questions of community participation, accountability, and control. A simple image that represents the participation waveform
of a design project might be used in design criticism to analyse case studies, or it
might be used by design justice practitioners to think through concrete community
accountability and control mechanisms in projects that we work on.

Design Justice as Community Organising

Design justice practitioners must also engage with fundamental questions about
the definition of *community*. It is possible to criticise simplistic conceptions of
community and representation without throwing up our hands and accepting
the Thatcherite position that 'there is no such thing as society' (Thatcher 1987).
The question of what a community is and how we can know what it wants is the
domain of democratic theory and political philosophy. It is also a key question
for fields including urban planning, participatory action research (PAR), development studies, and PD, among others (Fals-Borda 1987; White 1996). Design
justice practitioners choose to work in solidarity with and amplify the power of
community-based organisations. This is unlike many other approaches to PD, in
which designers partner with a community but tend to retain power in the process:
power to convene and structure the work, to make choices about who participates,
and, usually, to make key decisions at each point. Analysis of political power in
the design process—who sits at the table, who holds power over the project, what
decision-making process is used—will be fundamental to the successful future
articulation of design justice in theory and practice.

Ultimately, at its best, a design justice process is a form of community organising.
Design justice practitioners, such as community organisers, approach the question of who gets to speak for the community from a community asset perspective
(Mathie and Cunningham 2003). This is rooted in the principle that wherever people face challenges, they are always already working to deal with those challenges;
wherever a community is oppressed, they are always already developing strategies

---

[10] For a humorously framed sampling of design process diagrams, see https://
designfuckingthinking.tumblr.com.

to resist oppression. This principle underpins what Black feminist author adrienne maree brown calls *emergent strategy* (brown 2017). Emergent strategy grounds design justice practitioners' commitment to work with community-based organisations that are led by, and have strong accountability mechanisms to, people from marginalised communities.

Disability Justice and Queer Crip Design

So far, this chapter has explored PD as one pathway toward community accountability and control. It turns now to additional lessons from the disability justice movement. Disability rights and disability justice activists popularised the phrase 'nothing about us without us' in the 1980s and 1990s (Charlton 1998). These linked movements have had an extensive impact on the design of everything from the built environment to human-computer interfaces, from international architectural standards to the technical requirements of broadcast media and the internet, and much more. For example, Gerard Goggin and Christopher Newell explore the ways that disability is constructed in new media spaces, as well as how Disabled people have organised to shape those spaces over time (Goggin and Newell 2003). Elizabeth Ellcessor's recent scholarship considers the importance of these movements to the development of media technologies, from closed captioning to the Web Content Accessibility Guidelines, and from the implications of copyright for accessible content transformation to the possibility of collaborative futures designed through coalitional politics (Ellcessor 2016).

Over time, disability rights and justice scholars and activists pushed for a shift from the medical model of disability, which locates disability within individual 'dysfunctional' bodies, toward the social-relational model: that is, an analysis of how disability is constructed by culture, institutions, and the built environment, which are all organised in ways that privilege some bodies and minds over others. For example, the medical model might seek 'solutions' for wheelchair users that would help them stop using wheelchairs, whereas the social-relational model might seek to ensure that buildings, streets, and bathrooms are all constructed to allow mobility for both wheelchair users and non-wheelchair-users (Kafer 2013). Disability justice work, developed by queer and trans∗ people of colour (QTPOC), has also developed an analysis of the interlocking nature of able-bodied supremacy, racial capitalism, settler colonialism, and other systems of oppression. According to Patty Berne, cofounder and executive director of QTPOC performance collective Sins Invalid, a disability justice analysis recognises that

> the very understanding of disability experience itself is being shaped by race, gender, class, gender expression, historical moment, relationship to colonization, and more. ... We don't believe human worth is dependent on what and how much a

person can produce. We critique a concept of 'labor' as defined by able-bodied supremacy, white supremacy, and gender normativity. ... We value our people as they are, for who they are.

(From '10 Principles of Disability Justice', by Patty Berne on behalf of Sins Invalid, quoted in Piepzna-Samarasinha 2019, pp.26–28).

Scholars, activists, and cultural workers like Patty Berne, the Sins Invalid collective, Alison Kafer, Leah Lakshmi Piepzna-Samarasinha, Aimi Hamraie, and many others have extensively documented this history and have developed tools for intersectional feminist, queer, and Crip analysis and practice (Kafer 2013; Piepzna-Samarasinha 2018; Hamraie 2013).[11]

Another lesson from disability activism is that involving members of the community that is most directly affected by a design process is crucial, both because justice demands it and also because the tacit and experiential knowledge of community members is sure to produce ideas, approaches, and innovations that a nonmember of the community would be extremely unlikely to come up with.

A third key lesson is that it is entirely possible to create formal community accountability and control mechanisms in design processes, and that these can in part be institutionalised. Institutionalisation of disability activists' victories proceeded through a combination of grassroots action, lawsuits,[12] policymaking (the Americans with Disabilities Act), and lobbying standards-setting bodies to create and enforce accessibility standards. For these activists, it was important to pressure multiple actors, including lawmakers, government agencies, universities, and private sector firms, to change research and design practices, adopt new approaches, and implement new standards of care (Shepard and Hayduk 2002). Although these victories are only partial and there is an enormous amount of work to do to deepen the gains that have been secured, disability justice must be a key component of design justice theory and practice.

#MoreThanCode: Findings from the Technology for Social Justice Project

The final section of this chapter explores key findings about community-led technology design practices from #MoreThanCode. #MoreThanCode is a PAR report by the Tech for Social Justice Project (T4SJ), meant to amplify the voices of diverse technology practitioners in the United States who speak about their career paths, visions of how technology can be used to support social justice, and experiences of key barriers and supports along the way. After talking with designers, developers, researchers, community organisers, funders, and other practitioners around

---

[11] see https://www.sinsinvalid.org
[12] For example, see https://www.d.umn.edu/~lcarlson/atteam/lawsuits.html.

the country, the T4SJ Project synthesised hundreds of concrete suggestions for community accountability into the following recommendations:

Adopt codesign methods. This means spending time with a community partner, in their space, learning about needs, and working together through all stages of design. Usually, no new tech development is necessary to address the most pressing issues. Codesign methods have a growing practitioner base, but they could be better documented.

Develop specific, concrete mechanisms for community accountability. Nearly all interviewees said that the people most affected by an issue have to be involved throughout all stages of any tech project meant to address that issue. All actors in this field need to move past stating this as a goal and toward implementing specific, concrete accountability mechanisms. For example: funders should require concrete community accountability mechanisms from their grantees, and educators should centre community accountability in education programmes.

Centre community needs over tools. Community needs and priorities must drive technology design and development, and technology is most useful when priorities are set by those who are not technologists. Be humble and respect community knowledge. Process and solution should be driven by the community; do not make community members token participants.

Invest in education (both formal and informal) that teaches codesign methods to more practitioners. Support existing efforts in this space, create new ones, and push existing educational programmes and institutions to adopt codesign perspectives and practices.

Create tech clinics, modelled on legal clinics. Public interest law and legal services work are client-oriented, and lawyers doing this work are constantly interacting with people who need to navigate larger unequal systems. This is considered part of their legal education. Tech can learn from this model, where the services provided by white collar workers are directed by the communities that they serve (see also Browne, Chapter 19 in this volume).

Avoid 'parachuting' technologists into communities. In general, parachuting is a failed model. Do not do it. Stop parachuting technologists into organisations or focusing on isolated 'social good' technology projects, devoid of context, when the real need is capacity building. We are not saying 'never bring someone in from outside a community'. We do think that it is worthwhile to develop better models for sharing local knowledge with national groups and for national groups to share their perspectives with local groups in such a way that all parties can benefit.

Stop reinventing the wheel! Well-meaning technologists often reinvent the wheel, without researching existing solutions. Designers, developers, and project leads, no matter what sector they are in, should begin projects by researching existing projects and organisations. This also stems from competitive, rather than collaborative, mindsets ('ours will be better, so we'll just compete'). It is important to work together to develop shared tools and platforms, instead of only competing for scarce technology resources.

Support maintenance, not just 'innovation'. Significant resources are necessary to maintain and improve existing movement tech, but most focus is on the creation of new projects. We need more resources to update, improve, and maintain already proven tools (Costanza-Chock et al. 2018).

References

Asaro, P. (2014) Participatory Design. In *The Encyclopedia of Science, Technology and Ethics*, 2nd edn, ed. Carl Mitcham, pp. 345–347. New York: Macmillon. https://peterasaro.org/writing/Asaro%20Participatory_Design.pdf.

Asaro, Peter M. (2000) 'Transforming Society by Transforming Technology: The Science and Politics of Participatory Design'. *Accounting, Management and Information Technologies* 10(4): 257–290.

Bannon, Liam, Jeffrey Bardzell, and Susanne Bødker. (2019) 'Reimagining Participatory Design'. *Interactions* 26(1): 26–32.

Bar, François, Matthew S. Weber, and Francis Pisani. (2016) 'Mobile Technology Appropriation in a Distant Mirror: Baroquization, Creolization, and Cannibalism'. *New Media & Society* 18(4): 617–636. https://doi.org/10.1177/1461444816629474.

Bardzell, Shaowen. (2010) Feminist HCI: Taking Stock and Outlining an Agenda for Design. In *Proceedings of the SIGCHI Conference on Human Factors in Computing Systems*, pp. 1301–1310. New York: ACM.

Bezdek, Barbara L. (2013) 'Citizen Engagement in the Shrinking City: Toward Development Justice in an Era of Growing Inequality'. *St. Louis University Public Law Review* 33: 2014–2020.

brown, adrienne maree. (2017) *Emergent Strategy: Shaping Change, Changing Worlds*. Chico, CA, USA: AK Press.

Byrne, E., and P. M. Alexander. (2006) Questions of Ethics: Participatory Information Systems Research in Community Settings. In *Proceedings of the 2006 Annual Research Conference of the South African Institute of Computer Scientists and Information Technologists on IT Research in Developing Countries*, eds Judith Bishop and Derrick Kourie, pp. 117–126. Republic of South Africa: South African Institute for Computer Scientists and Information Technologists.

Charlton, James I. (1998) *Nothing about Us without Us: Disability Oppression and Empowerment*. Berkeley and Los Angeles, CA, USA: University of California Press.

Costanza-Chock, Sasha, Maya Wagoner, Berhan Taye, Caroline Rivas, Chris Schweidler, Georgia Bullen, and the T4SJ Project. (2018) *#MoreThanCode: Practitioners Reimagine the Landscape of Technology for Justice and Equity*. Research Action Design & Open Technology Institute. morethancode.cc.

Dunn, Christine E. (2007) 'Participatory GIS: A People's GIS?' *Progress in Human Geography* 31(5): 616–637.

Eglash, Ron. (2004) Appropriating Technology: An Introduction. In *Appropriating Technology: Vernacular Science and Social Power*, eds Ron Eglash, Jennifer L. Croissant, Giovanna Di Chiro, and Rayvon Fouché, pp. vii–xxi. Minneapolis: University of Minnesota Press.

Ellcessor, Elizabeth. (2016) *Restricted Access: Media, Disability, and the Politics of Participation*. New York: NYU Press.

Fals-Borda, Orlando. (1987) 'The Application of Participatory Action Research in Latin America'. *International Sociology* 2(4)(December): 329–347.

Flower, Ashley, Matthew K. Burns, and Nicole A. Bottsford-Miller. (2007) 'Meta-Analysis of Disability Simulation Research'. *Remedial and Special Education* 28(2): 72–79.

Goggin, Gerard, and Christopher Newell. (2003) *Digital Disability: The Social Construction of Disability in New Media*. Lanham, MD: Roman & Littlefield.

Gregory, Judith. (2003) 'Scandinavian Approaches to Participatory Design'. *International Journal of Engineering Education* 19(1): 62–74.

Guo, Frank Y., Sanjay Shamdasani, and Bruce Randall. (2011) Creating Effective Personas for Product Design: Insights from a Case Study. In *Proceedings of the 4th International Conference on Internationalization, Design and Global Development*, pp. 37–46.

Hamraie, Aimi. (2013) 'Designing Collective Access: A Feminist Disability Theory of Universal Design'. *Disability Studies Quarterly* 33: 1041–5718.

Harkinson, Josh. (2014) 'Silicon Valley Firms Are Even Whiter and More Male Than You Thought'. *Mother Jones* 29 May, https://www.motherjones.com/media/2014/05/google-diversity-labor-gender-race-gap-workers-silicon-valley.

Herring, Cedric. (2009) 'Does Diversity Pay? Race, Gender, and the Business Case for Diversity'. *American Sociological Review* 74(2): 208–224.

Irani, Lilly. (2015) 'Hackathons and the Making of Entrepreneurial Citizenship'. *Science, Technology, & Human Values* 40(5): 799–824.

Kafer, Alison. (2013) *Feminist, Queer, Crip*. Bloomington, IN, USA: Indiana University Press.

Long, Frank. (2009) Real or Imaginary: The Effectiveness of Using Personas in Product Design. In *Proceedings of the Irish Ergonomics Society Annual Conference*. May, pp. 1–10. Dublin: Irish Ergonomics Society. https://s3.amazonaws.com/media.loft.io/attachments/Long%20%20Real%20or%20Imaginary.pdf.

Mathie, Alison, and Gord Cunningham. (2003) 'From Clients to Citizens: Asset-Based Community Development as a Strategy for Community-Driven Development'. *Development in Practice* 13(5): 474–486.

McCann, Laurenellen. (2015) *Experimental Modes of Civic Engagement in Civic Tech: Meeting People Where They Are*. Chicago: Smart Chicago Collaborative.

Melendez, Steven. (2014) 'Contratados: A Yelp to Help Migrant Workers Fight Fraud'. *Fast Company*, 9 October. https://www.fastcompany.com/3036812/contratados-is-a-yelp-that-fights-fraud-for-migrant-workers.

Miller, Meg. (2017) 'AirBnB Debuts a Toolkit for Inclusive Design'. *Fast Company*, 1 August. https://www.fastcodesign.com/90135013/airbnb-debuts-a-toolkit-for-inclusive-design.

Muller, Michael J. (2003) 'Participatory Design: The Third Space in HCI'. *Human-Computer Interaction: Development Process* 4235: 165–185.

Nielsen, Lene. (2012) *Personas: User Focused Design*. London: Springer Science & Business Media.

Norman, Donald A. (2006) *The Design of Everyday Things*. New York: Basic Books.

O'Neil, Daniel X. (2013) Building a Smarter Chicago. In *Beyond Transparency: Open Data and the Future of Civic Innovation*, eds Brett Goldstein and Lauren Dyson, pp. 27–38. San Francisco: Code for America Press.

O'Neil, Daniel X. (2016) 'An Education in Civic Technology'. *Civicist*, 1 June. http://civichall.org/civicist/an-education-in-community-technology.

Papanek, Victor. (1974) *Design for the Real World*. St. Albans: Paladin.

Piepzna-Samarasinha, Leah Lakshmi. (2018) *Care Work: Dreaming Disability Justice*. Vancouver: Arsenal Pulp Press.

Prashad, Vijay. (2013) *The Poorer Nations: A Possible History of the Global South*. London: Verso Books.

Pursell, Carroll. (1993) 'The Rise and Fall of the Appropriate Technology Movement in the United States, 1965–1985'. *Technology and Culture* 34(3)(July): 629–637. https://doi.org/10.2307/3106707.

Ries, Eric. (2011) *The Lean Startup: How Today's Entrepreneurs Use Continuous Innovation to Create Radically Successful Businesses*. New York: Crown Business.

Sanders, Elizabeth B-N., and Pieter Jan Stappers. (2008) 'Co-Creation and the New Landscapes of Design'. *CoDesign* 4(1): 5–18.

Sanoff, Henry. (2008) 'Multiple Views of Participatory Design'. *International Journal of Architectural Research* 2(1): 57–69.

Schuler, Douglas, and Aki Namioka (eds) (1993) *Participatory Design: Principles and Practices*. Boca Raton, FL, USA: CRC Press.

Schumacher, Ernst Friedrich. (1999) *Small Is Beautiful: Economics as if People Mattered; 25 Years Later ... With Commentaries*. Point Roberts, WA, USA: Hartley & Marks.

Shepard, Benjamin, and Ronald Hayduk (eds) (2002) *From ACT UP to the WTO: Urban Protest and Community Building in the Era of Globalization*. London: Verso.

Silbey, Susan S. (2018) '#MeToo at MIT: Harassment and Systemic Gender Subordination'. *MIT Faculty Newsletter* January/February. http://web.mit.edu/fnl/volume/303/silbey.html.

Srinivasan, Ramesh. (2017) *Whose Global Village? Rethinking How Technology Shapes Our World*. New York: NYU Press.

Steen, Marc. (2011) 'Tensions in Human-Centred Design'. *CoDesign* 7(1): 45–60.

Swift, Mike. (2010) 'Blacks, Latinos, and Women Lose Ground at Silicon Valley Tech Companies'. *San Jose Mercury News* 11 February. http://www.mercurynews.com/ci_14383730?source=pkg.

Thatcher, Margaret. (1987) 'Aids, Education and the Year 2000!' Interview by Douglas Keay. *Woman's Own* 31 October.

Turner, Fred. (2010) *From Counterculture to Cyberculture: Stewart Brand, the Whole Earth Network, and the Rise of Digital Utopianism*. Chicago, IL, USA: University of Chicago Press.

Von Hippel, Eric. (2005) *Democratizing Innovation*. Cambridge, MA, USA: MIT Press.

White, Sarah C. (1996) 'Depoliticising Development: The Uses and Abuses of Participation'. *Development in Practice* 6(1): 6–15.

Willoughby, Kelvin W. (1990) *Technology Choice: A Critique of the Appropriate Technology Movement*. Boulder, CO, USA: Westview Press.

Wittkower, Dylan E. (2016) *Principles of Anti-Discriminatory Design*. Philosophy Faculty Publications, no. 28. https://digitalcommons.odu.edu/philosophy_fac_pubs/28.

Index